R for Data Science

Import, Tidy, Transform, Visualize, and Model Data

Hadley Wickham and Garrett Grolemund

Beijing • Boston • Farnham • Sebastopol • Tokyo

R for Data Science

by Hadley Wickham and Garrett Grolemund

Printed in the United States of America.

Published by O'Reilly Media, Inc., 1005 Gravenstein Highway North, Sebastopol, CA 95472.

O'Reilly books may be purchased for educational, business, or sales promotional use. Online editions are also available for most titles (*http://oreilly.com/safari*). For more information, contact our corporate/institutional sales department: 800-998-9938 or *corporate@oreilly.com*.

Editors: Marie Beaugureau and Mike Loukides
Production Editor: Nicholas Adams
Copyeditor: Kim Cofer
Proofreader: Charles Roumeliotis
Indexer: Wendy Catalano
Interior Designer: David Futato
Cover Designer: Karen Montgomery
Illustrator: Rebecca Demarest

December 2016: First Edition

Revision History for the First Edition

2016-12-06: First Release
2016-12-22: Second Release

See *http://oreilly.com/catalog/errata.csp?isbn=9781491910399* for release details.

978-1-491-91039-9

[LSI]

Table of Contents

Part III. Program

Part IV. Model

Part V. Communicate

Preface

Data science is an exciting discipline that allows you to turn raw data into understanding, insight, and knowledge. The goal of *R for Data Science* is to help you learn the most important tools in R that will allow you to do data science. After reading this book, you'll have the tools to tackle a wide variety of data science challenges, using the best parts of R.

What You Will Learn

Data science is a huge field, and there's no way you can master it by reading a single book. The goal of this book is to give you a solid foundation in the most important tools. Our model of the tools needed in a typical data science project looks something like this:

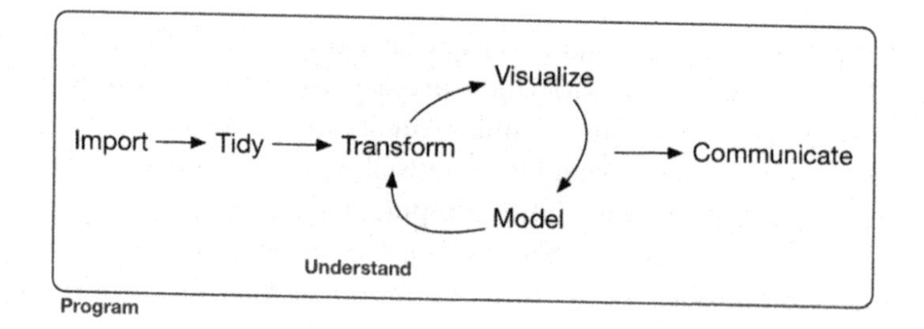

First you must *import* your data into R. This typically means that you take data stored in a file, database, or web API, and load it into a data frame in R. If you can't get your data into R, you can't do data science on it!

Once you've imported your data, it is a good idea to *tidy* it. Tidying your data means storing it in a consistent form that matches the semantics of the dataset with the way it is stored. In brief, when your data is tidy, each column is a variable, and each row is an observation. Tidy data is important because the consistent structure lets you focus your struggle on questions about the data, not fighting to get the data into the right form for different functions.

Once you have tidy data, a common first step is to *transform* it. Transformation includes narrowing in on observations of interest (like all people in one city, or all data from the last year), creating new variables that are functions of existing variables (like computing velocity from speed and time), and calculating a set of summary statistics (like counts or means). Together, tidying and transforming are called *wrangling*, because getting your data in a form that's natural to work with often feels like a fight!

Once you have tidy data with the variables you need, there are two main engines of knowledge generation: visualization and modeling. These have complementary strengths and weaknesses so any real analysis will iterate between them many times.

Visualization is a fundamentally human activity. A good visualization will show you things that you did not expect, or raise new questions about the data. A good visualization might also hint that you're asking the wrong question, or you need to collect different data. Visualizations can surprise you, but don't scale particularly well because they require a human to interpret them.

Models are complementary tools to visualization. Once you have made your questions sufficiently precise, you can use a model to answer them. Models are a fundamentally mathematical or computational tool, so they generally scale well. Even when they don't, it's usually cheaper to buy more computers than it is to buy more brains! But every model makes assumptions, and by its very nature a model cannot question its own assumptions. That means a model cannot fundamentally surprise you.

The last step of data science is *communication*, an absolutely critical part of any data analysis project. It doesn't matter how well your models and visualization have led you to understand the data unless you can also communicate your results to others.

Surrounding all these tools is *programming*. Programming is a cross-cutting tool that you use in every part of the project. You don't need to be an expert programmer to be a data scientist, but learning more about programming pays off because becoming a better programmer allows you to automate common tasks, and solve new problems with greater ease.

You'll use these tools in every data science project, but for most projects they're not enough. There's a rough 80-20 rule at play; you can tackle about 80% of every project using the tools that you'll learn in this book, but you'll need other tools to tackle the remaining 20%. Throughout this book we'll point you to resources where you can learn more.

How This Book Is Organized

The previous description of the tools of data science is organized roughly according to the order in which you use them in an analysis (although of course you'll iterate through them multiple times). In our experience, however, this is not the best way to learn them:

- Starting with data ingest and tidying is suboptimal because 80% of the time it's routine and boring, and the other 20% of the time it's weird and frustrating. That's a bad place to start learning a new subject! Instead, we'll start with visualization and transformation of data that's already been imported and tidied. That way, when you ingest and tidy your own data, your motivation will stay high because you know the pain is worth it.
- Some topics are best explained with other tools. For example, we believe that it's easier to understand how models work if you already know about visualization, tidy data, and programming.
- Programming tools are not necessarily interesting in their own right, but do allow you to tackle considerably more challenging problems. We'll give you a selection of programming tools in the middle of the book, and then you'll see they can combine with the data science tools to tackle interesting modeling problems.

Within each chapter, we try to stick to a similar pattern: start with some motivating examples so you can see the bigger picture, and then dive into the details. Each section of the book is paired with exercises to help you practice what you've learned. While it's tempting to skip the exercises, there's no better way to learn than practicing on real problems.

What You Won't Learn

There are some important topics that this book doesn't cover. We believe it's important to stay ruthlessly focused on the essentials so you can get up and running as quickly as possible. That means this book can't cover every important topic.

Big Data

This book proudly focuses on small, in-memory datasets. This is the right place to start because you can't tackle big data unless you have experience with small data. The tools you learn in this book will easily handle hundreds of megabytes of data, and with a little care you can typically use them to work with 1–2 Gb of data. If you're routinely working with larger data (10–100 Gb, say), you should learn more about data.table (*http://bit.ly/Rdatatable*). This book doesn't teach data.table because it has a very concise interface, which makes it harder to learn since it offers fewer linguistic cues. But if you're working with large data, the performance payoff is worth the extra effort required to learn it.

If your data is bigger than this, carefully consider if your big data problem might actually be a small data problem in disguise. While the complete data might be big, often the data needed to answer a specific question is small. You might be able to find a subset, subsample, or summary that fits in memory and still allows you to answer the question that you're interested in. The challenge here is finding the right small data, which often requires a lot of iteration.

Another possibility is that your big data problem is actually a large number of small data problems. Each individual problem might fit in memory, but you have millions of them. For example, you might want to fit a model to each person in your dataset. That would be trivial if you had just 10 or 100 people, but instead you have a million. Fortunately each problem is independent of the others (a setup that is sometimes called embarrassingly parallel), so you just need a system (like Hadoop or Spark) that allows you to send different datasets to different computers for processing. Once you've figured out how to answer the question for a single subset using the tools described in this book, you learn new tools like sparklyr, rhipe, and ddr to solve it for the full dataset.

Python, Julia, and Friends

In this book, you won't learn anything about Python, Julia, or any other programming language useful for data science. This isn't because we think these tools are bad. They're not! And in practice, most data science teams use a mix of languages, often at least R and Python.

However, we strongly believe that it's best to master one tool at a time. You will get better faster if you dive deep, rather than spreading yourself thinly over many topics. This doesn't mean you should only know one thing, just that you'll generally learn faster if you stick to one thing at a time. You should strive to learn new things throughout your career, but make sure your understanding is solid before you move on to the next interesting thing.

We think R is a great place to start your data science journey because it is an environment designed from the ground up to support data science. R is not just a programming language, but it is also an interactive environment for doing data science. To support interaction, R is a much more flexible language than many of its peers. This flexibility comes with its downsides, but the big upside is how easy it is to evolve tailored grammars for specific parts of the data science process. These mini languages help you think about problems as a data scientist, while supporting fluent interaction between your brain and the computer.

Nonrectangular Data

This book focuses exclusively on rectangular data: collections of values that are each associated with a variable and an observation. There are lots of datasets that do not naturally fit in this paradigm: including images, sounds, trees, and text. But rectangular data frames are extremely common in science and industry, and we believe that they're a great place to start your data science journey.

Hypothesis Confirmation

It's possible to divide data analysis into two camps: hypothesis generation and hypothesis confirmation (sometimes called confirmatory analysis). The focus of this book is unabashedly on hypothesis generation, or data exploration. Here you'll look deeply at the data and, in combination with your subject knowledge, generate many interesting hypotheses to help explain why the data behaves the way it does. You evaluate the hypotheses informally, using your skepticism to challenge the data in multiple ways.

The complement of hypothesis generation is hypothesis confirmation. Hypothesis confirmation is hard for two reasons:

- You need a precise mathematical model in order to generate falsifiable predictions. This often requires considerable statistical sophistication.
- You can only use an observation once to confirm a hypothesis. As soon as you use it more than once you're back to doing exploratory analysis. This means to do hypothesis confirmation you need to "preregister" (write out in advance) your analysis plan, and not deviate from it even when you have seen the data. We'll talk a little about some strategies you can use to make this easier in Part IV.

It's common to think about modeling as a tool for hypothesis confirmation, and visualization as a tool for hypothesis generation. But that's a false dichotomy: models are often used for exploration, and with a little care you can use visualization for confirmation. The key difference is how often you look at each observation: if you look only once, it's confirmation; if you look more than once, it's exploration.

Prerequisites

We've made a few assumptions about what you already know in order to get the most out of this book. You should be generally numerically literate, and it's helpful if you have some programming experience already. If you've never programmed before, you might find *Hands-On Programming with R* by Garrett to be a useful adjunct to this book.

There are four things you need to run the code in this book: R, RStudio, a collection of R packages called the *tidyverse*, and a handful of other packages. Packages are the fundamental units of reproducible R code. They include reusable functions, the documentation that describes how to use them, and sample data.

R

To download R, go to CRAN, the *comprehensive R archive network*. CRAN is composed of a set of mirror servers distributed around the world and is used to distribute R and R packages. Don't try and pick a mirror that's close to you: instead use the cloud mirror, *https://cloud.r-project.org*, which automatically figures it out for you.

A new major version of R comes out once a year, and there are 2–3 minor releases each year. It's a good idea to update regularly. Upgrading can be a bit of a hassle, especially for major versions, which require you to reinstall all your packages, but putting it off only makes it worse.

RStudio

RStudio is an integrated development environment, or IDE, for R programming. Download and install it from *http://www.rstudio.com/download*. RStudio is updated a couple of times a year. When a new version is available, RStudio will let you know. It's a good idea to upgrade regularly so you can take advantage of the latest and greatest features. For this book, make sure you have RStudio 1.0.0.

When you start RStudio, you'll see two key regions in the interface:

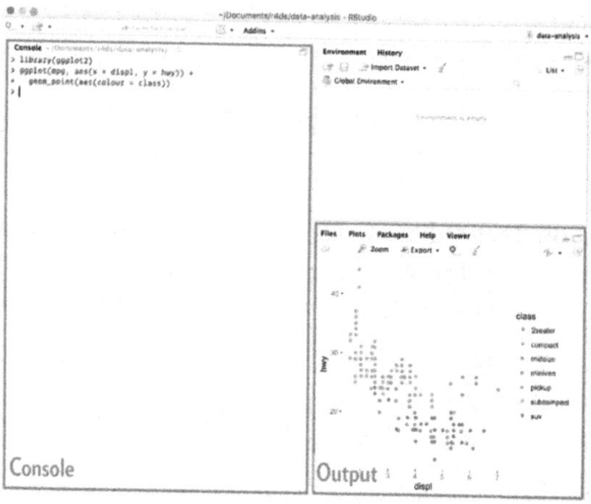

For now, all you need to know is that you type R code in the console pane, and press Enter to run it. You'll learn more as we go along!

The Tidyverse

You'll also need to install some R packages. An R *package* is a collection of functions, data, and documentation that extends the capabilities of base R. Using packages is key to the successful use of R. The majority of the packages that you will learn in this book are part of the so-called tidyverse. The packages in the tidyverse share a common philosophy of data and R programming, and are designed to work together naturally.

You can install the complete tidyverse with a single line of code:

```
install.packages("tidyverse")
```

On your own computer, type that line of code in the console, and then press Enter to run it. R will download the packages from CRAN and install them onto your computer. If you have problems installing, make sure that you are connected to the internet, and that *https://cloud.r-project.org/* isn't blocked by your firewall or proxy.

You will not be able to use the functions, objects, and help files in a package until you load it with `library()`. Once you have installed a package, you can load it with the `library()` function:

```
library(tidyverse)
#> Loading tidyverse: ggplot2
#> Loading tidyverse: tibble
#> Loading tidyverse: tidyr
#> Loading tidyverse: readr
#> Loading tidyverse: purrr
#> Loading tidyverse: dplyr
#> Conflicts with tidy packages ----------------------------
#> filter(): dplyr, stats
#> lag():    dplyr, stats
```

This tells you that tidyverse is loading the **ggplot2**, **tibble**, **tidyr**, **readr**, **purrr**, and **dplyr** packages. These are considered to be the *core* of the tidyverse because you'll use them in almost every analysis.

Packages in the tidyverse change fairly frequently. You can see if updates are available, and optionally install them, by running `tidyverse_update()`.

Other Packages

There are many other excellent packages that are not part of the tidyverse, because they solve problems in a different domain, or are designed with a different set of underlying principles. This doesn't make them better or worse, just different. In other words, the complement to the tidyverse is not the messyverse, but many other universes of interrelated packages. As you tackle more data science projects with R, you'll learn new packages and new ways of thinking about data.

In this book we'll use three data packages from outside the tidyverse:

```
install.packages(c("nycflights13", "gapminder", "Lahman"))
```

These packages provide data on airline flights, world development, and baseball that we'll use to illustrate key data science ideas.

Running R Code

The previous section showed you a couple of examples of running R code. Code in the book looks like this:

```
1 + 2
#> [1] 3
```

If you run the same code in your local console, it will look like this:

```
> 1 + 2
[1] 3
```

There are two main differences. In your console, you type after the `>`, called the *prompt*; we don't show the prompt in the book. In the book, output is commented out with `#>`; in your console it appears directly after your code. These two differences mean that if you're working with an electronic version of the book, you can easily copy code out of the book and into the console.

Throughout the book we use a consistent set of conventions to refer to code:

- Functions are in a code font and followed by parentheses, like `sum()` or `mean()`.
- Other R objects (like data or function arguments) are in a code font, without parentheses, like `flights` or `x`.

- If we want to make it clear what package an object comes from, we'll use the package name followed by two colons, like `dplyr::mutate()` or `nycflights13::flights`. This is also valid R code.

Getting Help and Learning More

This book is not an island; there is no single resource that will allow you to master R. As you start to apply the techniques described in this book to your own data you will soon find questions that I do not answer. This section describes a few tips on how to get help, and to help you keep learning.

If you get stuck, start with Google. Typically, adding "R" to a query is enough to restrict it to relevant results: if the search isn't useful, it often means that there aren't any R-specific results available. Google is particularly useful for error messages. If you get an error message and you have no idea what it means, try googling it! Chances are that someone else has been confused by it in the past, and there will be help somewhere on the web. (If the error message isn't in English, run `Sys.setenv(LANGUAGE = "en")` and re-run the code; you're more likely to find help for English error messages.)

If Google doesn't help, try stackoverflow (*http://stackoverflow.com*). Start by spending a little time searching for an existing answer; including `[R]` restricts your search to questions and answers that use R. If you don't find anything useful, prepare a minimal reproducible example or **reprex**. A good reprex makes it easier for other people to help you, and often you'll figure out the problem yourself in the course of making it.

There are three things you need to include to make your example reproducible: required packages, data, and code:

- *Packages* should be loaded at the top of the script, so it's easy to see which ones the example needs. This is a good time to check that you're using the latest version of each package; it's possible you've discovered a bug that's been fixed since you installed the package. For packages in the tidyverse, the easiest way to check is to run `tidyverse_update()`.

- The easiest way to include *data* in a question is to use `dput()` to generate the R code to re-create it. For example, to re-create the `mtcars` dataset in R, I'd perform the following steps:

  1. Run `dput(mtcars)` in R.
  2. Copy the output.
  3. In my reproducible script, type `mtcars <-` then paste.

  Try and find the smallest subset of your data that still reveals the problem.

- Spend a little bit of time ensuring that your *code* is easy for others to read:
  - Make sure you've used spaces and your variable names are concise, yet informative.
  - Use comments to indicate where your problem lies.
  - Do your best to remove everything that is not related to the problem.

  The shorter your code is, the easier it is to understand, and the easier it is to fix.

Finish by checking that you have actually made a reproducible example by starting a fresh R session and copying and pasting your script in.

You should also spend some time preparing yourself to solve problems before they occur. Investing a little time in learning R each day will pay off handsomely in the long run. One way is to follow what Hadley, Garrett, and everyone else at RStudio are doing on the RStudio blog (*https://blog.rstudio.org*). This is where we post announcements about new packages, new IDE features, and in-person courses. You might also want to follow Hadley (@hadleywickham (*https://twitter.com/hadleywickham*)) or Garrett (@statgarrett (*https://twitter.com/statgarrett*)) on Twitter, or follow @rstudiotips (*https://twitter.com/rstudiotips*) to keep up with new features in the IDE.

To keep up with the R community more broadly, we recommend reading *http://www.r-bloggers.com:* it aggregates over 500 blogs about R from around the world. If you're an active Twitter user, follow the #rstats hashtag. Twitter is one of the key tools that Hadley uses to keep up with new developments in the community.

Acknowledgments

This book isn't just the product of Hadley and Garrett, but is the result of many conversations (in person and online) that we've had with the many people in the R community. There are a few people we'd like to thank in particular, because they have spent many hours answering our dumb questions and helping us to better think about data science:

- Jenny Bryan and Lionel Henry for many helpful discussions around working with lists and list-columns.
- The three chapters on workflow were adapted (with permission) from "R basics, workspace and working directory, RStudio projects" (*http://bit.ly/Rbasicsworkflow*) by Jenny Bryan.
- Genevera Allen for discussions about models, modeling, the statistical learning perspective, and the difference between hypothesis generation and hypothesis confirmation.
- Yihui Xie for his work on the **bookdown** (*https://github.com/rstudio/bookdown*) package, and for tirelessly responding to my feature requests.
- Bill Behrman for his thoughtful reading of the entire book, and for trying it out with his data science class at Stanford.
- The #rstats twitter community who reviewed all of the draft chapters and provided tons of useful feedback.
- Tal Galili for augmenting his **dendextend** package to support a section on clustering that did not make it into the final draft.

This book was written in the open, and many people contributed pull requests to fix minor problems. Special thanks goes to everyone who contributed via GitHub (listed in alphabetical order): adi pradhan, Ahmed ElGabbas, Ajay Deonarine, @Alex, Andrew Landgraf, @batpigandme, @behrman, Ben Marwick, Bill Behrman, Brandon Greenwell, Brett Klamer, Christian G. Warden, Christian Mongeau, Colin Gillespie, Cooper Morris, Curtis Alexander, Daniel Gromer, David Clark, Derwin McGeary, Devin Pastoor, Dylan Cashman, Earl Brown, Eric Watt, Etienne B. Racine, Flemming Villalona, Gregory Jefferis, @harrismcgehee, Hengni Cai, Ian Lyttle, Ian Sealy, Jakub

Nowosad, Jennifer (Jenny) Bryan, @jennybc, Jeroen Janssens, Jim Hester, @jjchern, Joanne Jang, John Sears, Jon Calder, Jonathan Page, @jonathanflint, Julia Stewart Lowndes, Julian During, Justinas Petuchovas, Kara Woo, @kdpsingh, Kenny Darrell, Kirill Sevastyanenko, @koalabearski, @KyleHumphrey, Lawrence Wu, Matthew Sedaghatfar, Mine Cetinkaya-Rundel, @MJMarshall, Mustafa Ascha, @nate-d-olson, Nelson Areal, Nick Clark, @nickelas, @nwaff, @OaCantona, Patrick Kennedy, Peter Hurford, Rademeyer Vermaak, Radu Grosu, @rlzijdeman, Robert Schuessler, @robinlovelace, @robinsones, S'busiso Mkhondwane, @seamus-mckinsey, @seanpwilliams, Shannon Ellis, @shoili, @sibusiso16, @spirgel, Steve Mortimer, @svenski, Terence Teo, Thomas Klebel, TJ Mahr, Tom Prior, Will Beasley, Yihui Xie.

Online Version

An online version of this book is available at *http://r4ds.had.co.nz*. It will continue to evolve in between reprints of the physical book. The source of the book is available at *https://github.com/hadley/r4ds*. The book is powered by *https://bookdown.org*, which makes it easy to turn R markdown files into HTML, PDF, and EPUB.

This book was built with:

```
devtools::session_info(c("tidyverse"))
#> Session info -------------------------------------------------------
#>  setting  value
#>  version  R version 3.3.1 (2016-06-21)
#>  system   x86_64, darwin13.4.0
#>  ui       X11
#>  language (EN)
#>  collate  en_US.UTF-8
#>  tz       America/Los_Angeles
#>  date     2016-10-10
#> Packages -----------------------------------------------------------
#>  package      * version   date       source
#>  assertthat     0.1       2013-12-06 CRAN (R 3.3.0)
#>  BH             1.60.0-2  2016-05-07 CRAN (R 3.3.0)
#>  broom          0.4.1     2016-06-24 CRAN (R 3.3.0)
#>  colorspace     1.2-6     2015-03-11 CRAN (R 3.3.0)
#>  curl           2.1       2016-09-22 CRAN (R 3.3.0)
#>  DBI            0.5-1     2016-09-10 CRAN (R 3.3.0)
#>  dichromat      2.0-0     2013-01-24 CRAN (R 3.3.0)
#>  digest         0.6.10    2016-08-02 CRAN (R 3.3.0)
#>  dplyr        * 0.5.0     2016-06-24 CRAN (R 3.3.0)
#>  forcats        0.1.1     2016-09-16 CRAN (R 3.3.0)
#>  foreign        0.8-67    2016-09-13 CRAN (R 3.3.0)
```

```
#>  ggplot2      * 2.1.0.9001 2016-10-06 local
#>  gtable         0.2.0      2016-02-26 CRAN (R 3.3.0)
#>  haven          1.0.0      2016-09-30 local
#>  hms            0.2-1      2016-07-28 CRAN (R 3.3.1)
#>  httr           1.2.1      2016-07-03 cran (@1.2.1)
#>  jsonlite       1.1        2016-09-14 CRAN (R 3.3.0)
#>  labeling       0.3        2014-08-23 CRAN (R 3.3.0)
#>  lattice        0.20-34    2016-09-06 CRAN (R 3.3.0)
#>  lazyeval       0.2.0      2016-06-12 CRAN (R 3.3.0)
#>  lubridate      1.6.0      2016-09-13 CRAN (R 3.3.0)
#>  magrittr       1.5        2014-11-22 CRAN (R 3.3.0)
#>  MASS           7.3-45     2016-04-21 CRAN (R 3.3.1)
#>  mime           0.5        2016-07-07 cran (@0.5)
#>  mnormt         1.5-4      2016-03-09 CRAN (R 3.3.0)
#>  modelr         0.1.0      2016-08-31 CRAN (R 3.3.0)
#>  munsell        0.4.3      2016-02-13 CRAN (R 3.3.0)
#>  nlme           3.1-128    2016-05-10 CRAN (R 3.3.1)
#>  openssl        0.9.4      2016-05-25 cran (@0.9.4)
#>  plyr           1.8.4      2016-06-08 cran (@1.8.4)
#>  psych          1.6.9      2016-09-17 CRAN (R 3.3.0)
#>  purrr        * 0.2.2      2016-06-18 CRAN (R 3.3.0)
#>  R6             2.1.3      2016-08-19 CRAN (R 3.3.0)
#>  RColorBrewer   1.1-2      2014-12-07 CRAN (R 3.3.0)
#>  Rcpp           0.12.7     2016-09-05 CRAN (R 3.3.0)
#>  readr        * 1.0.0      2016-08-03 CRAN (R 3.3.0)
#>  readxl         0.1.1      2016-03-28 CRAN (R 3.3.0)
#>  reshape2       1.4.1      2014-12-06 CRAN (R 3.3.0)
#>  rvest          0.3.2      2016-06-17 CRAN (R 3.3.0)
#>  scales         0.4.0.9003 2016-10-06 local
#>  selectr        0.3-0      2016-08-30 CRAN (R 3.3.0)
#>  stringi        1.1.2      2016-10-01 CRAN (R 3.3.1)
#>  stringr        1.1.0      2016-08-19 cran (@1.1.0)
#>  tibble       * 1.2        2016-08-26 CRAN (R 3.3.0)
#>  tidyr        * 0.6.0      2016-08-12 CRAN (R 3.3.0)
#>  tidyverse    * 1.0.0      2016-09-09 CRAN (R 3.3.0)
#>  xml2           1.0.0.9001 2016-09-30 local
```

Conventions Used in This Book

The following typographical conventions are used in this book:

Italic
: Indicates new terms, URLs, email addresses, filenames, and file extensions.

Bold
: Indicates the names of R packages.

`Constant width`
: Used for program listings, as well as within paragraphs to refer to program elements such as variable or function names, databases, data types, environment variables, statements, and keywords.

`Constant width bold`
: Shows commands or other text that should be typed literally by the user.

`Constant width italic`
: Shows text that should be replaced with user-supplied values or by values determined by context.

This element signifies a tip or suggestion.

Using Code Examples

Source code is available for download at *https://github.com/hadley/r4ds*.

This book is here to help you get your job done. In general, if example code is offered with this book, you may use it in your programs and documentation. You do not need to contact us for permission unless you're reproducing a significant portion of the code. For example, writing a program that uses several chunks of code from this book does not require permission. Selling or distributing a CD-ROM of examples from O'Reilly books does require permission. Answering a question by citing this book and quoting example code does not require permission. Incorporating a significant amount of example code from this book into your product's documentation does require permission.

We appreciate, but do not require, attribution. An attribution usually includes the title, author, publisher, and ISBN. For example: "*R for Data Science* by Hadley Wickham and Garrett Grolemund (O'Reilly). Copyright 2017 Garrett Grolemund, Hadley Wickham, 978-1-491-91039-9."

If you feel your use of code examples falls outside fair use or the permission given above, feel free to contact us at *permissions@oreilly.com*.

O'Reilly Safari

Safari Safari (formerly Safari Books Online) is a membership-based training and reference platform for enterprise, government, educators, and individuals.

Members have access to thousands of books, training videos, Learning Paths, interactive tutorials, and curated playlists from over 250 publishers, including O'Reilly Media, Harvard Business Review, Prentice Hall Professional, Addison-Wesley Professional, Microsoft Press, Sams, Que, Peachpit Press, Adobe, Focal Press, Cisco Press, John Wiley & Sons, Syngress, Morgan Kaufmann, IBM Redbooks, Packt, Adobe Press, FT Press, Apress, Manning, New Riders, McGraw-Hill, Jones & Bartlett, and Course Technology, among others.

For more information, please visit *http://oreilly.com/safari*.

How to Contact Us

Please address comments and questions concerning this book to the publisher:

> O'Reilly Media, Inc.
> 1005 Gravenstein Highway North
> Sebastopol, CA 95472
> 800-998-9938 (in the United States or Canada)
> 707-829-0515 (international or local)
> 707-829-0104 (fax)

We have a web page for this book, where we list errata, examples, and any additional information. You can access this page at *http://bit.ly/r-for-data-science*.

To comment or ask technical questions about this book, send email to *bookquestions@oreilly.com*.

For more information about our books, courses, conferences, and news, see our website at *http://www.oreilly.com*.

Find us on Facebook: *http://facebook.com/oreilly*

Follow us on Twitter: *http://twitter.com/oreillymedia*

Watch us on YouTube: *http://www.youtube.com/oreillymedia*

PART I
Explore

The goal of the first part of this book is to get you up to speed with the basic tools of *data exploration* as quickly as possible. Data exploration is the art of looking at your data, rapidly generating hypotheses, quickly testing them, then repeating again and again and again. The goal of data exploration is to generate many promising leads that you can later explore in more depth.

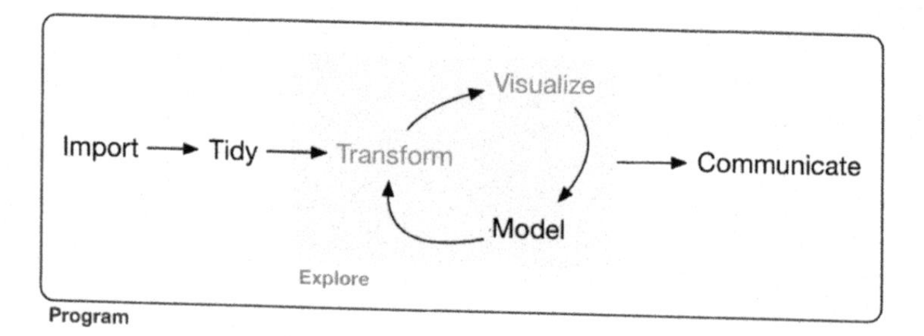

In this part of the book you will learn some useful tools that have an immediate payoff:

- Visualization is a great place to start with R programming, because the payoff is so clear: you get to make elegant and informative plots that help you understand data. In Chapter 1 you'll dive into visualization, learning the basic structure of a **ggplot2** plot, and powerful techniques for turning data into plots.

- Visualization alone is typically not enough, so in Chapter 3 you'll learn the key verbs that allow you to select important variables, filter out key observations, create new variables, and compute summaries.
- Finally, in Chapter 5, you'll combine visualization and transformation with your curiosity and skepticism to ask and answer interesting questions about data.

Modeling is an important part of the exploratory process, but you don't have the skills to effectively learn or apply it yet. We'll come back to it in Part IV, once you're better equipped with more data wrangling and programming tools.

Nestled among these three chapters that teach you the tools of exploration are three chapters that focus on your R workflow. In Chapter 2, Chapter 4, and Chapter 6 you'll learn good practices for writing and organizing your R code. These will set you up for success in the long run, as they'll give you the tools to stay organized when you tackle real projects.

CHAPTER 1

Data Visualization with ggplot2

Introduction

> The simple graph has brought more information to the data analyst's mind than any other device.
>
> —John Tukey

This chapter will teach you how to visualize your data using **ggplot2**. R has several systems for making graphs, but **ggplot2** is one of the most elegant and most versatile. **ggplot2** implements the *grammar of graphics*, a coherent system for describing and building graphs. With **ggplot2**, you can do more faster by learning one system and applying it in many places.

If you'd like to learn more about the theoretical underpinnings of **ggplot2** before you start, I'd recommend reading "A Layered Grammar of Graphics" (*http://vita.had.co.nz/papers/layered-grammar.pdf*).

Prerequisites

This chapter focuses on **ggplot2**, one of the core members of the tidyverse. To access the datasets, help pages, and functions that we will use in this chapter, load the tidyverse by running this code:

```
library(tidyverse)
#> Loading tidyverse: ggplot2
#> Loading tidyverse: tibble
#> Loading tidyverse: tidyr
#> Loading tidyverse: readr
#> Loading tidyverse: purrr
```

```
#> Loading tidyverse: dplyr
#> Conflicts with tidy packages ----------------------------
#> filter(): dplyr, stats
#> lag():    dplyr, stats
```

That one line of code loads the core tidyverse, packages that you will use in almost every data analysis. It also tells you which functions from the tidyverse conflict with functions in base R (or from other packages you might have loaded).

If you run this code and get the error message "there is no package called 'tidyverse'," you'll need to first install it, then run `library()` once again:

```
install.packages("tidyverse")
library(tidyverse)
```

You only need to install a package once, but you need to reload it every time you start a new session.

If we need to be explicit about where a function (or dataset) comes from, we'll use the special form `package::function()`. For example, `ggplot2::ggplot()` tells you explicitly that we're using the `ggplot()` function from the **ggplot2** package.

First Steps

Let's use our first graph to answer a question: do cars with big engines use more fuel than cars with small engines? You probably already have an answer, but try to make your answer precise. What does the relationship between engine size and fuel efficiency look like? Is it positive? Negative? Linear? Nonlinear?

The mpg Data Frame

You can test your answer with the `mpg` data frame found in **ggplot2** (aka `ggplot2::mpg`). A *data frame* is a rectangular collection of variables (in the columns) and observations (in the rows). `mpg` contains observations collected by the US Environment Protection Agency on 38 models of cars:

```
mpg
#> # A tibble: 234 × 11
#>   manufacturer model displ  year   cyl      trans   drv
#>          <chr> <chr> <dbl> <int> <int>      <chr> <chr>
#> 1         audi    a4   1.8  1999     4   auto(l5)     f
#> 2         audi    a4   1.8  1999     4 manual(m5)     f
```

```
#> 3          audi     a4   2.0  2008     4 manual(m6)     f
#> 4          audi     a4   2.0  2008     4   auto(av)     f
#> 5          audi     a4   2.8  1999     6   auto(l5)     f
#> 6          audi     a4   2.8  1999     6 manual(m5)     f
#> # ... with 228 more rows, and 4 more variables:
#> # cty <int>, hwy <int>, fl <chr>, class <chr>
```

Among the variables in `mpg` are:

- `displ`, a car's engine size, in liters.
- `hwy`, a car's fuel efficiency on the highway, in miles per gallon (mpg). A car with a low fuel efficiency consumes more fuel than a car with a high fuel efficiency when they travel the same distance.

To learn more about `mpg`, open its help page by running `?mpg`.

Creating a ggplot

To plot `mpg`, run this code to put `displ` on the x-axis and `hwy` on the y-axis:

```
ggplot(data = mpg) +
  geom_point(mapping = aes(x = displ, y = hwy))
```

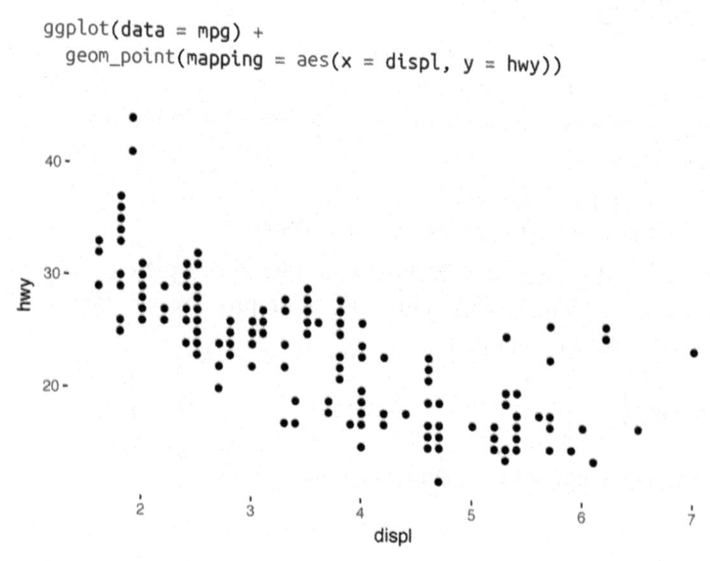

The plot shows a negative relationship between engine size (`displ`) and fuel efficiency (`hwy`). In other words, cars with big engines use more fuel. Does this confirm or refute your hypothesis about fuel efficiency and engine size?

With **ggplot2**, you begin a plot with the function `ggplot()`. `ggplot()` creates a coordinate system that you can add layers to. The first argument of `ggplot()` is the dataset to use in the graph. So `ggplot(data = mpg)` creates an empty graph, but it's not very interesting so I'm not going to show it here.

You complete your graph by adding one or more layers to `ggplot()`. The function `geom_point()` adds a layer of points to your plot, which creates a scatterplot. **ggplot2** comes with many geom functions that each add a different type of layer to a plot. You'll learn a whole bunch of them throughout this chapter.

Each geom function in **ggplot2** takes a `mapping` argument. This defines how variables in your dataset are mapped to visual properties. The `mapping` argument is always paired with `aes()`, and the `x` and `y` arguments of `aes()` specify which variables to map to the x- and y-axes. **ggplot2** looks for the mapped variable in the `data` argument, in this case, `mpg`.

A Graphing Template

Let's turn this code into a reusable template for making graphs with **ggplot2**. To make a graph, replace the bracketed sections in the following code with a dataset, a geom function, or a collection of mappings:

```
ggplot(data = <DATA>) +
  <GEOM_FUNCTION>(mapping = aes(<MAPPINGS>))
```

The rest of this chapter will show you how to complete and extend this template to make different types of graphs. We will begin with the `<MAPPINGS>` component.

Exercises

1. Run `ggplot(data = mpg)`. What do you see?
2. How many rows are in `mtcars`? How many columns?
3. What does the `drv` variable describe? Read the help for `?mpg` to find out.
4. Make a scatterplot of `hwy` versus `cyl`.

5. What happens if you make a scatterplot of `class` versus `drv`? Why is the plot not useful?

Aesthetic Mappings

> The greatest value of a picture is when it forces us to notice what we never expected to see.
>
> —John Tukey

In the following plot, one group of points (highlighted in red) seems to fall outside of the linear trend. These cars have a higher mileage than you might expect. How can you explain these cars?

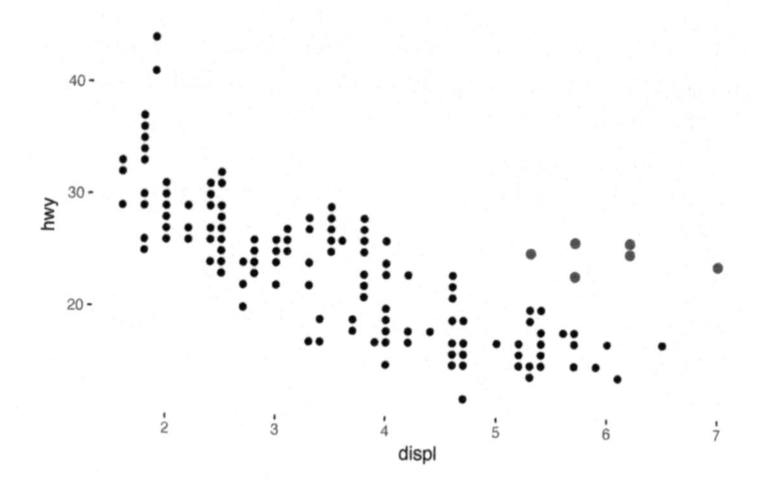

Let's hypothesize that the cars are hybrids. One way to test this hypothesis is to look at the `class` value for each car. The `class` variable of the `mpg` dataset classifies cars into groups such as compact, midsize, and SUV. If the outlying points are hybrids, they should be classified as compact cars or, perhaps, subcompact cars (keep in mind that this data was collected before hybrid trucks and SUVs became popular).

You can add a third variable, like `class`, to a two-dimensional scatterplot by mapping it to an *aesthetic*. An aesthetic is a visual property of the objects in your plot. Aesthetics include things like the size, the shape, or the color of your points. You can display a point (like the one shown next) in different ways by changing the values of its aesthetic properties. Since we already use the word "value" to describe data, let's use the word "level" to describe aesthetic properties. Here we change the levels of a point's size, shape, and color to make the point small, triangular, or blue:

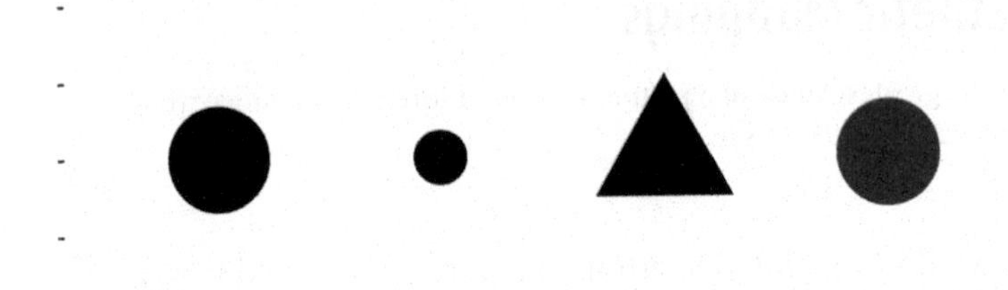

You can convey information about your data by mapping the aesthetics in your plot to the variables in your dataset. For example, you can map the colors of your points to the `class` variable to reveal the class of each car:

```
ggplot(data = mpg) +
  geom_point(mapping = aes(x = displ, y = hwy, color = class))
```

(If you prefer British English, like Hadley, you can use `colour` instead of `color`.)

To map an aesthetic to a variable, associate the name of the aesthetic to the name of the variable inside `aes()`. **ggplot2** will automatically assign a unique level of the aesthetic (here a unique color) to each unique value of the variable, a process known as *scaling*. **ggplot2** will also add a legend that explains which levels correspond to which values.

The colors reveal that many of the unusual points are two-seater cars. These cars don't seem like hybrids, and are, in fact, sports cars! Sports cars have large engines like SUVs and pickup trucks, but small bodies like midsize and compact cars, which improves their gas mileage. In hindsight, these cars were unlikely to be hybrids since they have large engines.

In the preceding example, we mapped `class` to the color aesthetic, but we could have mapped `class` to the size aesthetic in the same way. In this case, the exact size of each point would reveal its class affiliation. We get a *warning* here, because mapping an unordered variable (`class`) to an ordered aesthetic (`size`) is not a good idea:

```
ggplot(data = mpg) +
  geom_point(mapping = aes(x = displ, y = hwy, size = class))
#> Warning: Using size for a discrete variable is not advised.
```

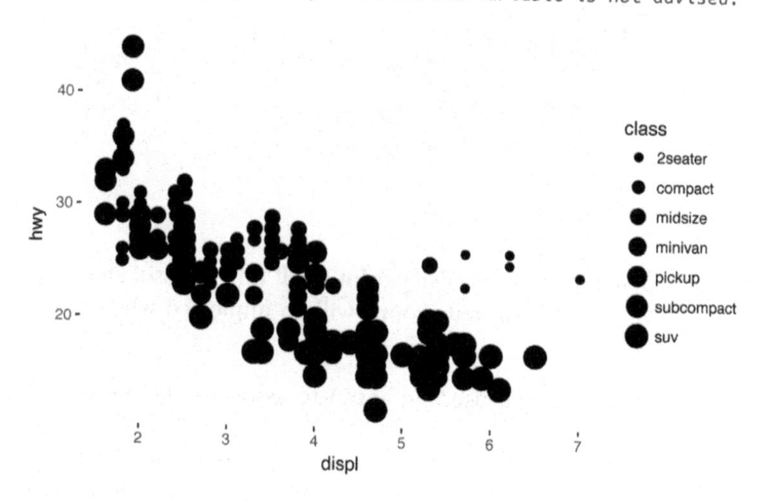

Or we could have mapped `class` to the *alpha* aesthetic, which controls the transparency of the points, or the shape of the points:

```
# Top
ggplot(data = mpg) +
  geom_point(mapping = aes(x = displ, y = hwy, alpha = class))

# Bottom
ggplot(data = mpg) +
  geom_point(mapping = aes(x = displ, y = hwy, shape = class))
```

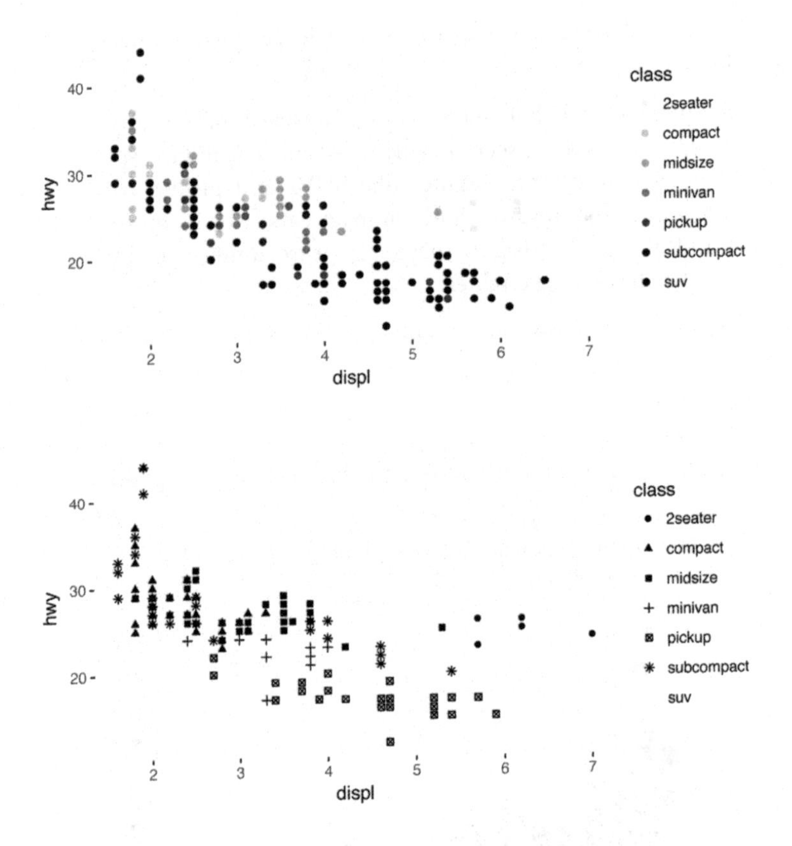

What happened to the SUVs? **ggplot2** will only use six shapes at a time. By default, additional groups will go unplotted when you use this aesthetic.

For each aesthetic you use, the `aes()` to associate the name of the aesthetic with a variable to display. The `aes()` function gathers together each of the aesthetic mappings used by a layer and passes them to the layer's mapping argument. The syntax highlights a useful insight about `x` and `y`: the x and y locations of a point are themselves aesthetics, visual properties that you can map to variables to display information about the data.

Once you map an aesthetic, **ggplot2** takes care of the rest. It selects a reasonable scale to use with the aesthetic, and it constructs a legend that explains the mapping between levels and values. For x and y aesthetics, **ggplot2** does not create a legend, but it creates an axis line with tick marks and a label. The axis line acts as a legend; it explains the mapping between locations and values.

You can also *set* the aesthetic properties of your geom manually. For example, we can make all of the points in our plot blue:

```
ggplot(data = mpg) +
  geom_point(mapping = aes(x = displ, y = hwy), color = "blue")
```

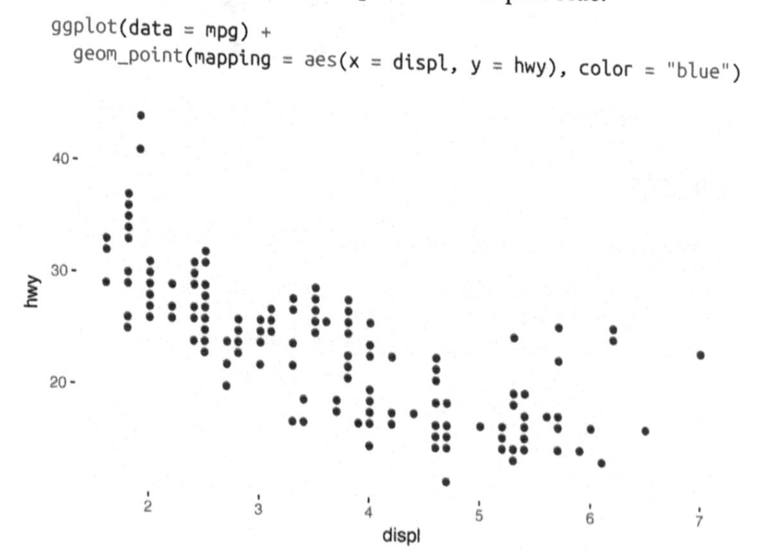

Here, the color doesn't convey information about a variable, but only changes the appearance of the plot. To set an aesthetic manually, set the aesthetic by name as an argument of your geom function; i.e., it goes *outside* of `aes()`. You'll need to pick a value that makes sense for that aesthetic:

- The name of a color as a character string.
- The size of a point in mm.
- The shape of a point as a number, as shown in Figure 1-1. There are some seeming duplicates: for example, 0, 15, and 22 are all squares. The difference comes from the interaction of the `color` and `fill` aesthetics. The hollow shapes (0–14) have a border determined by `color`; the solid shapes (15–18) are filled with `color`; and the filled shapes (21–24) have a border of `color` and are filled with `fill`.

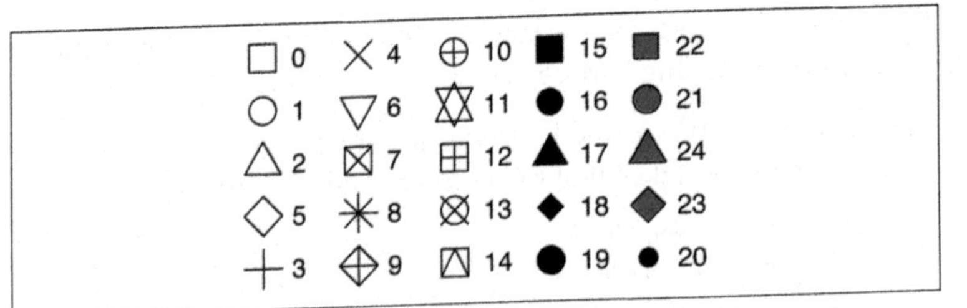

Figure 1-1. R has 25 built-in shapes that are identified by numbers

Exercises

1. What's gone wrong with this code? Why are the points not blue?

```
ggplot(data = mpg) +
  geom_point(
    mapping = aes(x = displ, y = hwy, color = "blue")
  )
```

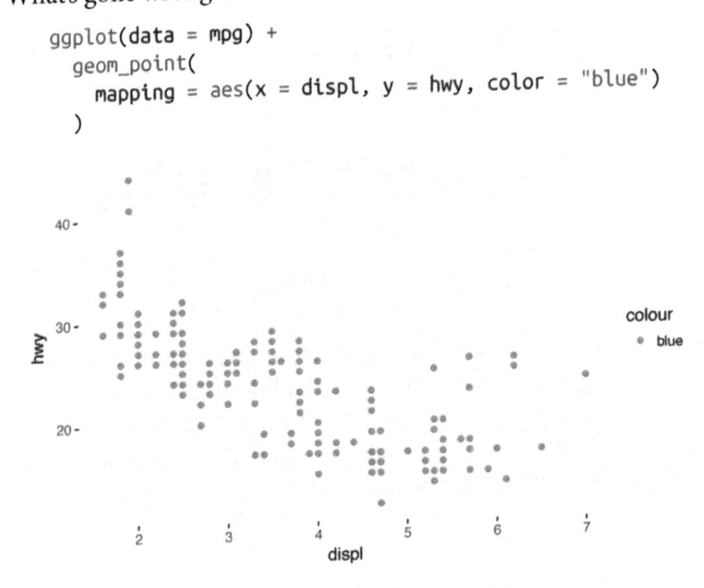

2. Which variables in `mpg` are categorical? Which variables are continuous? (Hint: type `?mpg` to read the documentation for the dataset.) How can you see this information when you run `mpg`?
3. Map a continuous variable to `color`, `size`, and `shape`. How do these aesthetics behave differently for categorical versus continuous variables?
4. What happens if you map the same variable to multiple aesthetics?
5. What does the `stroke` aesthetic do? What shapes does it work with? (Hint: use `?geom_point`.)

6. What happens if you map an aesthetic to something other than a variable name, like `aes(color = displ < 5)`?

Common Problems

As you start to run R code, you're likely to run into problems. Don't worry—it happens to everyone. I have been writing R code for years, and every day I still write code that doesn't work!

Start by carefully comparing the code that you're running to the code in the book. R is extremely picky, and a misplaced character can make all the difference. Make sure that every `(` is matched with a `)` and every `"` is paired with another `"`. Sometimes you'll run the code and nothing happens. Check the left-hand side of your console: if it's a `+`, it means that R doesn't think you've typed a complete expression and it's waiting for you to finish it. In this case, it's usually easy to start from scratch again by pressing Esc to abort processing the current command.

One common problem when creating **ggplot2** graphics is to put the `+` in the wrong place: it has to come at the end of the line, not the start. In other words, make sure you haven't accidentally written code like this:

```
ggplot(data = mpg)
+ geom_point(mapping = aes(x = displ, y = hwy))
```

If you're still stuck, try the help. You can get help about any R function by running `?function_name` in the console, or selecting the function name and pressing F1 in RStudio. Don't worry if the help doesn't seem that helpful—instead skip down to the examples and look for code that matches what you're trying to do.

If that doesn't help, carefully read the error message. Sometimes the answer will be buried there! But when you're new to R, the answer might be in the error message but you don't yet know how to understand it. Another great tool is Google: trying googling the error message, as it's likely someone else has had the same problem, and has received help online.

Facets

One way to add additional variables is with aesthetics. Another way, particularly useful for categorical variables, is to split your plot into *facets*, subplots that each display one subset of the data.

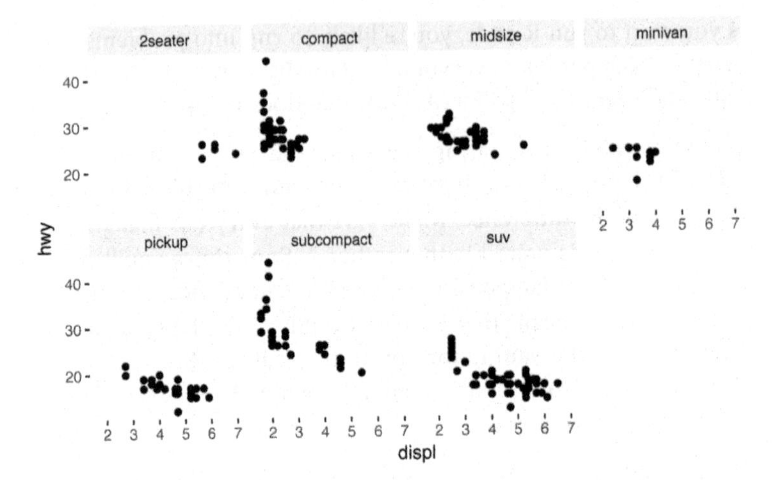

To facet your plot by a single variable, use `facet_wrap()`. The first argument of `facet_wrap()` should be a formula, which you create with ~ followed by a variable name (here "formula" is the name of a data structure in R, not a synonym for "equation"). The variable that you pass to `facet_wrap()` should be discrete:

```
ggplot(data = mpg) +
  geom_point(mapping = aes(x = displ, y = hwy)) +
  facet_wrap(~ class, nrow = 2)
```

To facet your plot on the combination of two variables, add `facet_grid()` to your plot call. The first argument of `facet_grid()` is also a formula. This time the formula should contain two variable names separated by a ~:

```
ggplot(data = mpg) +
  geom_point(mapping = aes(x = displ, y = hwy)) +
  facet_grid(drv ~ cyl)
```

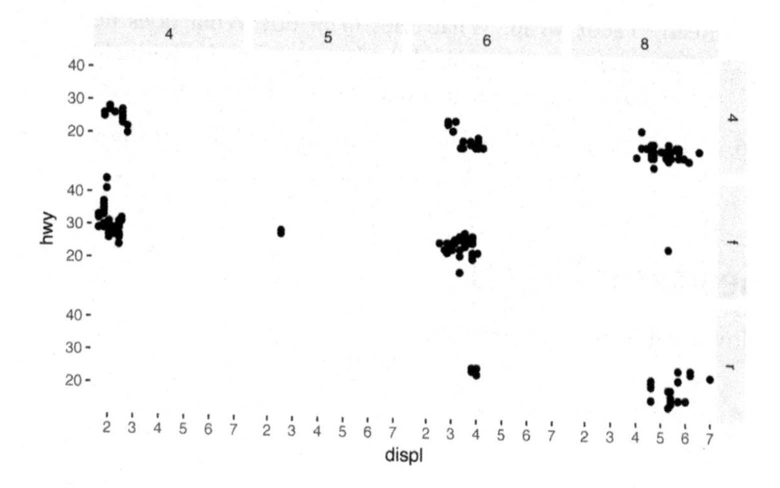

If you prefer to not facet in the rows or columns dimension, use a . instead of a variable name, e.g., + `facet_grid(. ~ cyl)`.

Exercises

1. What happens if you facet on a continuous variable?
2. What do the empty cells in a plot with `facet_grid(drv ~ cyl)` mean? How do they relate to this plot?

   ```
   ggplot(data = mpg) +
     geom_point(mapping = aes(x = drv, y = cyl))
   ```

3. What plots does the following code make? What does `.` do?

   ```
   ggplot(data = mpg) +
     geom_point(mapping = aes(x = displ, y = hwy)) +
     facet_grid(drv ~ .)

   ggplot(data = mpg) +
     geom_point(mapping = aes(x = displ, y = hwy)) +
     facet_grid(. ~ cyl)
   ```

4. Take the first faceted plot in this section:

   ```
   ggplot(data = mpg) +
     geom_point(mapping = aes(x = displ, y = hwy)) +
     facet_wrap(~ class, nrow = 2)
   ```

   What are the advantages to using faceting instead of the color aesthetic? What are the disadvantages? How might the balance change if you had a larger dataset?

5. Read `?facet_wrap`. What does `nrow` do? What does `ncol` do? What other options control the layout of the individual panels? Why doesn't `facet_grid()` have `nrow` and `ncol` variables?

6. When using `facet_grid()` you should usually put the variable with more unique levels in the columns. Why?

Geometric Objects

How are these two plots similar?

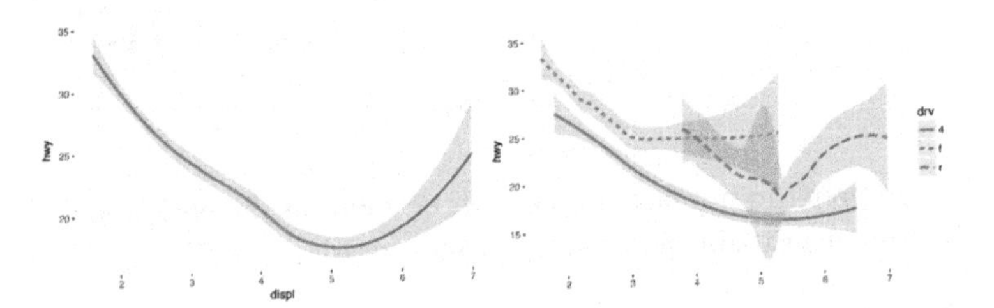

Both plots contain the same `x` variable and the same `y` variable, and both describe the same data. But the plots are not identical. Each plot uses a different visual object to represent the data. In **ggplot2** syntax, we say that they use different *geoms*.

A *geom* is the geometrical object that a plot uses to represent data. People often describe plots by the type of geom that the plot uses. For example, bar charts use bar geoms, line charts use line geoms, boxplots use boxplot geoms, and so on. Scatterplots break the trend; they use the point geom. As we see in the preceding plots, you can use different geoms to plot the same data. The plot on the left uses the point geom, and the plot on the right uses the smooth geom, a smooth line fitted to the data.

To change the geom in your plot, change the geom function that you add to `ggplot()`. For instance, to make the preceding plots, you can use this code:

```
# left
ggplot(data = mpg) +
  geom_point(mapping = aes(x = displ, y = hwy))

# right
ggplot(data = mpg) +
  geom_smooth(mapping = aes(x = displ, y = hwy))
```

Every geom function in **ggplot2** takes a `mapping` argument. However, not every aesthetic works with every geom. You could set the shape of a point, but you couldn't set the "shape" of a line. On the other hand, you *could* set the linetype of a line. `geom_smooth()` will draw a different line, with a different linetype, for each unique value of the variable that you map to linetype:

```
ggplot(data = mpg) +
  geom_smooth(mapping = aes(x = displ, y = hwy, linetype = drv))
```

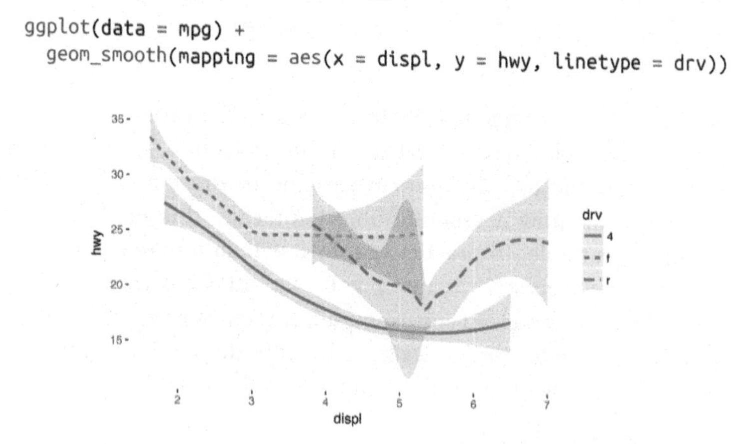

Here `geom_smooth()` separates the cars into three lines based on their `drv` value, which describes a car's drivetrain. One line describes all of the points with a `4` value, one line describes all of the points with an `f` value, and one line describes all of the points with an `r` value. Here, `4` stands for four-wheel drive, `f` for front-wheel drive, and `r` for rear-wheel drive.

If this sounds strange, we can make it more clear by overlaying the lines on top of the raw data and then coloring everything according to `drv`.

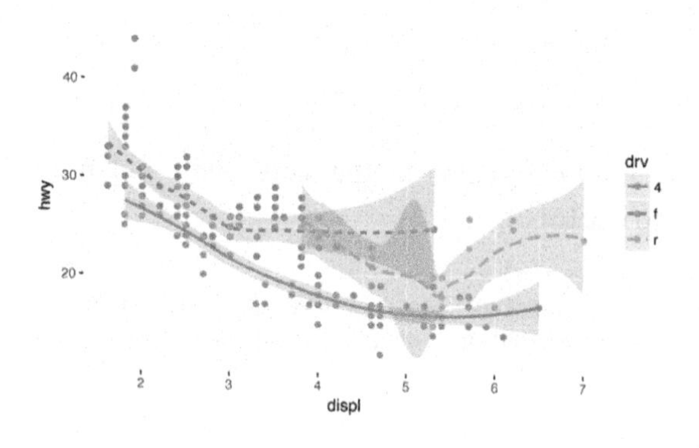

Notice that this plot contains two geoms in the same graph! If this makes you excited, buckle up. In the next section, we will learn how to place multiple geoms in the same plot.

ggplot2 provides over 30 geoms, and extension packages provide even more (see *https://www.ggplot2-exts.org* for a sampling). The best way to get a comprehensive overview is the **ggplot2** cheatsheet, which you can find at *http://rstudio.com/cheatsheets*. To learn more about any single geom, use help: `?geom_smooth`.

Many geoms, like `geom_smooth()`, use a single geometric object to display multiple rows of data. For these geoms, you can set the `group` aesthetic to a categorical variable to draw multiple objects. **ggplot2** will draw a separate object for each unique value of the grouping variable. In practice, **ggplot2** will automatically group the data for these geoms whenever you map an aesthetic to a discrete variable (as in the `linetype` example). It is convenient to rely on this feature because the group aesthetic by itself does not add a legend or distinguishing features to the geoms:

```
ggplot(data = mpg) +
  geom_smooth(mapping = aes(x = displ, y = hwy))

ggplot(data = mpg) +
  geom_smooth(mapping = aes(x = displ, y = hwy, group = drv))

ggplot(data = mpg) +
  geom_smooth(
    mapping = aes(x = displ, y = hwy, color = drv),
    show.legend = FALSE
  )
```

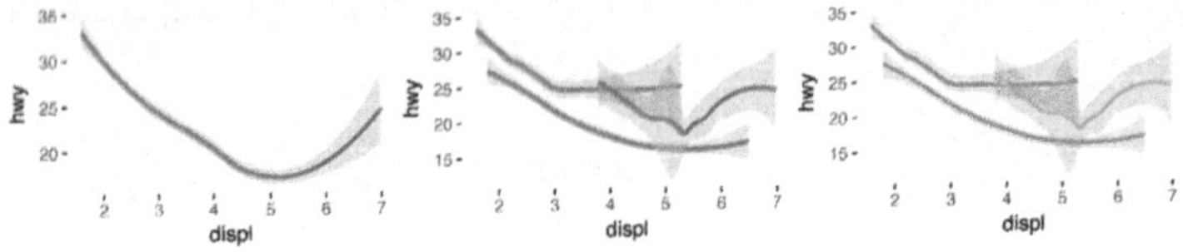

To display multiple geoms in the same plot, add multiple geom functions to `ggplot()`:

```
ggplot(data = mpg) +
  geom_point(mapping = aes(x = displ, y = hwy)) +
  geom_smooth(mapping = aes(x = displ, y = hwy))
```

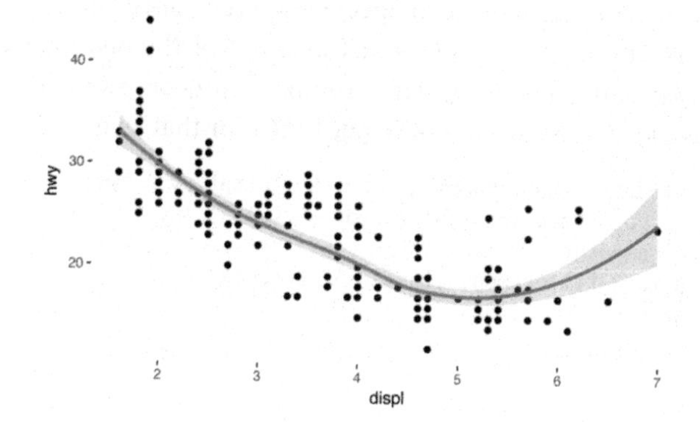

This, however, introduces some duplication in our code. Imagine if you wanted to change the y-axis to display cty instead of hwy. You'd need to change the variable in two places, and you might forget to update one. You can avoid this type of repetition by passing a set of mappings to ggplot(). **ggplot2** will treat these mappings as global mappings that apply to each geom in the graph. In other words, this code will produce the same plot as the previous code:

```
ggplot(data = mpg, mapping = aes(x = displ, y = hwy)) +
  geom_point() +
  geom_smooth()
```

If you place mappings in a geom function, **ggplot2** will treat them as local mappings for the layer. It will use these mappings to extend or overwrite the global mappings *for that layer only*. This makes it possible to display different aesthetics in different layers:

```
ggplot(data = mpg, mapping = aes(x = displ, y = hwy)) +
  geom_point(mapping = aes(color = class)) +
  geom_smooth()
```

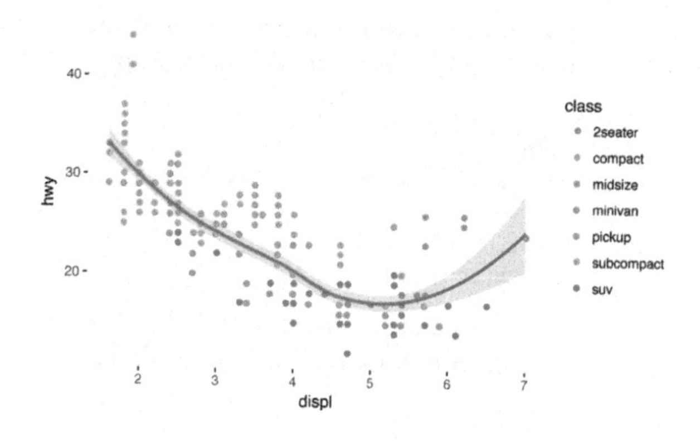

You can use the same idea to specify different `data` for each layer. Here, our smooth line displays just a subset of the `mpg` dataset, the subcompact cars. The local data argument in `geom_smooth()` overrides the global data argument in `ggplot()` for that layer only:

```
ggplot(data = mpg, mapping = aes(x = displ, y = hwy)) +
  geom_point(mapping = aes(color = class)) +
  geom_smooth(
    data = filter(mpg, class == "subcompact"),
    se = FALSE
  )
```

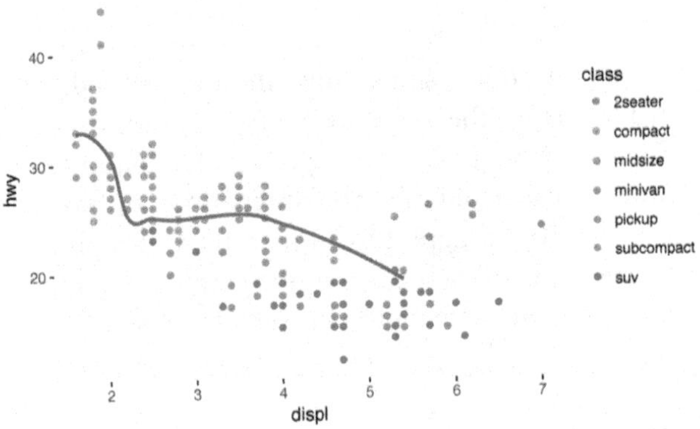

(You'll learn how `filter()` works in the next chapter: for now, just know that this command selects only the subcompact cars.)

Exercises

1. What geom would you use to draw a line chart? A boxplot? A histogram? An area chart?
2. Run this code in your head and predict what the output will look like. Then, run the code in R and check your predictions:

```
ggplot(
  data = mpg,
  mapping = aes(x = displ, y = hwy, color = drv)
) +
  geom_point() +
  geom_smooth(se = FALSE)
```

3. What does `show.legend = FALSE` do? What happens if you remove it? Why do you think I used it earlier in the chapter?
4. What does the `se` argument to `geom_smooth()` do?

5. Will these two graphs look different? Why/why not?

```
ggplot(data = mpg, mapping = aes(x = displ, y = hwy)) +
  geom_point() +
  geom_smooth()

ggplot() +
  geom_point(
    data = mpg,
    mapping = aes(x = displ, y = hwy)
  ) +
  geom_smooth(
    data = mpg,
    mapping = aes(x = displ, y = hwy)
  )
```

6. Re-create the R code necessary to generate the following graphs.

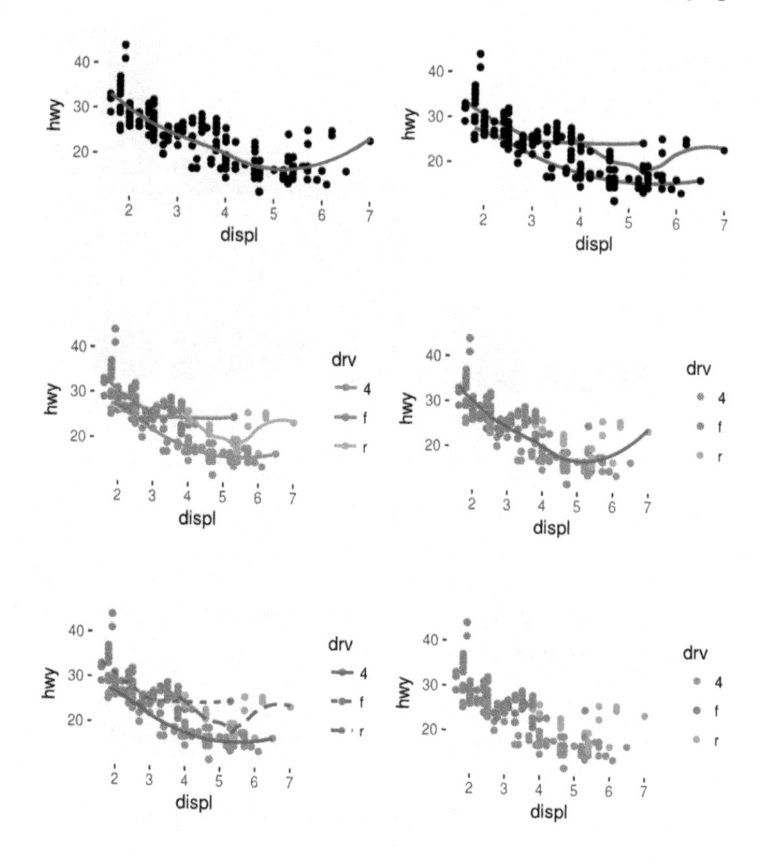

Statistical Transformations

Next, let's take a look at a bar chart. Bar charts seem simple, but they are interesting because they reveal something subtle about plots. Consider a basic bar chart, as drawn with `geom_bar()`. The following chart displays the total number of diamonds in the `diamonds` dataset, grouped by `cut`. The `diamonds` dataset comes in **ggplot2** and contains information about ~54,000 diamonds, including the `price`, `carat`, `color`, `clarity`, and `cut` of each diamond. The chart shows that more diamonds are available with high-quality cuts than with low quality cuts:

```
ggplot(data = diamonds) +
  geom_bar(mapping = aes(x = cut))
```

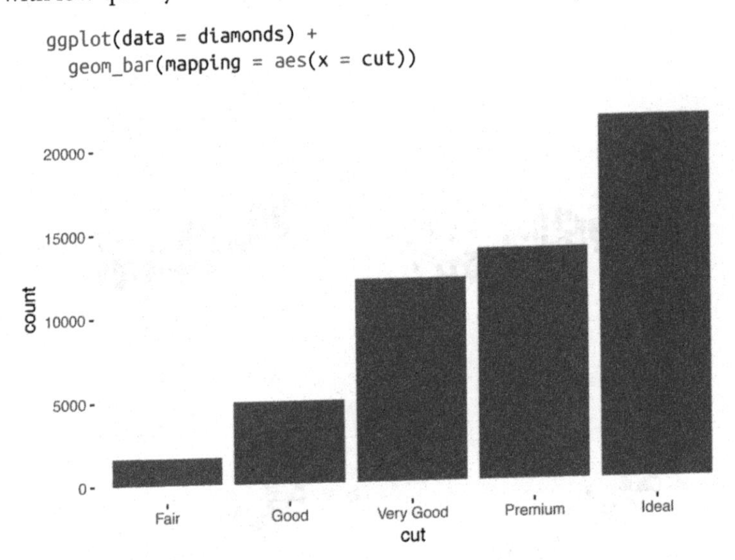

On the x-axis, the chart displays `cut`, a 'variable from `diamonds`. On the y-axis, it displays `count`, but `count` is not a variable in `diamonds`! Where does `count` come from? Many graphs, like scatterplots, plot the raw values of your dataset. Other graphs, like bar charts, calculate new values to plot:

- Bar charts, histograms, and frequency polygons bin your data and then plot bin counts, the number of points that fall in each bin.
- Smoothers fit a model to your data and then plot predictions from the model.

- Boxplots compute a robust summary of the distribution and display a specially formatted box.

The algorithm used to calculate new values for a graph is called a *stat*, short for statistical transformation. The following figure describes how this process works with `geom_bar()`.

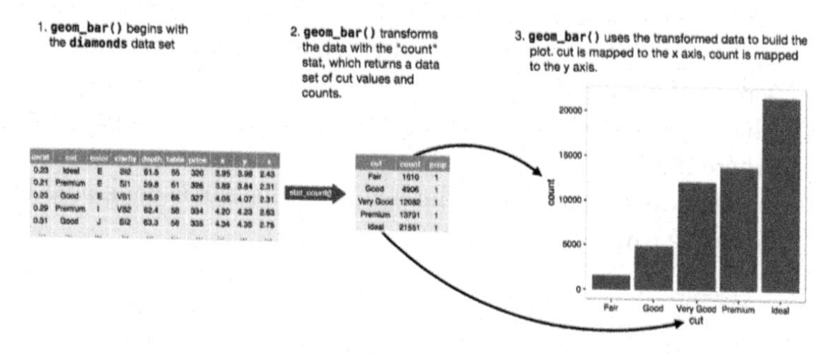

You can learn which stat a geom uses by inspecting the default value for the `stat` argument. For example, `?geom_bar` shows the default value for `stat` is "count," which means that `geom_bar()` uses `stat_count()`. `stat_count()` is documented on the same page as `geom_bar()`, and if you scroll down you can find a section called "Computed variables." That tells that it computes two new variables: `count` and `prop`.

You can generally use geoms and stats interchangeably. For example, you can re-create the previous plot using `stat_count()` instead of `geom_bar()`:

```
ggplot(data = diamonds) +
  stat_count(mapping = aes(x = cut))
```

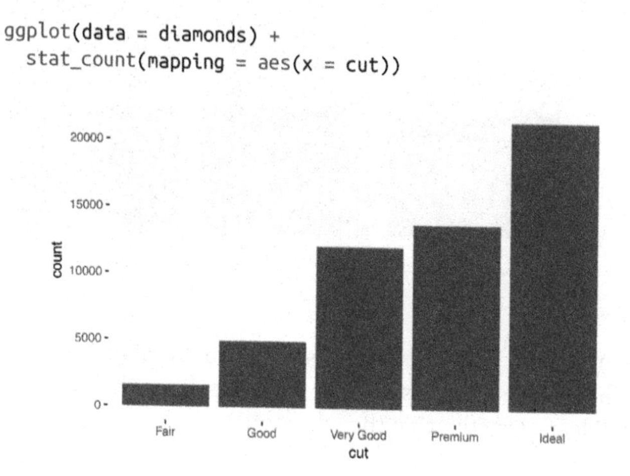

This works because every geom has a default stat, and every stat has a default geom. This means that you can typically use geoms without worrying about the underlying statistical transformation. There are three reasons you might need to use a stat explicitly:

- You might want to override the default stat. In the following code, I change the stat of `geom_bar()` from count (the default) to identity. This lets me map the height of the bars to the raw values of a *y* variable. Unfortunately when people talk about bar charts casually, they might be referring to this type of bar chart, where the height of the bar is already present in the data, or the previous bar chart where the height of the bar is generated by counting rows.

```
demo <- tribble(
  ~a,      ~b,
  "bar_1", 20,
  "bar_2", 30,
  "bar_3", 40
)

ggplot(data = demo) +
  geom_bar(
    mapping = aes(x = a, y = b), stat = "identity"
  )
```

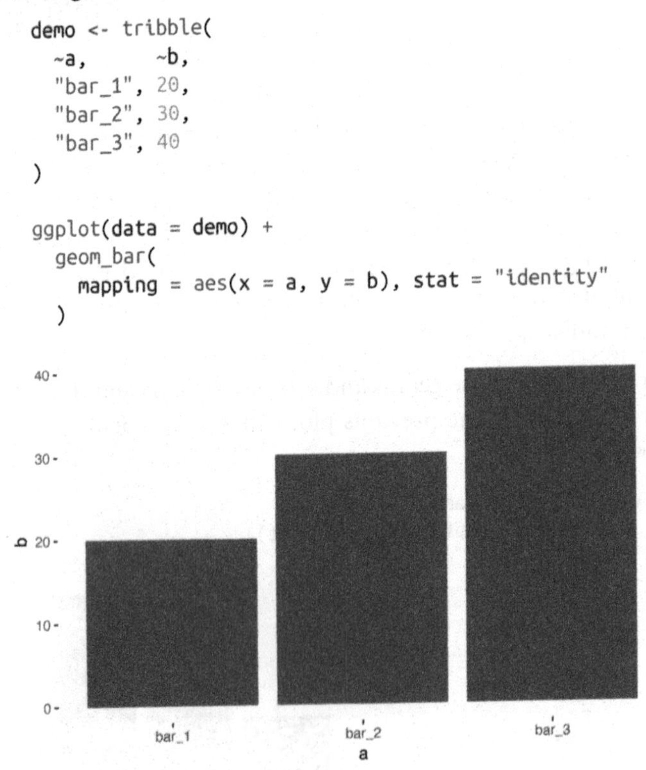

(Don't worry that you haven't seen `<-` or `tibble()` before. You might be able to guess at their meaning from the context, and you'll learn exactly what they do soon!)

- You might want to override the default mapping from transformed variables to aesthetics. For example, you might want to display a bar chart of proportion, rather than count:

```
ggplot(data = diamonds) +
  geom_bar(
    mapping = aes(x = cut, y = ..prop.., group = 1)
  )
```

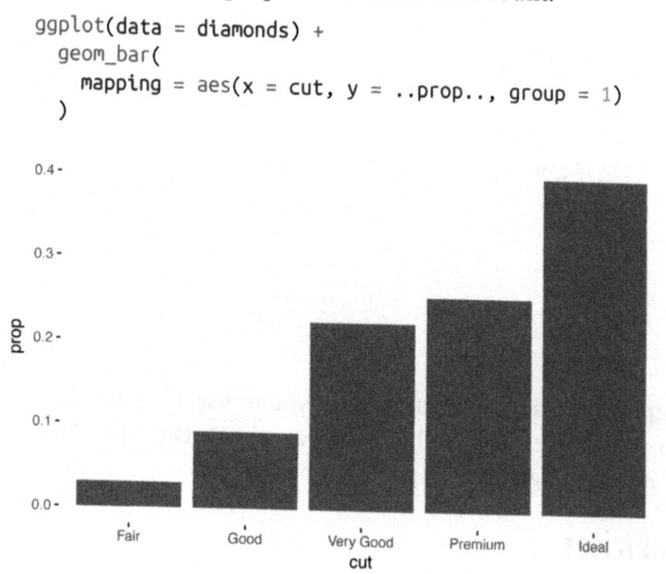

To find the variables computed by the stat, look for the help section titled "Computed variables."

- You might want to draw greater attention to the statistical transformation in your code. For example, you might use `stat_summary()`, which summarizes the y values for each unique x value, to draw attention to the summary that you're computing:

```
ggplot(data = diamonds) +
  stat_summary(
    mapping = aes(x = cut, y = depth),
    fun.ymin = min,
    fun.ymax = max,
    fun.y = median
  )
```

ggplot2 provides over 20 stats for you to use. Each stat is a function, so you can get help in the usual way, e.g., `?stat_bin`. To see a complete list of stats, try the **ggplot2** cheatsheet.

Exercises

1. What is the default geom associated with `stat_summary()`? How could you rewrite the previous plot to use that geom function instead of the stat function?
2. What does `geom_col()` do? How is it different to `geom_bar()`?
3. Most geoms and stats come in pairs that are almost always used in concert. Read through the documentation and make a list of all the pairs. What do they have in common?
4. What variables does `stat_smooth()` compute? What parameters control its behavior?
5. In our proportion bar chart, we need to set `group = 1`. Why? In other words what is the problem with these two graphs?

```
ggplot(data = diamonds) +
  geom_bar(mapping = aes(x = cut, y = ..prop..))
ggplot(data = diamonds) +
  geom_bar(
    mapping = aes(x = cut, fill = color, y = ..prop..)
  )
```

Position Adjustments

There's one more piece of magic associated with bar charts. You can color a bar chart using either the `color` aesthetic, or more usefully, `fill`:

```
ggplot(data = diamonds) +
  geom_bar(mapping = aes(x = cut, color = cut))
ggplot(data = diamonds) +
  geom_bar(mapping = aes(x = cut, fill = cut))
```

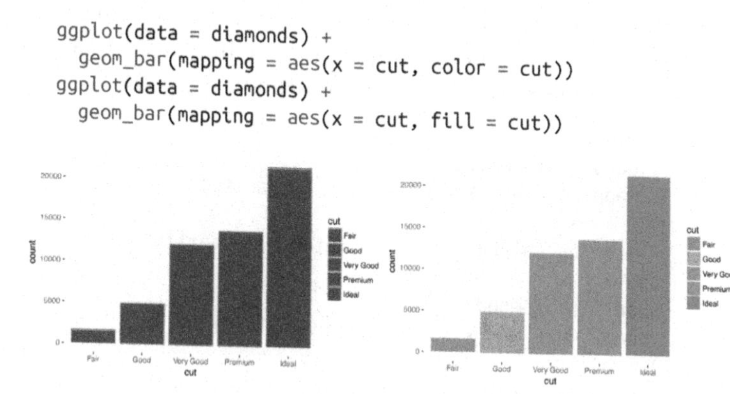

Note what happens if you map the `fill` aesthetic to another variable, like `clarity`: the bars are automatically stacked. Each colored rectangle represents a combination of `cut` and `clarity`:

```
ggplot(data = diamonds) +
  geom_bar(mapping = aes(x = cut, fill = clarity))
```

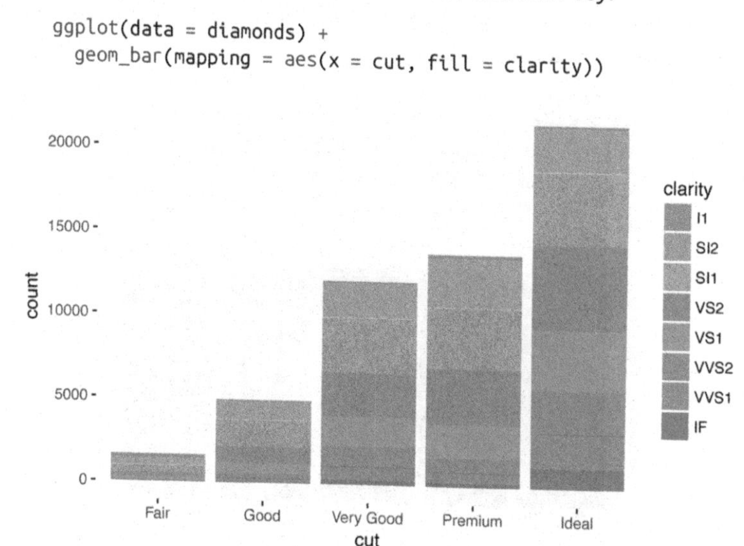

The stacking is performed automatically by the *position adjustment* specified by the `position` argument. If you don't want a stacked bar chart, you can use one of three other options: `"identity"`, `"dodge"` or `"fill"`:

- `position = "identity"` will place each object exactly where it falls in the context of the graph. This is not very useful for bars, because it overlaps them. To see that overlapping we either need to make the bars slightly transparent by setting `alpha` to a small value, or completely transparent by setting `fill = NA`:

```
ggplot(
  data = diamonds,
  mapping = aes(x = cut, fill = clarity)
) +
  geom_bar(alpha = 1/5, position = "identity")
ggplot(
  data = diamonds,
  mapping = aes(x = cut, color = clarity)
) +
  geom_bar(fill = NA, position = "identity")
```

  The identity position adjustment is more useful for 2D geoms, like points, where it is the default.

- `position = "fill"` works like stacking, but makes each set of stacked bars the same height. This makes it easier to compare proportions across groups:

```
ggplot(data = diamonds) +
  geom_bar(
    mapping = aes(x = cut, fill = clarity),
    position = "fill"
  )
```

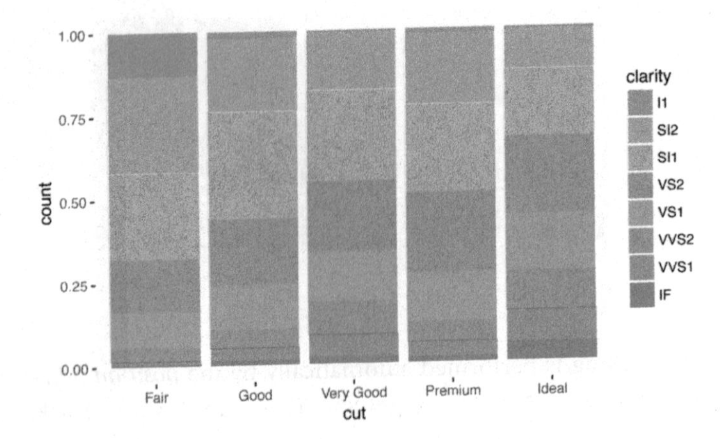

- `position = "dodge"` places overlapping objects directly *beside* one another. This makes it easier to compare individual values:

```
ggplot(data = diamonds) +
  geom_bar(
    mapping = aes(x = cut, fill = clarity),
    position = "dodge"
  )
```

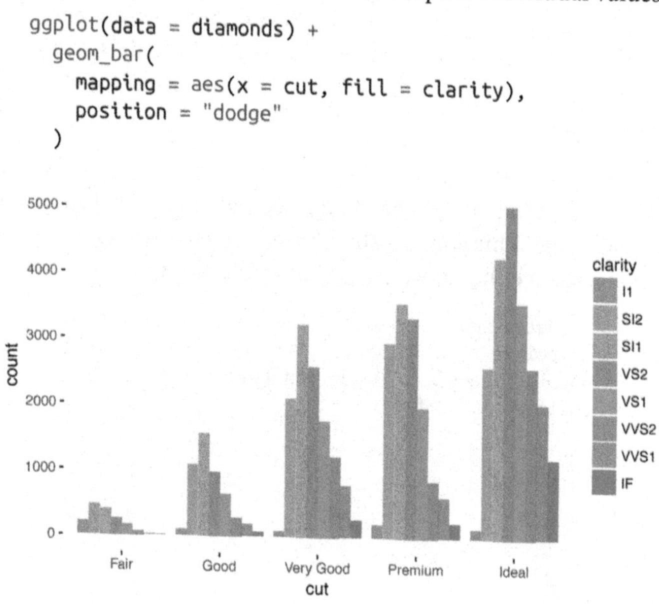

There's one other type of adjustment that's not useful for bar charts, but it can be very useful for scatterplots. Recall our first scatterplot. Did you notice that the plot displays only 126 points, even though there are 234 observations in the dataset?

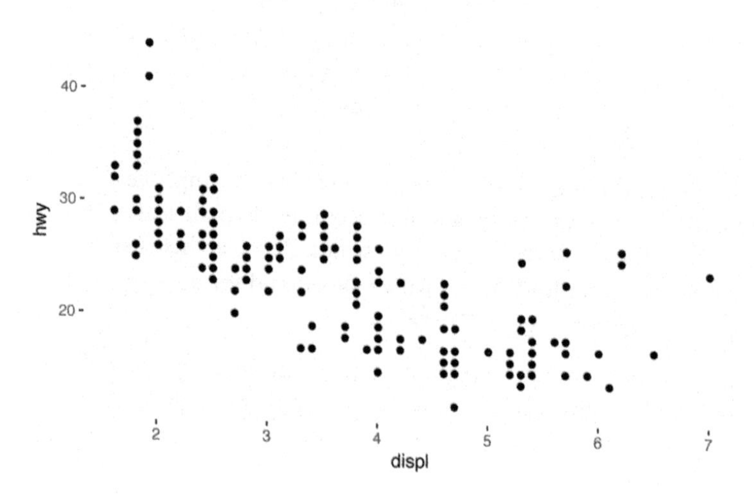

The values of `hwy` and `displ` are rounded so the points appear on a grid and many points overlap each other. This problem is known as *overplotting*. This arrangement makes it hard to see where the mass of the data is. Are the data points spread equally throughout the graph, or is there one special combination of `hwy` and `displ` that contains 109 values?

You can avoid this gridding by setting the position adjustment to "jitter." `position = "jitter"` adds a small amount of random noise to each point. This spreads the points out because no two points are likely to receive the same amount of random noise:

```
ggplot(data = mpg) +
  geom_point(
    mapping = aes(x = displ, y = hwy),
    position = "jitter"
  )
```

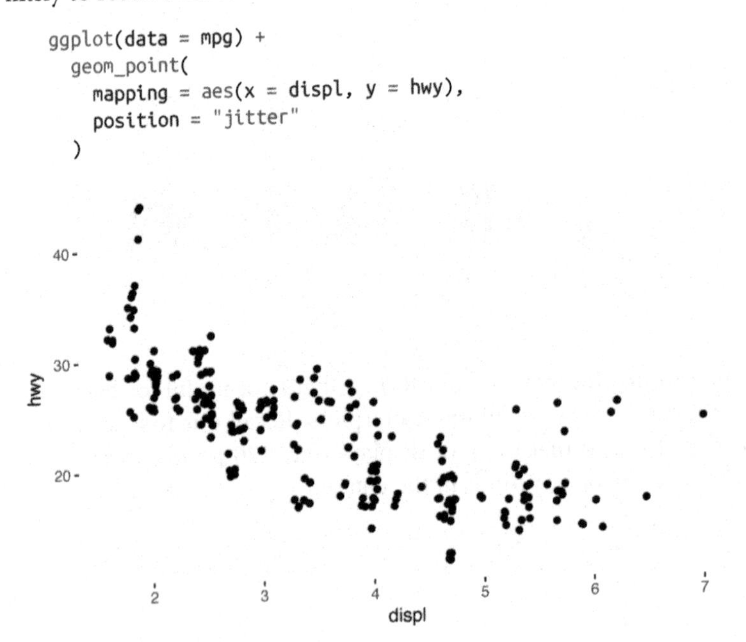

Adding randomness seems like a strange way to improve your plot, but while it makes your graph less accurate at small scales, it makes your graph *more* revealing at large scales. Because this is such a useful operation, **ggplot2** comes with a shorthand for `geom_point(position = "jitter")`: `geom_jitter()`.

To learn more about a position adjustment, look up the help page associated with each adjustment: `?position_dodge`, `?position_fill`, `?position_identity`, `?position_jitter`, and `?position_stack`.

Exercises

1. What is the problem with this plot? How could you improve it?

```
ggplot(data = mpg, mapping = aes(x = cty, y = hwy)) +
  geom_point()
```

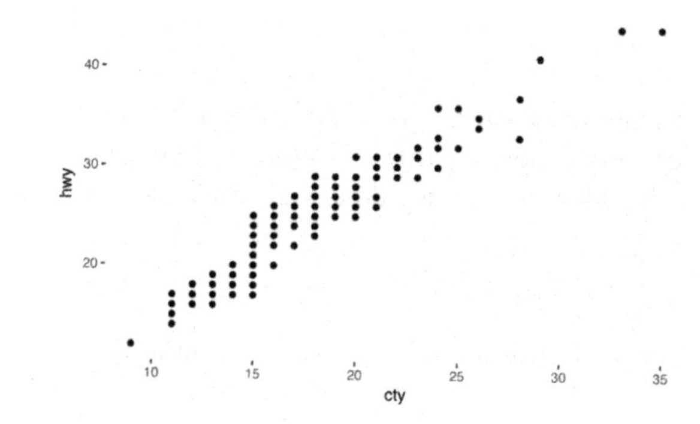

2. What parameters to `geom_jitter()` control the amount of jittering?
3. Compare and contrast `geom_jitter()` with `geom_count()`.
4. What's the default position adjustment for `geom_boxplot()`? Create a visualization of the `mpg` dataset that demonstrates it.

Coordinate Systems

Coordinate systems are probably the most complicated part of **ggplot2**. The default coordinate system is the Cartesian coordinate system where the x and y position act independently to find the location of each point. There are a number of other coordinate systems that are occasionally helpful:

- `coord_flip()` switches the x- and y-axes. This is useful (for example) if you want horizontal boxplots. It's also useful for long labels—it's hard to get them to fit without overlapping on the x-axis:

```
ggplot(data = mpg, mapping = aes(x = class, y = hwy)) +
  geom_boxplot()
ggplot(data = mpg, mapping = aes(x = class, y = hwy)) +
  geom_boxplot() +
  coord_flip()
```

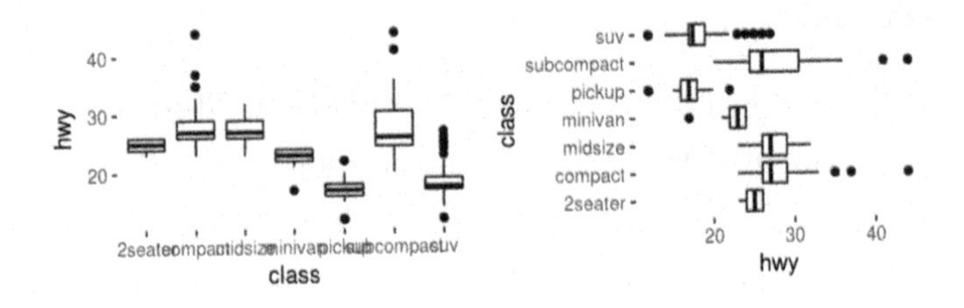

- `coord_quickmap()` sets the aspect ratio correctly for maps. This is very important if you're plotting spatial data with **ggplot2** (which unfortunately we don't have the space to cover in this book):

```
nz <- map_data("nz")

ggplot(nz, aes(long, lat, group = group)) +
  geom_polygon(fill = "white", color = "black")

ggplot(nz, aes(long, lat, group = group)) +
  geom_polygon(fill = "white", color = "black") +
  coord_quickmap()
```

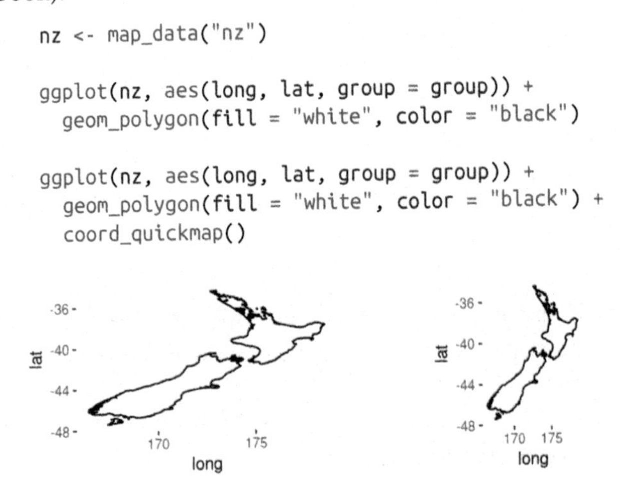

- `coord_polar()` uses polar coordinates. Polar coordinates reveal an interesting connection between a bar chart and a Coxcomb chart:

```
bar <- ggplot(data = diamonds) +
  geom_bar(
    mapping = aes(x = cut, fill = cut),
    show.legend = FALSE,
    width = 1
  ) +
  theme(aspect.ratio = 1) +
  labs(x = NULL, y = NULL)

bar + coord_flip()
bar + coord_polar()
```

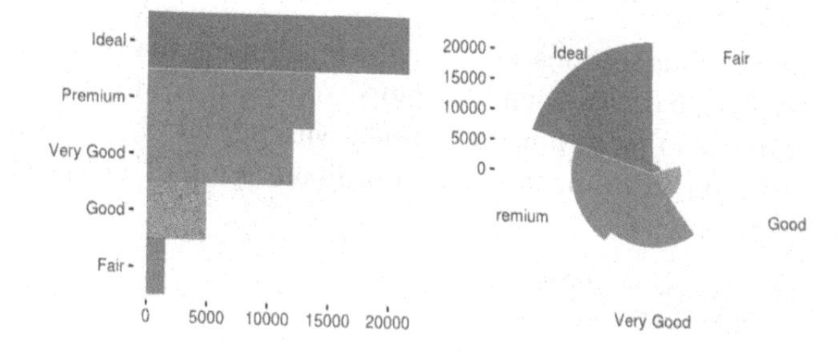

Exercises

1. Turn a stacked bar chart into a pie chart using `coord_polar()`.
2. What does `labs()` do? Read the documentation.
3. What's the difference between `coord_quickmap()` and `coord_map()`?
4. What does the following plot tell you about the relationship between city and highway mpg? Why is `coord_fixed()` important? What does `geom_abline()` do?

```
ggplot(data = mpg, mapping = aes(x = cty, y = hwy)) +
  geom_point() +
  geom_abline() +
  coord_fixed()
```

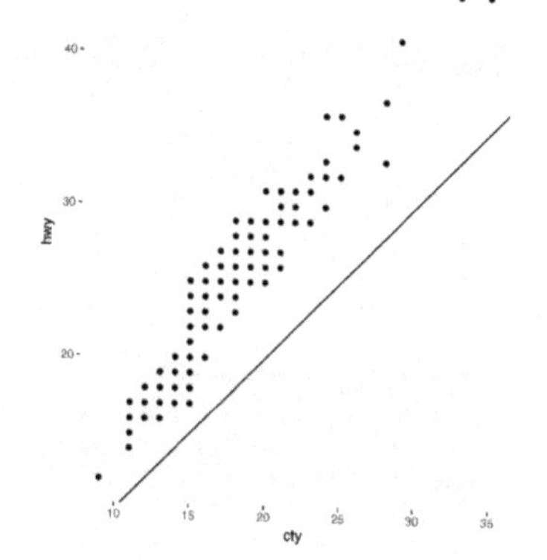

The Layered Grammar of Graphics

In the previous sections, you learned much more than how to make scatterplots, bar charts, and boxplots. You learned a foundation that you can use to make *any* type of plot with **ggplot2**. To see this, let's add position adjustments, stats, coordinate systems, and faceting to our code template:

```
ggplot(data = <DATA>) +
  <GEOM_FUNCTION>(
     mapping = aes(<MAPPINGS>),
     stat = <STAT>,
     position = <POSITION>
  ) +
  <COORDINATE_FUNCTION> +
  <FACET_FUNCTION>
```

Our new template takes seven parameters, the bracketed words that appear in the template. In practice, you rarely need to supply all seven parameters to make a graph because **ggplot2** will provide useful defaults for everything except the data, the mappings, and the geom function.

The seven parameters in the template compose the grammar of graphics, a formal system for building plots. The grammar of graphics is based on the insight that you can uniquely describe *any* plot as a combination of a dataset, a geom, a set of mappings, a stat, a position adjustment, a coordinate system, and a faceting scheme.

To see how this works, consider how you could build a basic plot from scratch: you could start with a dataset and then transform it into the information that you want to display (with a stat):

1. Begin with the **diamonds** data set

2. Compute counts for each cut value with **stat_count()**.

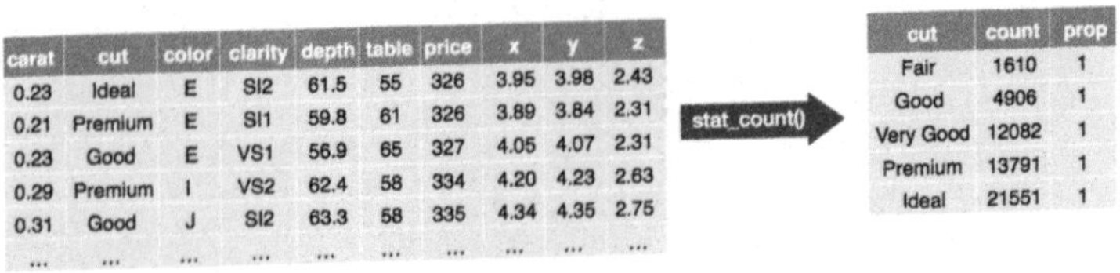

carat	cut	color	clarity	depth	table	price	x	y	z
0.23	Ideal	E	SI2	61.5	55	326	3.95	3.98	2.43
0.21	Premium	E	SI1	59.8	61	326	3.89	3.84	2.31
0.23	Good	E	VS1	56.9	65	327	4.05	4.07	2.31
0.29	Premium	I	VS2	62.4	58	334	4.20	4.23	2.63
0.31	Good	J	SI2	63.3	58	335	4.34	4.35	2.75
...	...	...	...	...	...	...	...	...	...

stat_count()

cut	count	prop
Fair	1610	1
Good	4906	1
Very Good	12082	1
Premium	13791	1
Ideal	21551	1

Next, you could choose a geometric object to represent each observation in the transformed data. You could then use the aesthetic properties of the geoms to represent variables in the data. You would map the values of each variable to the levels of an aesthetic:

3. Represent each observation with a bar.

4. Map the `fill` of each bar to the `..count..` variable.

carat	cut	color	clarity	depth	table	price	x	y	z
0.23	Ideal	E	SI2	61.5	55	326	3.95	3.98	2.43
0.21	Premium	E	SI1	59.8	61	326	3.89	3.84	2.31
0.23	Good	E	VS1	56.9	65	327	4.05	4.07	2.31
0.29	Premium	I	VS2	62.4	58	334	4.20	4.23	2.63
0.31	Good	J	SI2	63.3	58	335	4.34	4.35	2.75
...	...	...	...	...	...	...	...	...	...

stat_count()

cut	count	prop
Fair	1610	1
Good	4906	1
Very Good	12082	1
Premium	13791	1
Ideal	21551	1

You'd then select a coordinate system to place the geoms into. You'd use the location of the objects (which is itself an aesthetic property) to display the values of the x and y variables. At that point, you would have a complete graph, but you could further adjust the positions of the geoms within the coordinate system (a position adjustment) or split the graph into subplots (faceting). You could also extend the plot by adding one or more additional layers, where each additional layer uses a dataset, a geom, a set of mappings, a stat, and a position adjustment:

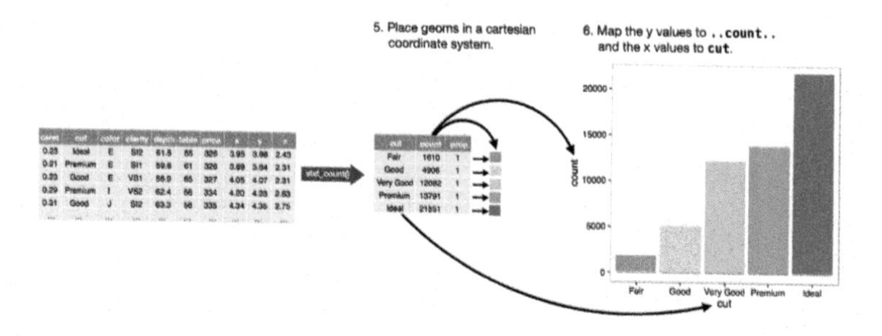

You could use this method to build *any* plot that you imagine. In other words, you can use the code template that you've learned in this chapter to build hundreds of thousands of unique plots.

CHAPTER 2
Workflow: Basics

You now have some experience running R code. I didn't give you many details, but you've obviously figured out the basics, or you would've thrown this book away in frustration! Frustration is natural when you start programming in R, because it is such a stickler for punctuation, and even one character out of place will cause it to complain. But while you should expect to be a little frustrated, take comfort in that it's both typical and temporary: it happens to everyone, and the only way to get over it is to keep trying.

Before we go any further, let's make sure you've got a solid foundation in running R code, and that you know about some of the most helpful RStudio features.

Coding Basics

Let's review some basics we've so far omitted in the interests of getting you plotting as quickly as possible. You can use R as a calculator:

```
1 / 200 * 30
#> [1] 0.15
(59 + 73 + 2) / 3
#> [1] 44.7
sin(pi / 2)
#> [1] 1
```

You can create new objects with `<-`:

```
x <- 3 * 4
```

All R statements where you create objects, *assignment* statements, have the same form:

```
object_name <- value
```

When reading that code say "object name gets value" in your head.

You will make lots of assignments and `<-` is a pain to type. Don't be lazy and use `=`: it will work, but it will cause confusion later. Instead, use RStudio's keyboard shortcut: Alt-- (the minus sign). Notice that RStudio automagically surrounds `<-` with spaces, which is a good code formatting practice. Code is miserable to read on a good day, so giveyoureyesabreak and use spaces.

What's in a Name?

Object names must start with a letter, and can only contain letters, numbers, `_`, and `.`. You want your object names to be descriptive, so you'll need a convention for multiple words. I recommend *snake_case* where you separate lowercase words with `_`:

```
i_use_snake_case
otherPeopleUseCamelCase
some.people.use.periods
And_aFew.People_RENOUNCEconvention
```

We'll come back to code style later, in Chapter 15.

You can inspect an object by typing its name:

```
x
#> [1] 12
```

Make another assignment:

```
this_is_a_really_long_name <- 2.5
```

To inspect this object, try out RStudio's completion facility: type "this," press Tab, add characters until you have a unique prefix, then press Return.

Oops, you made a mistake! `this_is_a_really_long_name` should have value 3.5 not 2.5. Use another keyboard shortcut to help you fix it. Type "this" then press Cmd/Ctrl-↑. That will list all the commands you've typed that start with those letters. Use the arrow keys to navigate, then press Enter to retype the command. Change 2.5 to 3.5 and rerun.

Make yet another assignment:

```
r_rocks <- 2 ^ 3
```

Let's try to inspect it:

```
r_rock
#> Error: object 'r_rock' not found
R_rocks
#> Error: object 'R_rocks' not found
```

There's an implied contract between you and R: it will do the tedious computation for you, but in return, you must be completely precise in your instructions. Typos matter. Case matters.

Calling Functions

R has a large collection of built-in functions that are called like this:

```
function_name(arg1 = val1, arg2 = val2, ...)
```

Let's try using `seq()`, which makes regular *seq*uences of numbers and, while we're at it, learn more helpful features of RStudio. Type `se` and hit Tab. A pop-up shows you possible completions. Specify `seq()` by typing more (a "q") to disambiguate, or by using ↑/↓ arrows to select. Notice the floating tooltip that pops up, reminding you of the function's arguments and purpose. If you want more help, press F1 to get all the details in the help tab in the lower-right pane.

Press Tab once more when you've selected the function you want. RStudio will add matching opening (`(`) and closing (`)`) parentheses for you. Type the arguments `1, 10` and hit Return:

```
seq(1, 10)
#>  [1]  1  2  3  4  5  6  7  8  9 10
```

Type this code and notice similar assistance help with the paired quotation marks:

```
x <- "hello world"
```

Quotation marks and parentheses must always come in a pair. RStudio does its best to help you, but it's still possible to mess up and end up with a mismatch. If this happens, R will show you the continuation character "+":

```
> x <- "hello
+
```

The + tells you that R is waiting for more input; it doesn't think you're done yet. Usually that means you've forgotten either a " or a). Either add the missing pair, or press Esc to abort the expression and try again.

If you make an assignment, you don't get to see the value. You're then tempted to immediately double-check the result:

```
y <- seq(1, 10, length.out = 5)
y
#> [1]  1.00  3.25  5.50  7.75 10.00
```

This common action can be shortened by surrounding the assignment with parentheses, which causes assignment and "print to screen" to happen:

```
(y <- seq(1, 10, length.out = 5))
#> [1]  1.00  3.25  5.50  7.75 10.00
```

Now look at your environment in the upper-right pane:

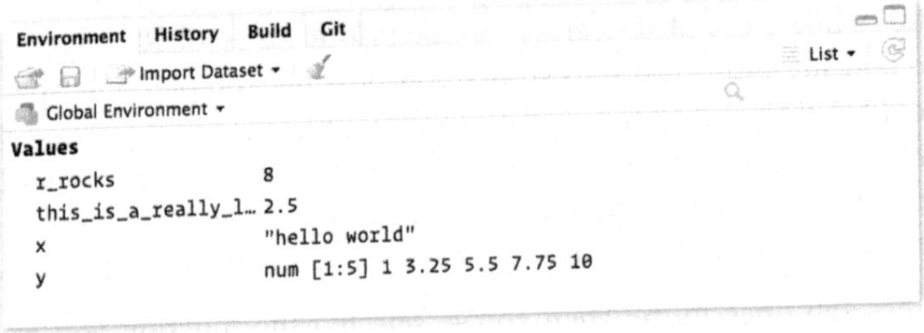

Here you can see all of the objects that you've created.

Exercises

1. Why does this code not work?

   ```
   my_variable <- 10
   my_varıable
   #> Error in eval(expr, envir, enclos):
   #> object 'my_varıable' not found
   ```

   Look carefully! (This may seem like an exercise in pointlessness, but training your brain to notice even the tiniest difference will pay off when programming.)

2. Tweak each of the following R commands so that they run correctly:

```
library(tidyverse)

ggplot(dota = mpg) +
  geom_point(mapping = aes(x = displ, y = hwy))

fliter(mpg, cyl = 8)
filter(diamond, carat > 3)
```

3. Press Alt-Shift-K. What happens? How can you get to the same place using the menus?

CHAPTER 3

Data Transformation with dplyr

Introduction

Visualization is an important tool for insight generation, but it is rare that you get the data in exactly the right form you need. Often you'll need to create some new variables or summaries, or maybe you just want to rename the variables or reorder the observations in order to make the data a little easier to work with. You'll learn how to do all that (and more!) in this chapter, which will teach you how to transform your data using the **dplyr** package and a new dataset on flights departing New York City in 2013.

Prerequisites

In this chapter we're going to focus on how to use the **dplyr** package, another core member of the tidyverse. We'll illustrate the key ideas using data from the **nycflights13** package, and use **ggplot2** to help us understand the data.

```
library(nycflights13)
library(tidyverse)
```

Take careful note of the conflicts message that's printed when you load the tidyverse. It tells you that **dplyr** overwrites some functions in base R. If you want to use the base version of these functions after loading **dplyr**, you'll need to use their full names: `stats::filter()` and `stats::lag()`.

nycflights13

To explore the basic data manipulation verbs of **dplyr**, we'll use `nycflights13::flights`. This data frame contains all 336,776 flights that departed from New York City in 2013. The data comes from the US Bureau of Transportation Statistics (*http://bit.ly/transstats*), and is documented in `?flights`:

```
flights
#> # A tibble: 336,776 × 19
#>     year month   day dep_time sched_dep_time dep_delay
#>    <int> <int> <int>    <int>          <int>     <dbl>
#> 1  2013     1     1      517            515         2
#> 2  2013     1     1      533            529         4
#> 3  2013     1     1      542            540         2
#> 4  2013     1     1      544            545        -1
#> 5  2013     1     1      554            600        -6
#> 6  2013     1     1      554            558        -4
#> # ... with 336,776 more rows, and 13 more variables:
#> #   arr_time <int>, sched_arr_time <int>, arr_delay <dbl>,
#> #   carrier <chr>, flight <int>, tailnum <chr>, origin <chr>,
#> #   dest <chr>, air_time <dbl>, distance <dbl>, hour <dbl>,
#> #   minute <dbl>, time_hour <dttm>
```

You might notice that this data frame prints a little differently from other data frames you might have used in the past: it only shows the first few rows and all the columns that fit on one screen. (To see the whole dataset, you can run `View(flights)`, which will open the dataset in the RStudio viewer.) It prints differently because it's a *tibble*. Tibbles are data frames, but slightly tweaked to work better in the tidyverse. For now, you don't need to worry about the differences; we'll come back to tibbles in more detail in Part II.

You might also have noticed the row of three- (or four-) letter abbreviations under the column names. These describe the type of each variable:

- `int` stands for integers.
- `dbl` stands for doubles, or real numbers.
- `chr` stands for character vectors, or strings.
- `dttm` stands for date-times (a date + a time).

There are three other common types of variables that aren't used in this dataset but you'll encounter later in the book:

- `lgl` stands for logical, vectors that contain only `TRUE` or `FALSE`.
- `fctr` stands for factors, which R uses to represent categorical variables with fixed possible values.
- `date` stands for dates.

dplyr Basics

In this chapter you are going to learn the five key **dplyr** functions that allow you to solve the vast majority of your data-manipulation challenges:

- Pick observations by their values (`filter()`).
- Reorder the rows (`arrange()`).
- Pick variables by their names (`select()`).
- Create new variables with functions of existing variables (`mutate()`).
- Collapse many values down to a single summary (`summarize()`).

These can all be used in conjunction with `group_by()`, which changes the scope of each function from operating on the entire dataset to operating on it group-by-group. These six functions provide the verbs for a language of data manipulation.

All verbs work similarly:

1. The first argument is a data frame.
2. The subsequent arguments describe what to do with the data frame, using the variable names (without quotes).
3. The result is a new data frame.

Together these properties make it easy to chain together multiple simple steps to achieve a complex result. Let's dive in and see how these verbs work.

Filter Rows with filter()

`filter()` allows you to subset observations based on their values. The first argument is the name of the data frame. The second and subsequent arguments are the expressions that filter the data frame. For example, we can select all flights on January 1st with:

```
filter(flights, month == 1, day == 1)
#> # A tibble: 842 × 19
#>    year month   day dep_time sched_dep_time dep_delay
#>   <int> <int> <int>    <int>          <int>     <dbl>
#> 1  2013     1     1      517            515         2
#> 2  2013     1     1      533            529         4
#> 3  2013     1     1      542            540         2
#> 4  2013     1     1      544            545        -1
#> 5  2013     1     1      554            600        -6
#> 6  2013     1     1      554            558        -4
#> # ... with 836 more rows, and 13 more variables:
#> #   arr_time <int>, sched_arr_time <int>, arr_delay <dbl>,
#> #   carrier <chr>, flight <int>, tailnum <chr>,origin <chr>,
#> #   dest <chr>, air_time <dbl>, distance <dbl>, hour <dbl>,
#> #   minute <dbl>, time_hour <dttm>
```

When you run that line of code, **dplyr** executes the filtering operation and returns a new data frame. **dplyr** functions never modify their inputs, so if you want to save the result, you'll need to use the assignment operator, <-:

```
jan1 <- filter(flights, month == 1, day == 1)
```

R either prints out the results, or saves them to a variable. If you want to do both, you can wrap the assignment in parentheses:

```
(dec25 <- filter(flights, month == 12, day == 25))
#> # A tibble: 719 × 19
#>    year month   day dep_time sched_dep_time dep_delay
#>   <int> <int> <int>    <int>          <int>     <dbl>
#> 1  2013    12    25      456            500        -4
#> 2  2013    12    25      524            515         9
#> 3  2013    12    25      542            540         2
#> 4  2013    12    25      546            550        -4
#> 5  2013    12    25      556            600        -4
#> 6  2013    12    25      557            600        -3
#> # ... with 713 more rows, and 13 more variables:
#> #   arr_time <int>, sched_arr_time <int>, arr_delay <dbl>,
#> #   carrier <chr>, flight <int>, tailnum <chr>,origin <chr>,
#> #   dest <chr>, air_time <dbl>, distance <dbl>, hour <dbl>,
#> #   minute <dbl>, time_hour <dttm>
```

Comparisons

To use filtering effectively, you have to know how to select the observations that you want using the comparison operators. R provides the standard suite: >, >=, <, <=, != (not equal), and == (equal).

When you're starting out with R, the easiest mistake to make is to use = instead of == when testing for equality. When this happens you'll get an informative error:

```
filter(flights, month = 1)
#> Error: filter() takes unnamed arguments. Do you need `==`?
```

There's another common problem you might encounter when using ==: floating-point numbers. These results might surprise you!

```
sqrt(2) ^ 2 == 2
#> [1] FALSE
1/49 * 49 == 1
#> [1] FALSE
```

Computers use finite precision arithmetic (they obviously can't store an infinite number of digits!) so remember that every number you see is an approximation. Instead of relying on ==, use `near()`:

```
near(sqrt(2) ^ 2,  2)
#> [1] TRUE
near(1 / 49 * 49, 1)
#> [1] TRUE
```

Logical Operators

Multiple arguments to `filter()` are combined with "and": every expression must be true in order for a row to be included in the output. For other types of combinations, you'll need to use Boolean operators yourself: `&` is "and," `|` is "or," and `!` is "not." The following figure shows the complete set of Boolean operations.

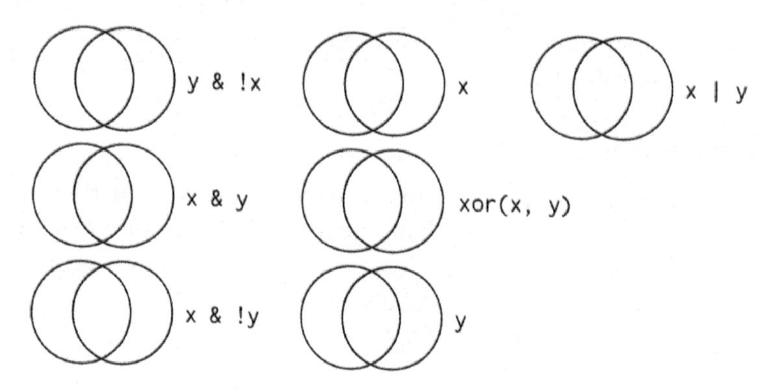

The following code finds all flights that departed in November or December:

```
filter(flights, month == 11 | month == 12)
```

The order of operations doesn't work like English. You can't write `filter(flights, month == 11 | 12)`, which you might literally translate into "finds all flights that departed in November or December." Instead it finds all months that equal `11 | 12`, an expression that evaluates to `TRUE`. In a numeric context (like here), `TRUE` becomes one, so this finds all flights in January, not November or December. This is quite confusing!

A useful shorthand for this problem is `x %in% y`. This will select every row where `x` is one of the values in `y`. We could use it to rewrite the preceding code:

```
nov_dec <- filter(flights, month %in% c(11, 12))
```

Sometimes you can simplify complicated subsetting by remembering De Morgan's law: `!(x & y)` is the same as `!x | !y`, and `!(x | y)` is the same as `!x & !y`. For example, if you wanted to find flights that weren't delayed (on arrival or departure) by more than two hours, you could use either of the following two filters:

```
filter(flights, !(arr_delay > 120 | dep_delay > 120))
filter(flights, arr_delay <= 120, dep_delay <= 120)
```

As well as `&` and `|`, R also has `&&` and `||`. Don't use them here! You'll learn when you should use them in "Conditional Execution" on page 276.

Whenever you start using complicated, multipart expressions in `filter()`, consider making them explicit variables instead. That makes it much easier to check your work. You'll learn how to create new variables shortly.

Missing Values

One important feature of R that can make comparison tricky is missing values, or `NA`s ("not availables"). `NA` represents an unknown value so missing values are "contagious"; almost any operation involving an unknown value will also be unknown:

```
NA > 5
#> [1] NA
10 == NA
#> [1] NA
NA + 10
#> [1] NA
```

```
NA / 2
#> [1] NA
```

The most confusing result is this one:

```
NA == NA
#> [1] NA
```

It's easiest to understand why this is true with a bit more context:

```
# Let x be Mary's age. We don't know how old she is.
x <- NA

# Let y be John's age. We don't know how old he is.
y <- NA

# Are John and Mary the same age?
x == y
#> [1] NA
# We don't know!
```

If you want to determine if a value is missing, use `is.na()`:

```
is.na(x)
#> [1] TRUE
```

`filter()` only includes rows where the condition is `TRUE`; it excludes both `FALSE` and `NA` values. If you want to preserve missing values, ask for them explicitly:

```
df <- tibble(x = c(1, NA, 3))
filter(df, x > 1)
#> # A tibble: 1 × 1
#>       x
#>   <dbl>
#> 1     3
filter(df, is.na(x) | x > 1)
#> # A tibble: 2 × 1
#>       x
#>   <dbl>
#> 1    NA
#> 2     3
```

Exercises

1. Find all flights that:
   a. Had an arrival delay of two or more hours
   b. Flew to Houston (`IAH` or `HOU`)
   c. Were operated by United, American, or Delta

   d. Departed in summer (July, August, and September)
   e. Arrived more than two hours late, but didn't leave late
   f. Were delayed by at least an hour, but made up over 30 minutes in flight
   g. Departed between midnight and 6 a.m. (inclusive)

2. Another useful **dplyr** filtering helper is `between()`. What does it do? Can you use it to simplify the code needed to answer the previous challenges?
3. How many flights have a missing `dep_time`? What other variables are missing? What might these rows represent?
4. Why is `NA ^ 0` not missing? Why is `NA | TRUE` not missing? Why is `FALSE & NA` not missing? Can you figure out the general rule? (`NA * 0` is a tricky counterexample!)

Arrange Rows with arrange()

`arrange()` works similarly to `filter()` except that instead of selecting rows, it changes their order. It takes a data frame and a set of column names (or more complicated expressions) to order by. If you provide more than one column name, each additional column will be used to break ties in the values of preceding columns:

```
arrange(flights, year, month, day)
#> # A tibble: 336,776 × 19
#>    year month   day dep_time sched_dep_time dep_delay
#>   <int> <int> <int>    <int>          <int>     <dbl>
#> 1  2013     1     1      517            515         2
#> 2  2013     1     1      533            529         4
#> 3  2013     1     1      542            540         2
#> 4  2013     1     1      544            545        -1
#> 5  2013     1     1      554            600        -6
#> 6  2013     1     1      554            558        -4
#> # ... with 3.368e+05 more rows, and 13 more variables:
#> #   arr_time <int>, sched_arr_time <int>, arr_delay <dbl>,
#> #   carrier <chr>, flight <int>, tailnum <chr>, origin <chr>,
#> #   dest <chr>, air_time <dbl>, distance <dbl>, hour <dbl>,
#> #   minute <dbl>, time_hour <dttm>
```

Use `desc()` to reorder by a column in descending order:

```
arrange(flights, desc(arr_delay))
#> # A tibble: 336,776 × 19
#>    year month   day dep_time sched_dep_time dep_delay
```

```
#>    <int> <int> <int>    <int>          <int>     <dbl>
#> 1  2013     1     9      641            900      1301
#> 2  2013     6    15     1432           1935      1137
#> 3  2013     1    10     1121           1635      1126
#> 4  2013     9    20     1139           1845      1014
#> 5  2013     7    22      845           1600      1005
#> 6  2013     4    10     1100           1900       960
#> # ... with 3.368e+05 more rows, and 13 more variables:
#> #   arr_time <int>, sched_arr_time <int>, arr_delay <dbl>,
#> #   carrier <chr>, flight <int>, tailnum <chr>, origin <chr>,
#> #   dest <chr>, air_time <dbl>, distance <dbl>, hour <dbl>,
#> #   minute <dbl>, time_hour <dttm>,
```

Missing values are always sorted at the end:

```
df <- tibble(x = c(5, 2, NA))
arrange(df, x)
#> # A tibble: 3 × 1
#>       x
#>   <dbl>
#> 1     2
#> 2     5
#> 3    NA
arrange(df, desc(x))
#> # A tibble: 3 × 1
#>       x
#>   <dbl>
#> 1     5
#> 2     2
#> 3    NA
```

Exercises

1. How could you use `arrange()` to sort all missing values to the start? (Hint: use `is.na()`.)
2. Sort `flights` to find the most delayed flights. Find the flights that left earliest.
3. Sort `flights` to find the fastest flights.
4. Which flights traveled the longest? Which traveled the shortest?

Select Columns with select()

It's not uncommon to get datasets with hundreds or even thousands of variables. In this case, the first challenge is often narrowing in on the variables you're actually interested in. `select()` allows you to rapidly zoom in on a useful subset using operations based on the names of the variables.

`select()` is not terribly useful with the flight data because we only have 19 variables, but you can still get the general idea:

```
# Select columns by name
select(flights, year, month, day)
#> # A tibble: 336,776 × 3
#>    year month   day
#>   <int> <int> <int>
#> 1  2013     1     1
#> 2  2013     1     1
#> 3  2013     1     1
#> 4  2013     1     1
#> 5  2013     1     1
#> 6  2013     1     1
#> # ... with 3.368e+05 more rows

# Select all columns between year and day (inclusive)
select(flights, year:day)
#> # A tibble: 336,776 × 3
#>    year month   day
#>   <int> <int> <int>
#> 1  2013     1     1
#> 2  2013     1     1
#> 3  2013     1     1
#> 4  2013     1     1
#> 5  2013     1     1
#> 6  2013     1     1
#> # ... with 3.368e+05 more rows

# Select all columns except those from year to day (inclusive)
select(flights, -(year:day))
#> # A tibble: 336,776 × 16
#>   dep_time sched_dep_time dep_delay arr_time sched_arr_time
#>      <int>          <int>     <dbl>    <int>          <int>
#> 1      517            515         2      830            819
#> 2      533            529         4      850            830
#> 3      542            540         2      923            850
#> 4      544            545        -1     1004           1022
#> 5      554            600        -6      812            837
#> 6      554            558        -4      740            728
#> # ... with 3.368e+05 more rows, and 12 more variables:
#> #   arr_delay <dbl>, carrier <chr>, flight <int>,
#> #   tailnum <chr>, origin <chr>, dest <chr>, air_time <dbl>,
#> #   distance <dbl>,  hour <dbl>, minute <dbl>,
#> #   time_hour <dttm>
```

There are a number of helper functions you can use within `select()`:

- `starts_with("abc")` matches names that begin with "abc".
- `ends_with("xyz")` matches names that end with "xyz".
- `contains("ijk")` matches names that contain "ijk".
- `matches("(.)\\1")` selects variables that match a regular expression. This one matches any variables that contain repeated characters. You'll learn more about regular expressions in Chapter 11.
- `num_range("x", 1:3)` matches `x1`, `x2`, and `x3`.

See `?select` for more details.

`select()` can be used to rename variables, but it's rarely useful because it drops all of the variables not explicitly mentioned. Instead, use `rename()`, which is a variant of `select()` that keeps all the variables that aren't explicitly mentioned:

```
rename(flights, tail_num = tailnum)
#> # A tibble: 336,776 × 19
#>    year month   day dep_time sched_dep_time dep_delay
#>   <int> <int> <int>    <int>          <int>     <dbl>
#> 1  2013     1     1      517            515         2
#> 2  2013     1     1      533            529         4
#> 3  2013     1     1      542            540         2
#> 4  2013     1     1      544            545        -1
#> 5  2013     1     1      554            600        -6
#> 6  2013     1     1      554            558        -4
#> # ... with 3.368e+05 more rows, and 13 more variables:
#> #   arr_time <int>, sched_arr_time <int>, arr_delay <dbl>,
#> #   carrier <chr>, flight <int>, tail_num <chr>,
#> #   origin <chr>, dest <chr>, air_time <dbl>,
#> #   distance <dbl>, hour <dbl>, minute <dbl>,
#> #   time_hour <dttm>
```

Another option is to use `select()` in conjunction with the `everything()` helper. This is useful if you have a handful of variables you'd like to move to the start of the data frame:

```
select(flights, time_hour, air_time, everything())
#> # A tibble: 336,776 × 19
#>             time_hour air_time  year month   day dep_time
#>                <dttm>    <dbl> <int> <int> <int>    <int>
#> 1 2013-01-01 05:00:00      227  2013     1     1      517
#> 2 2013-01-01 05:00:00      227  2013     1     1      533
#> 3 2013-01-01 05:00:00      160  2013     1     1      542
#> 4 2013-01-01 05:00:00      183  2013     1     1      544
#> 5 2013-01-01 06:00:00      116  2013     1     1      554
```

```
#> 6 2013-01-01 05:00:00       150  2013     1     1      554
#> # ... with 3.368e+05 more rows, and 13 more variables:
#> #   sched_dep_time <int>, dep_delay <dbl>, arr_time <int>,
#> #   sched_arr_time <int>, arr_delay <dbl>, carrier <chr>,
#> #   flight <int>, tailnum <chr>, origin <chr>, dest <chr>,
#> #   distance <dbl>, hour <dbl>, minute <dbl>
```

Exercises

1. Brainstorm as many ways as possible to select `dep_time`, `dep_delay`, `arr_time`, and `arr_delay` from `flights`.
2. What happens if you include the name of a variable multiple times in a `select()` call?
3. What does the `one_of()` function do? Why might it be helpful in conjunction with this vector?

   ```
   vars <- c(
     "year", "month", "day", "dep_delay", "arr_delay"
   )
   ```

4. Does the result of running the following code surprise you? How do the select helpers deal with case by default? How can you change that default?

   ```
   select(flights, contains("TIME"))
   ```

Add New Variables with mutate()

Besides selecting sets of existing columns, it's often useful to add new columns that are functions of existing columns. That's the job of `mutate()`.

`mutate()` always adds new columns at the end of your dataset so we'll start by creating a narrower dataset so we can see the new variables. Remember that when you're in RStudio, the easiest way to see all the columns is `View()`:

```
flights_sml <- select(flights,
  year:day,
  ends_with("delay"),
  distance,
  air_time
)
mutate(flights_sml,
  gain = arr_delay - dep_delay,
  speed = distance / air_time * 60
```

```
)
#> # A tibble: 336,776 × 9
#>    year month   day dep_delay arr_delay distance air_time
#>   <int> <int> <int>     <dbl>     <dbl>    <dbl>    <dbl>
#> 1  2013     1     1         2        11     1400      227
#> 2  2013     1     1         4        20     1416      227
#> 3  2013     1     1         2        33     1089      160
#> 4  2013     1     1        -1       -18     1576      183
#> 5  2013     1     1        -6       -25      762      116
#> 6  2013     1     1        -4        12      719      150
#> # ... with 3.368e+05 more rows, and 2 more variables:
#> #   gain <dbl>, speed <dbl>
```

Note that you can refer to columns that you've just created:

```
mutate(flights_sml,
  gain = arr_delay - dep_delay,
  hours = air_time / 60,
  gain_per_hour = gain / hours
)
#> # A tibble: 336,776 × 10
#>    year month   day dep_delay arr_delay distance air_time
#>   <int> <int> <int>     <dbl>     <dbl>    <dbl>    <dbl>
#> 1  2013     1     1         2        11     1400      227
#> 2  2013     1     1         4        20     1416      227
#> 3  2013     1     1         2        33     1089      160
#> 4  2013     1     1        -1       -18     1576      183
#> 5  2013     1     1        -6       -25      762      116
#> 6  2013     1     1        -4        12      719      150
#> # ... with 3.368e+05 more rows, and 3 more variables:
#> #   gain <dbl>, hours <dbl>, gain_per_hour <dbl>
```

If you only want to keep the new variables, use `transmute()`:

```
transmute(flights,
  gain = arr_delay - dep_delay,
  hours = air_time / 60,
  gain_per_hour = gain / hours
)
#> # A tibble: 336,776 × 3
#>    gain hours gain_per_hour
#>   <dbl> <dbl>         <dbl>
#> 1     9  3.78          2.38
#> 2    16  3.78          4.23
#> 3    31  2.67         11.62
#> 4   -17  3.05         -5.57
#> 5   -19  1.93         -9.83
#> 6    16  2.50          6.40
#> # ... with 3.368e+05 more rows
```

Useful Creation Functions

There are many functions for creating new variables that you can use with `mutate()`. The key property is that the function must be vectorized: it must take a vector of values as input, and return a vector with the same number of values as output. There's no way to list every possible function that you might use, but here's a selection of functions that are frequently useful:

Arithmetic operators +, -, *, /, ^
: These are all vectorized, using the so-called "recycling rules." If one parameter is shorter than the other, it will be automatically extended to be the same length. This is most useful when one of the arguments is a single number: `air_time / 60`, `hours * 60 + minute`, etc.

  Arithmetic operators are also useful in conjunction with the aggregate functions you'll learn about later. For example, `x / sum(x)` calculates the proportion of a total, and `y - mean(y)` computes the difference from the mean.

Modular arithmetic (%/% and %%)
: `%/%` (integer division) and `%%` (remainder), where `x == y * (x %/% y) + (x %% y)`. Modular arithmetic is a handy tool because it allows you to break integers into pieces. For example, in the flights dataset, you can compute `hour` and `minute` from `dep_time` with:

```
transmute(flights,
  dep_time,
  hour = dep_time %/% 100,
  minute = dep_time %% 100
)
#> # A tibble: 336,776 × 3
#>   dep_time  hour minute
#>      <int> <dbl>  <dbl>
#> 1      517     5     17
#> 2      533     5     33
#> 3      542     5     42
#> 4      544     5     44
#> 5      554     5     54
#> 6      554     5     54
#> # ... with 3.368e+05 more rows
```

Logs `log()`, `log2()`, `log10()`
: Logarithms are an incredibly useful transformation for dealing with data that ranges across multiple orders of magnitude. They also convert multiplicative relationships to additive, a feature we'll come back to in Part IV.

  All else being equal, I recommend using `log2()` because it's easy to interpret: a difference of 1 on the log scale corresponds to doubling on the original scale and a difference of –1 corresponds to halving.

Offsets
: `lead()` and `lag()` allow you to refer to leading or lagging values. This allows you to compute running differences (e.g., `x - lag(x)`) or find when values change (`x != lag(x)`). They are most useful in conjunction with `group_by()`, which you'll learn about shortly:

```
(x <- 1:10)
#>  [1]  1  2  3  4  5  6  7  8  9 10
lag(x)
#>  [1] NA  1  2  3  4  5  6  7  8  9
lead(x)
#>  [1]  2  3  4  5  6  7  8  9 10 NA
```

Cumulative and rolling aggregates
: R provides functions for running sums, products, mins, and maxes: `cumsum()`, `cumprod()`, `cummin()`, `cummax()`; and **dplyr** provides `cummean()` for cumulative means. If you need rolling aggregates (i.e., a sum computed over a rolling window), try the **RcppRoll** package:

```
x
#>  [1]  1  2  3  4  5  6  7  8  9 10
cumsum(x)
#>  [1]  1  3  6 10 15 21 28 36 45 55
cummean(x)
#>  [1] 1.0 1.5 2.0 2.5 3.0 3.5 4.0 4.5 5.0 5.5
```

Logical comparisons `<`, `<=`, `>`, `>=`, `!=`
: If you're doing a complex sequence of logical operations it's often a good idea to store the interim values in new variables so you can check that each step is working as expected.

Ranking

There are a number of ranking functions, but you should start with `min_rank()`. It does the most usual type of ranking (e.g., first, second, third, fourth). The default gives the smallest values the smallest ranks; use `desc(x)` to give the largest values the smallest ranks:

```
y <- c(1, 2, 2, NA, 3, 4)
min_rank(y)
#> [1]  1  2  2 NA  4  5
min_rank(desc(y))
#> [1]  5  3  3 NA  2  1
```

If `min_rank()` doesn't do what you need, look at the variants `row_number()`, `dense_rank()`, `percent_rank()`, `cume_dist()`, and `ntile()`. See their help pages for more details:

```
row_number(y)
#> [1]  1  2  3 NA  4  5
dense_rank(y)
#> [1]  1  2  2 NA  3  4
percent_rank(y)
#> [1] 0.00 0.25 0.25   NA 0.75 1.00
cume_dist(y)
#> [1] 0.2 0.6 0.6  NA 0.8 1.0
```

Exercises

1. Currently `dep_time` and `sched_dep_time` are convenient to look at, but hard to compute with because they're not really continuous numbers. Convert them to a more convenient representation of number of minutes since midnight.

2. Compare `air_time` with `arr_time - dep_time`. What do you expect to see? What do you see? What do you need to do to fix it?

3. Compare `dep_time`, `sched_dep_time`, and `dep_delay`. How would you expect those three numbers to be related?

4. Find the 10 most delayed flights using a ranking function. How do you want to handle ties? Carefully read the documentation for `min_rank()`.

5. What does `1:3 + 1:10` return? Why?

6. What trigonometric functions does R provide?

Grouped Summaries with summarize()

The last key verb is `summarize()`. It collapses a data frame to a single row:

```
summarize(flights, delay = mean(dep_delay, na.rm = TRUE))
#> # A tibble: 1 × 1
#>   delay
#>   <dbl>
#> 1  12.6
```

(We'll come back to what that `na.rm = TRUE` means very shortly.)

`summarize()` is not terribly useful unless we pair it with `group_by()`. This changes the unit of analysis from the complete dataset to individual groups. Then, when you use the **dplyr** verbs on a grouped data frame they'll be automatically applied "by group." For example, if we applied exactly the same code to a data frame grouped by date, we get the average delay per date:

```
by_day <- group_by(flights, year, month, day)
summarize(by_day, delay = mean(dep_delay, na.rm = TRUE))
#> Source: local data frame [365 x 4]
#> Groups: year, month [?]
#>
#>    year month   day delay
#>   <int> <int> <int> <dbl>
#> 1  2013     1     1 11.55
#> 2  2013     1     2 13.86
#> 3  2013     1     3 10.99
#> 4  2013     1     4  8.95
#> 5  2013     1     5  5.73
#> 6  2013     1     6  7.15
#> # ... with 359 more rows
```

Together `group_by()` and `summarize()` provide one of the tools that you'll use most commonly when working with **dplyr**: grouped summaries. But before we go any further with this, we need to introduce a powerful new idea: the pipe.

Combining Multiple Operations with the Pipe

Imagine that we want to explore the relationship between the distance and average delay for each location. Using what you know about **dplyr**, you might write code like this:

```
by_dest <- group_by(flights, dest)
delay <- summarize(by_dest,
  count = n(),
```

```
    dist = mean(distance, na.rm = TRUE),
    delay = mean(arr_delay, na.rm = TRUE)
  )
delay <- filter(delay, count > 20, dest != "HNL")

# It looks like delays increase with distance up to ~750 miles
# and then decrease. Maybe as flights get longer there's more
# ability to make up delays in the air?
ggplot(data = delay, mapping = aes(x = dist, y = delay)) +
  geom_point(aes(size = count), alpha = 1/3) +
  geom_smooth(se = FALSE)
#> `geom_smooth()` using method = 'loess'
```

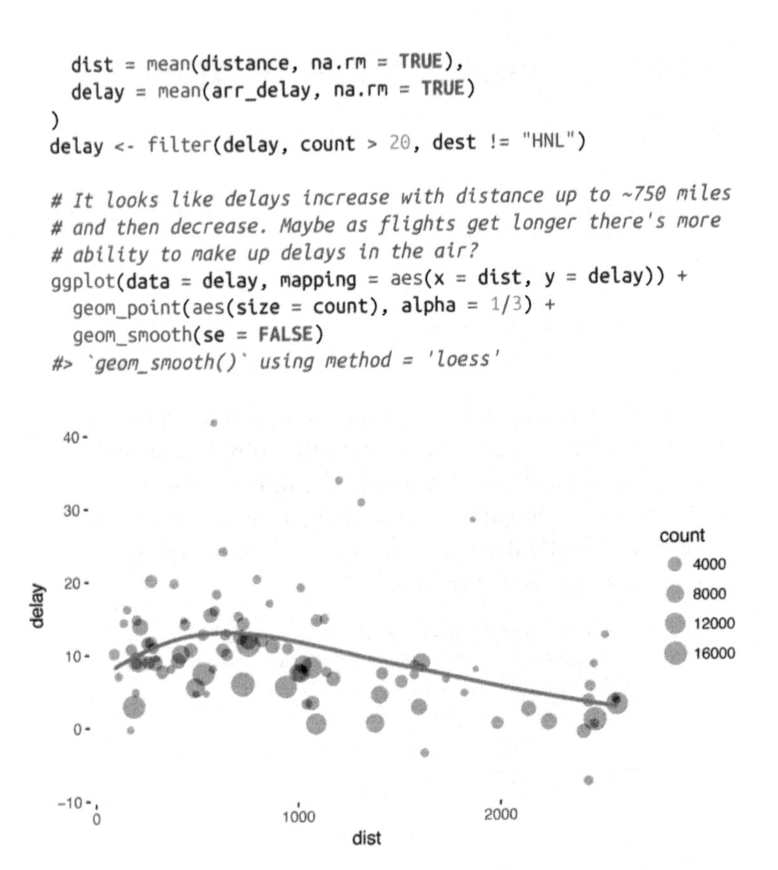

There are three steps to prepare this data:

1. Group flights by destination.
2. Summarize to compute distance, average delay, and number of flights.
3. Filter to remove noisy points and Honolulu airport, which is almost twice as far away as the next closest airport.

This code is a little frustrating to write because we have to give each intermediate data frame a name, even though we don't care about it. Naming things is hard, so this slows down our analysis.

There's another way to tackle the same problem with the pipe, `%>%`:

```
delays <- flights %>%
  group_by(dest) %>%
  summarize(
    count = n(),
```

```
    dist = mean(distance, na.rm = TRUE),
    delay = mean(arr_delay, na.rm = TRUE)
  ) %>%
  filter(count > 20, dest != "HNL")
```

This focuses on the transformations, not what's being transformed, which makes the code easier to read. You can read it as a series of imperative statements: group, then summarize, then filter. As suggested by this reading, a good way to pronounce `%>%` when reading code is "then."

Behind the scenes, `x %>% f(y)` turns into `f(x, y)`, and `x %>% f(y) %>% g(z)` turns into `g(f(x, y), z)`, and so on. You can use the pipe to rewrite multiple operations in a way that you can read left-to-right, top-to-bottom. We'll use piping frequently from now on because it considerably improves the readability of code, and we'll come back to it in more detail in Chapter 14.

Working with the pipe is one of the key criteria for belonging to the tidyverse. The only exception is **ggplot2**: it was written before the pipe was discovered. Unfortunately, the next iteration of **ggplot2**, **ggvis**, which does use the pipe, isn't ready for prime time yet.

Missing Values

You may have wondered about the `na.rm` argument we used earlier. What happens if we don't set it?

```
flights %>%
  group_by(year, month, day) %>%
  summarize(mean = mean(dep_delay))
#> Source: local data frame [365 x 4]
#> Groups: year, month [?]
#>
#>    year month   day  mean
#>   <int> <int> <int> <dbl>
#> 1  2013     1     1    NA
#> 2  2013     1     2    NA
#> 3  2013     1     3    NA
#> 4  2013     1     4    NA
#> 5  2013     1     5    NA
#> 6  2013     1     6    NA
#> # ... with 359 more rows
```

We get a lot of missing values! That's because aggregation functions obey the usual rule of missing values: if there's any missing value in the input, the output will be a missing value. Fortunately, all aggregation functions have an `na.rm` argument, which removes the missing values prior to computation:

```
flights %>%
  group_by(year, month, day) %>%
  summarize(mean = mean(dep_delay, na.rm = TRUE))
#> Source: local data frame [365 x 4]
#> Groups: year, month [?]
#>
#>    year month   day  mean
#>   <int> <int> <int> <dbl>
#> 1  2013     1     1 11.55
#> 2  2013     1     2 13.86
#> 3  2013     1     3 10.99
#> 4  2013     1     4  8.95
#> 5  2013     1     5  5.73
#> 6  2013     1     6  7.15
#> # ... with 359 more rows
```

In this case, where missing values represent cancelled flights, we could also tackle the problem by first removing the cancelled flights. We'll save this dataset so we can reuse it in the next few examples:

```
not_cancelled <- flights %>%
  filter(!is.na(dep_delay), !is.na(arr_delay))

not_cancelled %>%
  group_by(year, month, day) %>%
  summarize(mean = mean(dep_delay))
#> Source: local data frame [365 x 4]
#> Groups: year, month [?]
#>
#>    year month   day  mean
#>   <int> <int> <int> <dbl>
#> 1  2013     1     1 11.44
#> 2  2013     1     2 13.68
#> 3  2013     1     3 10.91
#> 4  2013     1     4  8.97
#> 5  2013     1     5  5.73
#> 6  2013     1     6  7.15
#> # ... with 359 more rows
```

Counts

Whenever you do any aggregation, it's always a good idea to include either a count (`n()`), or a count of nonmissing values (`sum(!is.na(x))`). That way you can check that you're not drawing conclusions based on very small amounts of data. For example, let's look at the planes (identified by their tail number) that have the highest average delays:

```
delays <- not_cancelled %>%
  group_by(tailnum) %>%
  summarize(
    delay = mean(arr_delay)
  )

ggplot(data = delays, mapping = aes(x = delay)) +
  geom_freqpoly(binwidth = 10)
```

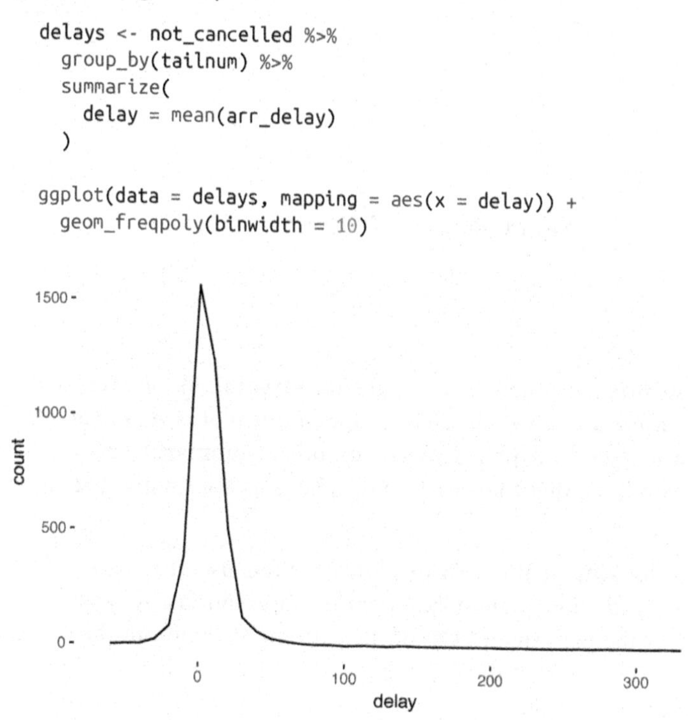

Wow, there are some planes that have an *average* delay of 5 hours (300 minutes)!

The story is actually a little more nuanced. We can get more insight if we draw a scatterplot of number of flights versus average delay:

```
delays <- not_cancelled %>%
  group_by(tailnum) %>%
  summarize(
    delay = mean(arr_delay, na.rm = TRUE),
    n = n()
  )

ggplot(data = delays, mapping = aes(x = n, y = delay)) +
  geom_point(alpha = 1/10)
```

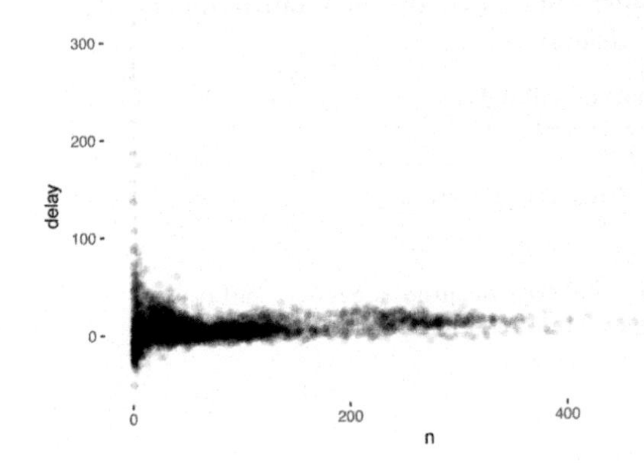

Not surprisingly, there is much greater variation in the average delay when there are few flights. The shape of this plot is very characteristic: whenever you plot a mean (or other summary) versus group size, you'll see that the variation decreases as the sample size increases.

When looking at this sort of plot, it's often useful to filter out the groups with the smallest numbers of observations, so you can see more of the pattern and less of the extreme variation in the smallest groups. This is what the following code does, as well as showing you a handy pattern for integrating **ggplot2** into **dplyr** flows. It's a bit painful that you have to switch from `%>%` to `+`, but once you get the hang of it, it's quite convenient:

```
delays %>%
  filter(n > 25) %>%
  ggplot(mapping = aes(x = n, y = delay)) +
    geom_point(alpha = 1/10)
```

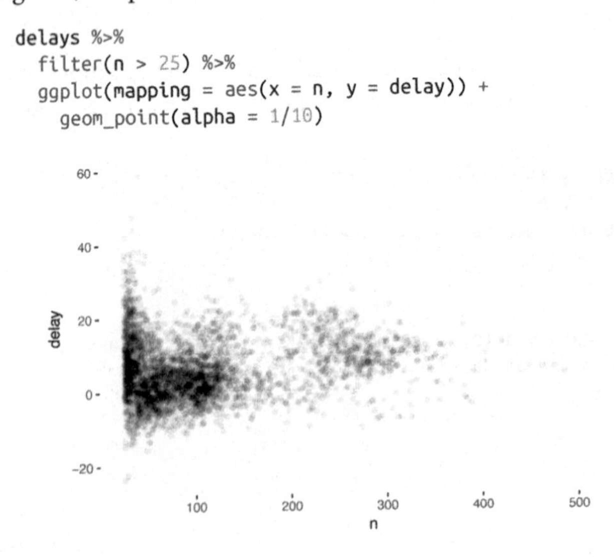

RStudio tip: a useful keyboard shortcut is Cmd/Ctrl-Shift-P. This resends the previously sent chunk from the editor to the console. This is very convenient when you're (e.g.) exploring the value of `n` in the preceding example. You send the whole block once with Cmd/Ctrl-Enter, then you modify the value of `n` and press Cmd/Ctrl-Shift-P to resend the complete block.

There's another common variation of this type of pattern. Let's look at how the average performance of batters in baseball is related to the number of times they're at bat. Here I use data from the **Lahman** package to compute the batting average (number of hits / number of attempts) of every major league baseball player.

When I plot the skill of the batter (measured by the batting average, `ba`) against the number of opportunities to hit the ball (measured by at bat, `ab`), you see two patterns:

- As above, the variation in our aggregate decreases as we get more data points.
- There's a positive correlation between skill (`ba`) and opportunities to hit the ball (`ab`). This is because teams control who gets to play, and obviously they'll pick their best players:

```
# Convert to a tibble so it prints nicely
batting <- as_tibble(Lahman::Batting)

batters <- batting %>%
  group_by(playerID) %>%
  summarize(
    ba = sum(H, na.rm = TRUE) / sum(AB, na.rm = TRUE),
    ab = sum(AB, na.rm = TRUE)
  )

batters %>%
  filter(ab > 100) %>%
  ggplot(mapping = aes(x = ab, y = ba)) +
    geom_point() +
    geom_smooth(se = FALSE)
#> `geom_smooth()` using method = 'gam'
```

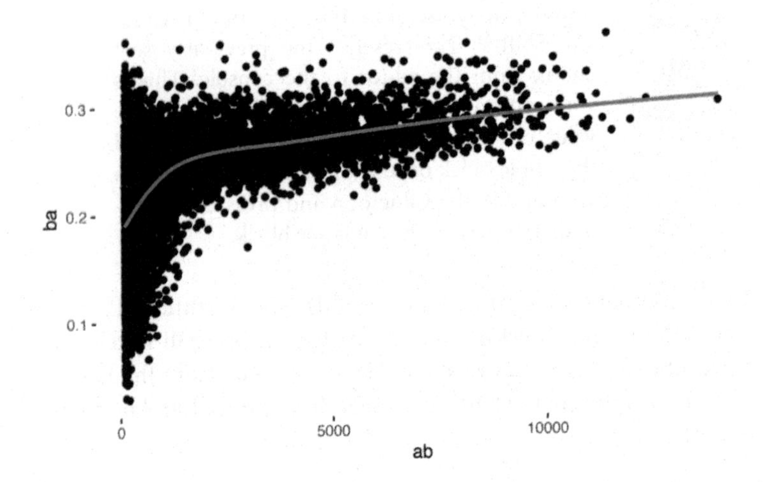

This also has important implications for ranking. If you naively sort on `desc(ba)`, the people with the best batting averages are clearly lucky, not skilled:

```
batters %>%
  arrange(desc(ba))
#> # A tibble: 18,659 × 3
#>    playerID    ba    ab
#>       <chr> <dbl> <int>
#> 1 abramge01     1     1
#> 2 banisje01     1     1
#> 3 bartocl01     1     1
#> 4  bassdo01     1     1
#> 5 birasst01     1     2
#> 6 bruneju01     1     1
#> # ... with 1.865e+04 more rows
```

You can find a good explanation of this problem at *http://bit.ly/Bayesbbal* and *http://bit.ly/notsortavg*.

Useful Summary Functions

Just using means, counts, and sum can get you a long way, but R provides many other useful summary functions:

Measures of location
: We've used `mean(x)`, but `median(x)` is also useful. The mean is the sum divided by the length; the median is a value where 50% of `x` is above it, and 50% is below it.

It's sometimes useful to combine aggregation with logical subsetting. We haven't talked about this sort of subsetting yet, but you'll learn more about it in "Subsetting" on page 304:

```
not_cancelled %>%
  group_by(year, month, day) %>%
  summarize(
    # average delay:
    avg_delay1 = mean(arr_delay),
    # average positive delay:
    avg_delay2 = mean(arr_delay[arr_delay > 0])
  )
#> Source: local data frame [365 x 5]
#> Groups: year, month [?]
#>
#>    year month   day avg_delay1 avg_delay2
#>   <int> <int> <int>      <dbl>      <dbl>
#> 1  2013     1     1      12.65       32.5
#> 2  2013     1     2      12.69       32.0
#> 3  2013     1     3       5.73       27.7
#> 4  2013     1     4      -1.93       28.3
#> 5  2013     1     5      -1.53       22.6
#> 6  2013     1     6       4.24       24.4
#> # ... with 359 more rows
```

Measures of spread `sd(x)`, `IQR(x)`, `mad(x)`

The mean squared deviation, or standard deviation or sd for short, is the standard measure of spread. The interquartile range `IQR()` and median absolute deviation `mad(x)` are robust equivalents that may be more useful if you have outliers:

```
# Why is distance to some destinations more variable
# than to others?
not_cancelled %>%
  group_by(dest) %>%
  summarize(distance_sd = sd(distance)) %>%
  arrange(desc(distance_sd))
#> # A tibble: 104 × 2
#>    dest distance_sd
#>   <chr>       <dbl>
#> 1   EGE       10.54
#> 2   SAN       10.35
#> 3   SFO       10.22
#> 4   HNL       10.00
#> 5   SEA        9.98
#> 6   LAS        9.91
#> # ... with 98 more rows
```

Measures of rank `min(x)`, `quantile(x, 0.25)`, `max(x)`
: Quantiles are a generalization of the median. For example, `quantile(x, 0.25)` will find a value of `x` that is greater than 25% of the values, and less than the remaining 75%:

```
# When do the first and last flights leave each day?
not_cancelled %>%
  group_by(year, month, day) %>%
  summarize(
    first = min(dep_time),
    last = max(dep_time)
  )
#> Source: local data frame [365 x 5]
#> Groups: year, month [?]
#>
#>    year month   day first  last
#>   <int> <int> <int> <int> <int>
#> 1  2013     1     1   517  2356
#> 2  2013     1     2    42  2354
#> 3  2013     1     3    32  2349
#> 4  2013     1     4    25  2358
#> 5  2013     1     5    14  2357
#> 6  2013     1     6    16  2355
#> # ... with 359 more rows
```

Measures of position `first(x)`, `nth(x, 2)`, `last(x)`
: These work similarly to `x[1]`, `x[2]`, and `x[length(x)]` but let you set a default value if that position does not exist (i.e., you're trying to get the third element from a group that only has two elements). For example, we can find the first and last departure for each day:

```
not_cancelled %>%
  group_by(year, month, day) %>%
  summarize(
    first_dep = first(dep_time),
    last_dep = last(dep_time)
  )
#> Source: local data frame [365 x 5]
#> Groups: year, month [?]
#>
#>    year month   day first_dep last_dep
#>   <int> <int> <int>     <int>    <int>
#> 1  2013     1     1       517     2356
#> 2  2013     1     2        42     2354
#> 3  2013     1     3        32     2349
#> 4  2013     1     4        25     2358
#> 5  2013     1     5        14     2357
```

```
#> 6  2013     1     6          16      2355
#> # ... with 359 more rows
```

These functions are complementary to filtering on ranks. Filtering gives you all variables, with each observation in a separate row:

```
not_cancelled %>%
  group_by(year, month, day) %>%
  mutate(r = min_rank(desc(dep_time))) %>%
  filter(r %in% range(r))
#> Source: local data frame [770 x 20]
#> Groups: year, month, day [365]
#>
#>     year month   day dep_time sched_dep_time dep_delay
#>    <int> <int> <int>    <int>          <int>     <dbl>
#> 1  2013     1     1      517            515         2
#> 2  2013     1     1     2356           2359        -3
#> 3  2013     1     2       42           2359        43
#> 4  2013     1     2     2354           2359        -5
#> 5  2013     1     3       32           2359        33
#> 6  2013     1     3     2349           2359       -10
#> # ... with 764 more rows, and 13 more variables:
#> #   arr_time <int>, sched_arr_time <int>,
#> #   arr_delay <dbl>, carrier <chr>, flight <int>,
#> #   tailnum <chr>, origin <chr>, dest <chr>,
#> #   air_time <dbl>, distance <dbl>, hour <dbl>,
#> #   minute <dbl>, time_hour <dttm>, r <int>
```

Counts

You've seen `n()`, which takes no arguments, and returns the size of the current group. To count the number of non-missing values, use `sum(!is.na(x))`. To count the number of distinct (unique) values, use `n_distinct(x)`:

```
# Which destinations have the most carriers?
not_cancelled %>%
  group_by(dest) %>%
  summarize(carriers = n_distinct(carrier)) %>%
  arrange(desc(carriers))
#> # A tibble: 104 × 2
#>     dest carriers
#>    <chr>    <int>
#> 1    ATL        7
#> 2    BOS        7
#> 3    CLT        7
#> 4    ORD        7
#> 5    TPA        7
```

```
#> 6   AUS        6
#> # ... with 98 more rows
```

Counts are so useful that **dplyr** provides a simple helper if all you want is a count:

```
not_cancelled %>%
  count(dest)
#> # A tibble: 104 × 2
#>     dest     n
#>    <chr> <int>
#> 1    ABQ   254
#> 2    ACK   264
#> 3    ALB   418
#> 4    ANC     8
#> 5    ATL 16837
#> 6    AUS  2411
#> # ... with 98 more rows
```

You can optionally provide a weight variable. For example, you could use this to "count" (sum) the total number of miles a plane flew:

```
not_cancelled %>%
  count(tailnum, wt = distance)
#> # A tibble: 4,037 × 2
#>   tailnum      n
#>     <chr>  <dbl>
#> 1  D942DN   3418
#> 2  N0EGMQ 239143
#> 3  N10156 109664
#> 4  N102UW  25722
#> 5  N103US  24619
#> 6  N104UW  24616
#> # ... with 4,031 more rows
```

Counts and proportions of logical values `sum(x > 10)`, `mean(y == 0)`
: When used with numeric functions, `TRUE` is converted to 1 and `FALSE` to 0. This makes `sum()` and `mean()` very useful: `sum(x)` gives the number of `TRUE`s in `x`, and `mean(x)` gives the proportion:

```
# How many flights left before 5am? (these usually
# indicate delayed flights from the previous day)
not_cancelled %>%
  group_by(year, month, day) %>%
  summarize(n_early = sum(dep_time < 500))
#> Source: local data frame [365 x 4]
#> Groups: year, month [?]
```

```
#>
#>    year month   day n_early
#>   <int> <int> <int>   <int>
#> 1  2013     1     1       0
#> 2  2013     1     2       3
#> 3  2013     1     3       4
#> 4  2013     1     4       3
#> 5  2013     1     5       3
#> 6  2013     1     6       2
#> # ... with 359 more rows

# What proportion of flights are delayed by more
# than an hour?
not_cancelled %>%
  group_by(year, month, day) %>%
  summarize(hour_perc = mean(arr_delay > 60))
#> Source: local data frame [365 x 4]
#> Groups: year, month [?]
#>
#>    year month   day hour_perc
#>   <int> <int> <int>     <dbl>
#> 1  2013     1     1    0.0722
#> 2  2013     1     2    0.0851
#> 3  2013     1     3    0.0567
#> 4  2013     1     4    0.0396
#> 5  2013     1     5    0.0349
#> 6  2013     1     6    0.0470
#> # ... with 359 more rows
```

Grouping by Multiple Variables

When you group by multiple variables, each summary peels off one level of the grouping. That makes it easy to progressively roll up a dataset:

```
daily <- group_by(flights, year, month, day)
(per_day   <- summarize(daily, flights = n()))
#> Source: local data frame [365 x 4]
#> Groups: year, month [?]
#>
#>    year month   day flights
#>   <int> <int> <int>   <int>
#> 1  2013     1     1     842
#> 2  2013     1     2     943
#> 3  2013     1     3     914
#> 4  2013     1     4     915
#> 5  2013     1     5     720
#> 6  2013     1     6     832
#> # ... with 359 more rows
```

```
(per_month <- summarize(per_day, flights = sum(flights)))
#> Source: local data frame [12 x 3]
#> Groups: year [?]
#>
#>    year month flights
#>   <int> <int>   <int>
#> 1  2013     1   27004
#> 2  2013     2   24951
#> 3  2013     3   28834
#> 4  2013     4   28330
#> 5  2013     5   28796
#> 6  2013     6   28243
#> # ... with 6 more rows

(per_year  <- summarize(per_month, flights = sum(flights)))
#> # A tibble: 1 × 2
#>    year flights
#>   <int>   <int>
#> 1  2013  336776
```

Be careful when progressively rolling up summaries: it's OK for sums and counts, but you need to think about weighting means and variances, and it's not possible to do it exactly for rank-based statistics like the median. In other words, the sum of groupwise sums is the overall sum, but the median of groupwise medians is not the overall median.

Ungrouping

If you need to remove grouping, and return to operations on ungrouped data, use `ungroup()`:

```
daily %>%
  ungroup() %>%             # no longer grouped by date
  summarize(flights = n())  # all flights
#> # A tibble: 1 × 1
#>   flights
#>     <int>
#> 1  336776
```

Exercises

1. Brainstorm at least five different ways to assess the typical delay characteristics of a group of flights. Consider the following scenarios:

- A flight is 15 minutes early 50% of the time, and 15 minutes late 50% of the time.
- A flight is always 10 minutes late.
- A flight is 30 minutes early 50% of the time, and 30 minutes late 50% of the time.
- 99% of the time a flight is on time. 1% of the time it's 2 hours late.

Which is more important: arrival delay or departure delay?

2. Come up with another approach that will give you the same output as `not_cancelled %>% count(dest)` and `not_cancelled %>% count(tailnum, wt = distance)` (without using `count()`).
3. Our definition of cancelled flights (`is.na(dep_delay) | is.na(arr_delay)`) is slightly suboptimal. Why? Which is the most important column?
4. Look at the number of cancelled flights per day. Is there a pattern? Is the proportion of cancelled flights related to the average delay?
5. Which carrier has the worst delays? Challenge: can you disentangle the effects of bad airports versus bad carriers? Why/why not? (Hint: think about `flights %>% group_by(carrier, dest) %>% summarize(n())`.)
6. For each plane, count the number of flights before the first delay of greater than 1 hour.
7. What does the `sort` argument to `count()` do? When might you use it?

Grouped Mutates (and Filters)

Grouping is most useful in conjunction with `summarize()`, but you can also do convenient operations with `mutate()` and `filter()`:

- Find the worst members of each group:

```
flights_sml %>%
  group_by(year, month, day) %>%
  filter(rank(desc(arr_delay)) < 10)
```

```
#> Source: local data frame [3,306 x 7]
#> Groups: year, month, day [365]
#>
#>     year month   day dep_delay arr_delay distance
#>    <int> <int> <int>     <dbl>     <dbl>    <dbl>
#> 1  2013     1     1       853       851      184
#> 2  2013     1     1       290       338     1134
#> 3  2013     1     1       260       263      266
#> 4  2013     1     1       157       174      213
#> 5  2013     1     1       216       222      708
#> 6  2013     1     1       255       250      589
#> # ... with 3,300 more rows, and 1 more variables:
#> #   air_time <dbl>
```

- Find all groups bigger than a threshold:

```
popular_dests <- flights %>%
  group_by(dest) %>%
  filter(n() > 365)
popular_dests
#> Source: local data frame [332,577 x 19]
#> Groups: dest [77]
#>
#>     year month   day dep_time sched_dep_time dep_delay
#>    <int> <int> <int>    <int>          <int>     <dbl>
#> 1  2013     1     1      517            515         2
#> 2  2013     1     1      533            529         4
#> 3  2013     1     1      542            540         2
#> 4  2013     1     1      544            545        -1
#> 5  2013     1     1      554            600        -6
#> 6  2013     1     1      554            558        -4
#> # ... with 3.326e+05 more rows, and 13 more variables:
#> #   arr_time <int>, sched_arr_time <int>,
#> #   arr_delay <dbl>, carrier <chr>, flight <int>,
#> #   tailnum <chr>, origin <chr>, dest <chr>,
#> #   air_time <dbl>, distance <dbl>, hour <dbl>,
#> #   minute <dbl>, time_hour <dttm>
```

- Standardize to compute per group metrics:

```
popular_dests %>%
  filter(arr_delay > 0) %>%
  mutate(prop_delay = arr_delay / sum(arr_delay)) %>%
  select(year:day, dest, arr_delay, prop_delay)
#> Source: local data frame [131,106 x 6]
#> Groups: dest [77]
#>
#>     year month   day  dest arr_delay prop_delay
#>    <int> <int> <int> <chr>     <dbl>      <dbl>
#> 1  2013     1     1   IAH        11   1.11e-04
```

```
#> 2  2013     1     1   IAH          20   2.01e-04
#> 3  2013     1     1   MIA          33   2.35e-04
#> 4  2013     1     1   ORD          12   4.24e-05
#> 5  2013     1     1   FLL          19   9.38e-05
#> 6  2013     1     1   ORD           8   2.83e-05
#> # ... with 1.311e+05 more rows
```

A grouped filter is a grouped mutate followed by an ungrouped filter. I generally avoid them except for quick-and-dirty manipulations: otherwise it's hard to check that you've done the manipulation correctly.

Functions that work most naturally in grouped mutates and filters are known as window functions (versus the summary functions used for summaries). You can learn more about useful window functions in the corresponding vignette: `vignette("window-functions")`.

Exercises

1. Refer back to the table of useful mutate and filtering functions. Describe how each operation changes when you combine it with grouping.
2. Which plane (`tailnum`) has the worst on-time record?
3. What time of day should you fly if you want to avoid delays as much as possible?
4. For each destination, compute the total minutes of delay. For each flight, compute the proportion of the total delay for its destination.
5. Delays are typically temporally correlated: even once the problem that caused the initial delay has been resolved, later flights are delayed to allow earlier flights to leave. Using `lag()` explores how the delay of a flight is related to the delay of the immediately preceding flight.
6. Look at each destination. Can you find flights that are suspiciously fast? (That is, flights that represent a potential data entry error.) Compute the air time of a flight relative to the shortest flight to that destination. Which flights were most delayed in the air?

7. Find all destinations that are flown by at least two carriers. Use that information to rank the carriers.

CHAPTER 4

Workflow: Scripts

So far you've been using the console to run code. That's a great place to start, but you'll find it gets cramped pretty quickly as you create more complex **ggplot2** graphics and **dplyr** pipes. To give yourself more room to work, it's a great idea to use the script editor. Open it up either by clicking the File menu and selecting New File, then R script, or using the keyboard shortcut Cmd/Ctrl-Shift-N. Now you'll see four panes:

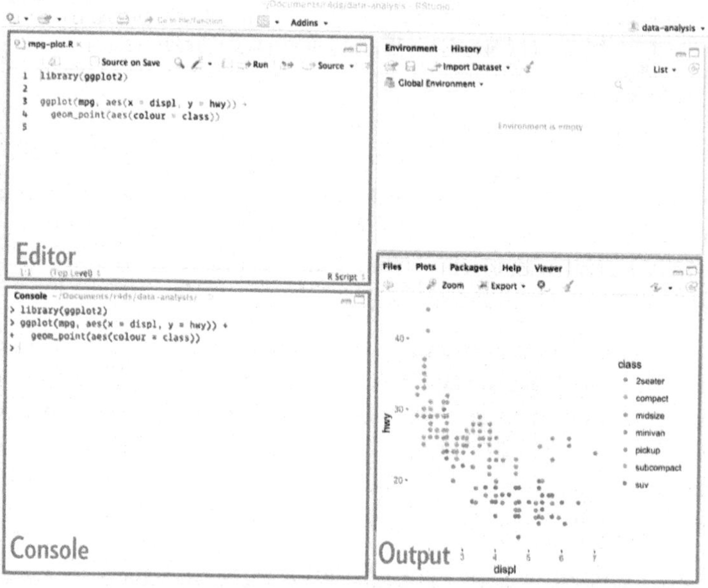

The script editor is a great place to put code you care about. Keep experimenting in the console, but once you have written code that works and does what you want, put it in the script editor. RStudio will automatically save the contents of the editor when you quit RStudio, and will automatically load it when you reopen. Nevertheless, it's a good idea to save your scripts regularly and to back them up.

Running Code

The script editor is also a great place to build up complex **ggplot2** plots or long sequences of **dplyr** manipulations. The key to using the script editor effectively is to memorize one of the most important keyboard shortcuts: Cmd/Ctrl-Enter. This executes the current R expression in the console. For example, take the following code. If your cursor is at █, pressing Cmd/Ctrl-Enter will run the complete command that generates `not_cancelled`. It will also move the cursor to the next statement (beginning with `not_cancelled %>%`). That makes it easy to run your complete script by repeatedly pressing Cmd/Ctrl-Enter:

```
library(dplyr)
library(nycflights13)

not_cancelled <- flights %>%
  filter(!is.na(dep_delay)█, !is.na(arr_delay))

not_cancelled %>%
  group_by(year, month, day) %>%
  summarize(mean = mean(dep_delay))
```

Instead of running expression-by-expression, you can also execute the complete script in one step: Cmd/Ctrl-Shift-S. Doing this regularly is a great way to check that you've captured all the important parts of your code in the script.

I recommend that you always start your script with the packages that you need. That way, if you share your code with others, they can easily see what packages they need to install. Note, however, that you should never include `install.packages()` or `setwd()` in a script that you share. It's very antisocial to change settings on someone else's computer!

When working through future chapters, I highly recommend starting in the editor and practicing your keyboard shortcuts. Over time, sending code to the console in this way will become so natural that you won't even think about it.

RStudio Diagnostics

The script editor will also highlight syntax errors with a red squiggly line and a cross in the sidebar:

```
3
4   x y <- 10
5
```

Hover over the cross to see what the problem is:

```
4   x y <- 10
unexpected token 'y'
unexpected token '<-'
```

RStudio will also let you know about potential problems:

```
17   3 == NA
use 'is.na' to check whether expression evaluates to
NA
```

Exercises

1. Go to the RStudio Tips twitter account at *@rstudiotips* (*https://twitter.com/rstudiotips*) and find one tip that looks interesting. Practice using it!
2. What other common mistakes will RStudio diagnostics report? Read *http://bit.ly/RStudiocodediag* to find out.

CHAPTER 5

Exploratory Data Analysis

Introduction

This chapter will show you how to use visualization and transformation to explore your data in a systematic way, a task that statisticians call exploratory data analysis, or EDA for short. EDA is an iterative cycle. You:

1. Generate questions about your data.
2. Search for answers by visualizing, transforming, and modeling your data.
3. Use what you learn to refine your questions and/or generate new questions.

EDA is not a formal process with a strict set of rules. More than anything, EDA is a state of mind. During the initial phases of EDA you should feel free to investigate every idea that occurs to you. Some of these ideas will pan out, and some will be dead ends. As your exploration continues, you will hone in on a few particularly productive areas that you'll eventually write up and communicate to others.

EDA is an important part of any data analysis, even if the questions are handed to you on a platter, because you always need to investigate the quality of your data. Data cleaning is just one application of EDA: you ask questions about whether or not your data meets your expectations. To do data cleaning, you'll need to deploy all the tools of EDA: visualization, transformation, and modeling.

Prerequisites

In this chapter we'll combine what you've learned about **dplyr** and **ggplot2** to interactively ask questions, answer them with data, and then ask new questions.

```
library(tidyverse)
```

Questions

> There are no routine statistical questions, only questionable statistical routines.
>
> —Sir David Cox

> Far better an approximate answer to the right question, which is often vague, than an exact answer to the wrong question, which can always be made precise.
>
> —John Tukey

Your goal during EDA is to develop an understanding of your data. The easiest way to do this is to use questions as tools to guide your investigation. When you ask a question, the question focuses your attention on a specific part of your dataset and helps you decide which graphs, models, or transformations to make.

EDA is fundamentally a creative process. And like most creative processes, the key to asking *quality* questions is to generate a large *quantity* of questions. It is difficult to ask revealing questions at the start of your analysis because you do not know what insights are contained in your dataset. On the other hand, each new question that you ask will expose you to a new aspect of your data and increase your chance of making a discovery. You can quickly drill down into the most interesting parts of your data—and develop a set of thought-provoking questions—if you follow up each question with a new question based on what you find.

There is no rule about which questions you should ask to guide your research. However, two types of questions will always be useful for making discoveries within your data. You can loosely word these questions as:

1. What type of variation occurs within my variables?
2. What type of covariation occurs between my variables?

The rest of this chapter will look at these two questions. I'll explain what variation and covariation are, and I'll show you several ways to answer each question. To make the discussion easier, let's define some terms:

- A *variable* is a quantity, quality, or property that you can measure.
- A *value* is the state of a variable when you measure it. The value of a variable may change from measurement to measurement.
- An *observation*, or a *case*, is a set of measurements made under similar conditions (you usually make all of the measurements in an observation at the same time and on the same object). An observation will contain several values, each associated with a different variable. I'll sometimes refer to an observation as a data point.
- *Tabular data* is a set of values, each associated with a variable and an observation. Tabular data is *tidy* if each value is placed in its own "cell," each variable in its own column, and each observation in its own row.

So far, all of the data that you've seen has been tidy. In real life, most data isn't tidy, so we'll come back to these ideas again in Chapter 9.

Variation

Variation is the tendency of the values of a variable to change from measurement to measurement. You can see variation easily in real life; if you measure any continuous variable twice, you will get two different results. This is true even if you measure quantities that are constant, like the speed of light. Each of your measurements will include a small amount of error that varies from measurement to measurement. Categorical variables can also vary if you measure across different subjects (e.g., the eye colors of different people), or different times (e.g., the energy levels of an electron at different moments). Every variable has its own pattern of variation, which can reveal interesting information. The best way to understand that pattern is to visualize the distribution of variables' values.

Visualizing Distributions

How you visualize the distribution of a variable will depend on whether the variable is categorical or continuous. A variable is *categorical* if it can only take one of a small set of values. In R, categorical variables are usually saved as factors or character vectors. To examine the distribution of a categorical variable, use a bar chart:

```
ggplot(data = diamonds) +
  geom_bar(mapping = aes(x = cut))
```

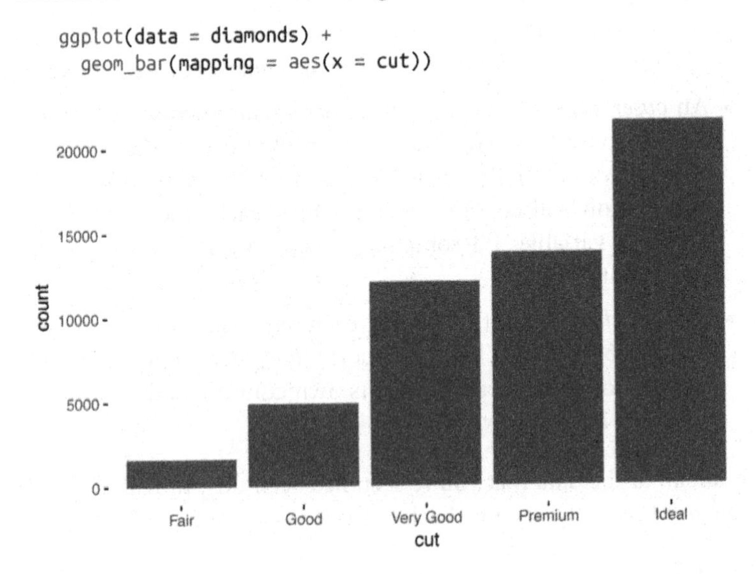

The height of the bars displays how many observations occurred with each x value. You can compute these values manually with `dplyr::count()`:

```
diamonds %>%
  count(cut)
#> # A tibble: 5 × 2
#>         cut     n
#>       <ord> <int>
#> 1      Fair  1610
#> 2      Good  4906
#> 3 Very Good 12082
#> 4   Premium 13791
#> 5     Ideal 21551
```

A variable is *continuous* if it can take any of an infinite set of ordered values. Numbers and date-times are two examples of continuous variables. To examine the distribution of a continuous variable, use a histogram:

```r
ggplot(data = diamonds) +
  geom_histogram(mapping = aes(x = carat), binwidth = 0.5)
```

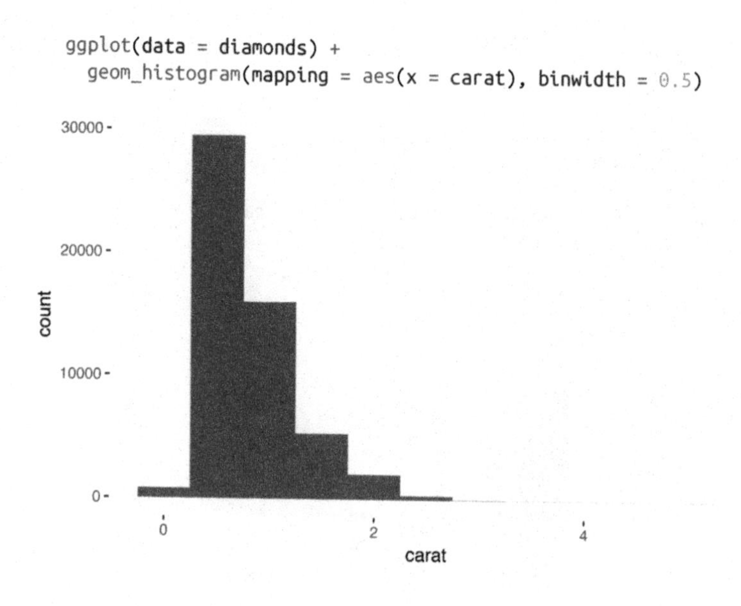

You can compute this by hand by combining `dplyr::count()` and `ggplot2::cut_width()`:

```r
diamonds %>%
  count(cut_width(carat, 0.5))
#> # A tibble: 11 × 2
#>   `cut_width(carat, 0.5)`     n
#>                    <fctr> <int>
#> 1            [-0.25,0.25]   785
#> 2             (0.25,0.75] 29498
#> 3             (0.75,1.25] 15977
#> 4             (1.25,1.75]  5313
#> 5             (1.75,2.25]  2002
#> 6             (2.25,2.75]   322
#> # ... with 5 more rows
```

A histogram divides the x-axis into equally spaced bins and then uses the height of each bar to display the number of observations that fall in each bin. In the preceding graph, the tallest bar shows that almost 30,000 observations have a `carat` value between 0.25 and 0.75, which are the left and right edges of the bar.

You can set the width of the intervals in a histogram with the `binwidth` argument, which is measured in the units of the `x` variable. You should always explore a variety of binwidths when working with histograms, as different binwidths can reveal different patterns. For example, here is how the preceding graph looks when we zoom into just the diamonds with a size of less than three carats and choose a smaller binwidth:

```
smaller <- diamonds %>%
  filter(carat < 3)

ggplot(data = smaller, mapping = aes(x = carat)) +
  geom_histogram(binwidth = 0.1)
```

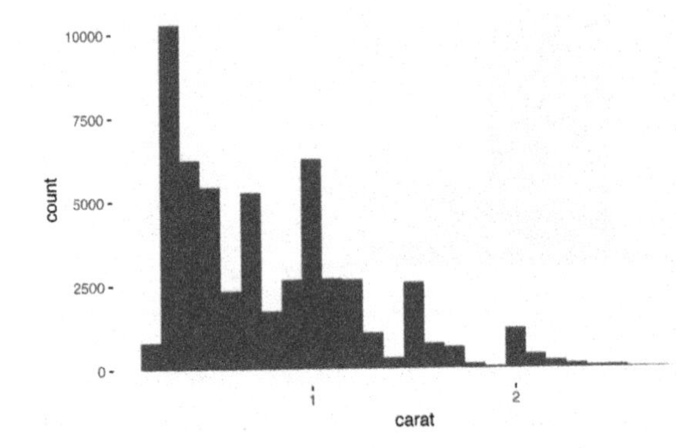

If you wish to overlay multiple histograms in the same plot, I recommend using `geom_freqpoly()` instead of `geom_histogram()`. `geom_freqpoly()` performs the same calculation as `geom_histogram()`, but instead of displaying the counts with bars, uses lines instead. It's much easier to understand overlapping lines than bars:

```
ggplot(data = smaller, mapping = aes(x = carat, color = cut)) +
  geom_freqpoly(binwidth = 0.1)
```

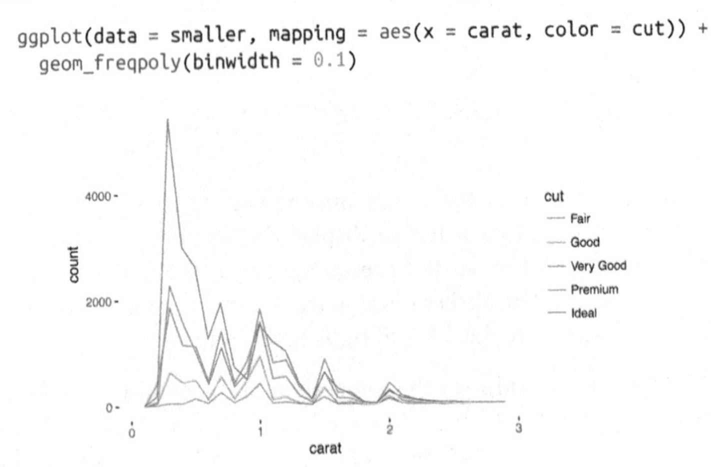

There are a few challenges with this type of plot, which we will come back to in "A Categorical and Continuous Variable" on page 93.

Typical Values

In both bar charts and histograms, tall bars show the common values of a variable, and shorter bars show less-common values. Places that do not have bars reveal values that were not seen in your data. To turn this information into useful questions, look for anything unexpected:

- Which values are the most common? Why?
- Which values are rare? Why? Does that match your expectations?
- Can you see any unusual patterns? What might explain them?

As an example, the following histogram suggests several interesting questions:

- Why are there more diamonds at whole carats and common fractions of carats?
- Why are there more diamonds slightly to the right of each peak than there are slightly to the left of each peak?
- Why are there no diamonds bigger than 3 carats?

```
ggplot(data = smaller, mapping = aes(x = carat)) +
  geom_histogram(binwidth = 0.01)
```

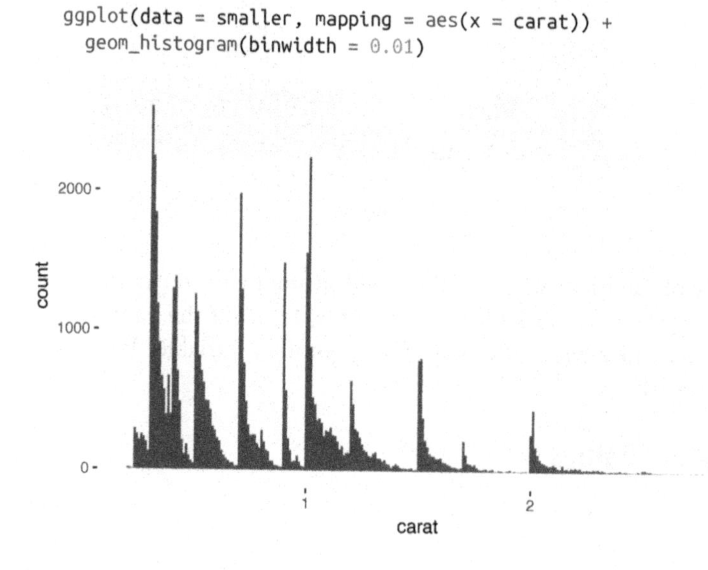

In general, clusters of similar values suggest that subgroups exist in your data. To understand the subgroups, ask:

- How are the observations within each cluster similar to each other?
- How are the observations in separate clusters different from each other?
- How can you explain or describe the clusters?
- Why might the appearance of clusters be misleading?

The following histogram shows the length (in minutes) of 272 eruptions of the Old Faithful Geyser in Yellowstone National Park. Eruption times appear to be clustered into two groups: there are short eruptions (of around 2 minutes) and long eruptions (4–5 minutes), but little in between:

```
ggplot(data = faithful, mapping = aes(x = eruptions)) +
  geom_histogram(binwidth = 0.25)
```

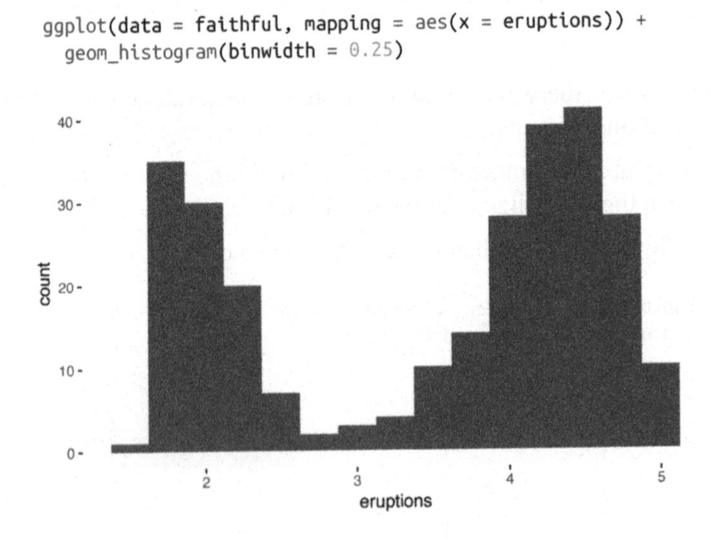

Many of the preceding questions will prompt you to explore a relationship *between* variables, for example, to see if the values of one variable can explain the behavior of another variable. We'll get to that shortly.

Unusual Values

Outliers are observations that are unusual; data points that don't seem to fit the pattern. Sometimes outliers are data entry errors; other times outliers suggest important new science. When you have a lot of data, outliers are sometimes difficult to see in a histogram. For example, take the distribution of the `y` variable from the diamonds dataset. The only evidence of outliers is the unusually wide limits on the y-axis:

```
ggplot(diamonds) +
  geom_histogram(mapping = aes(x = y), binwidth = 0.5)
```

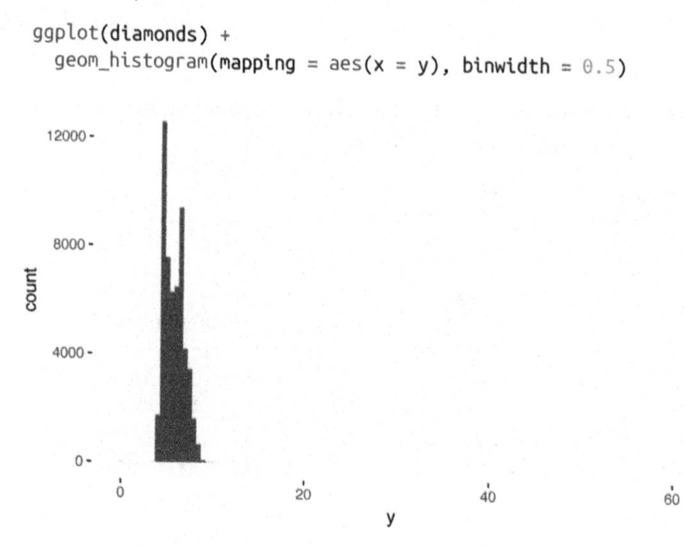

There are so many observations in the common bins that the rare bins are so short that you can't see them (although maybe if you stare intently at 0 you'll spot something). To make it easy to see the unusual values, we need to zoom in to small values of the y-axis with `coord_cartesian()`:

```
ggplot(diamonds) +
  geom_histogram(mapping = aes(x = y), binwidth = 0.5) +
  coord_cartesian(ylim = c(0, 50))
```

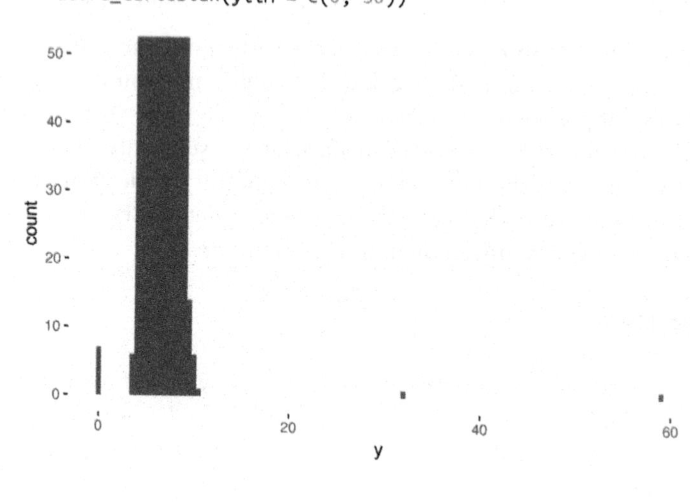

(coord_cartesian() also has an xlim() argument for when you need to zoom into the x-axis. **ggplot2** also has xlim() and ylim() functions that work slightly differently: they throw away the data outside the limits.)

This allows us to see that there are three unusual values: 0, ~30, and ~60. We pluck them out with **dplyr**:

```
unusual <- diamonds %>%
  filter(y < 3 | y > 20) %>%
  arrange(y)
unusual
#> # A tibble: 9 × 10
#>    carat       cut color clarity depth table price     x
#>    <dbl>     <ord> <ord>   <ord> <dbl> <dbl> <int> <dbl>
#> 1   1.00 Very Good     H     VS2  63.3    53  5139  0.00
#> 2   1.14      Fair     G     VS1  57.5    67  6381  0.00
#> 3   1.56     Ideal     G     VS2  62.2    54 12800  0.00
#> 4   1.20   Premium     D    VVS1  62.1    59 15686  0.00
#> 5   2.25   Premium     H     SI2  62.8    59 18034  0.00
#> 6   0.71      Good     F     SI2  64.1    60  2130  0.00
#> 7   0.71      Good     F     SI2  64.1    60  2130  0.00
#> 8   0.51     Ideal     E     VS1  61.8    55  2075  5.15
#> 9   2.00   Premium     H     SI2  58.9    57 12210  8.09
#> # ... with 2 more variables:
#> #   y <dbl>, z <dbl>
```

The y variable measures one of the three dimensions of these diamonds, in mm. We know that diamonds can't have a width of 0mm, so these values must be incorrect. We might also suspect that measurements of 32mm and 59mm are implausible: those diamonds are over an inch long, but don't cost hundreds of thousands of dollars!

It's good practice to repeat your analysis with and without the outliers. If they have minimal effect on the results, and you can't figure out why they're there, it's reasonable to replace them with missing values and move on. However, if they have a substantial effect on your results, you shouldn't drop them without justification. You'll need to figure out what caused them (e.g., a data entry error) and disclose that you removed them in your write-up.

Exercises

1. Explore the distribution of each of the x, y, and z variables in diamonds. What do you learn? Think about a diamond and how you might decide which dimension is the length, width, and depth.

2. Explore the distribution of `price`. Do you discover anything unusual or surprising? (Hint: carefully think about the `bin width` and make sure you try a wide range of values.)
3. How many diamonds are 0.99 carat? How many are 1 carat? What do you think is the cause of the difference?
4. Compare and contrast `coord_cartesian()` versus `xlim()` or `ylim()` when zooming in on a histogram. What happens if you leave `binwidth` unset? What happens if you try and zoom so only half a bar shows?

Missing Values

If you've encountered unusual values in your dataset, and simply want to move on to the rest of your analysis, you have two options:

- Drop the entire row with the strange values:

  ```
  diamonds2 <- diamonds %>%
    filter(between(y, 3, 20))
  ```

  I don't recommend this option as just because one measurement is invalid, doesn't mean all the measurements are. Additionally, if you have low-quality data, by time that you've applied this approach to every variable you might find that you don't have any data left!

- Instead, I recommend replacing the unusual values with missing values. The easiest way to do this is to use `mutate()` to replace the variable with a modified copy. You can use the `ifelse()` function to replace unusual values with `NA`:

  ```
  diamonds2 <- diamonds %>%
    mutate(y = ifelse(y < 3 | y > 20, NA, y))
  ```

`ifelse()` has three arguments. The first argument `test` should be a logical vector. The result will contain the value of the second argument, `yes`, when `test` is `TRUE`, and the value of the third argument, `no`, when it is false.

Like R, **ggplot2** subscribes to the philosophy that missing values should never silently go missing. It's not obvious where you should plot missing values, so **ggplot2** doesn't include them in the plot, but it does warn that they've been removed:

```
ggplot(data = diamonds2, mapping = aes(x = x, y = y)) +
  geom_point()
#> Warning: Removed 9 rows containing missing values
#> (geom_point).
```

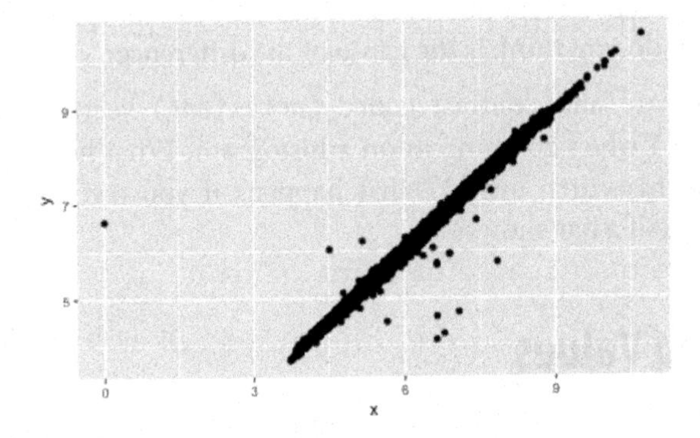

To suppress that warning, set `na.rm = TRUE`:

```
ggplot(data = diamonds2, mapping = aes(x = x, y = y)) +
  geom_point(na.rm = TRUE)
```

Other times you want to understand what makes observations with missing values different from observations with recorded values. For example, in `nycflights13::flights`, missing values in the `dep_time` variable indicate that the flight was cancelled. So you might want to compare the scheduled departure times for cancelled and noncancelled times. You can do this by making a new variable with `is.na()`:

```
nycflights13::flights %>%
  mutate(
    cancelled = is.na(dep_time),
    sched_hour = sched_dep_time %/% 100,
    sched_min = sched_dep_time %% 100,
    sched_dep_time = sched_hour + sched_min / 60
  ) %>%
  ggplot(mapping = aes(sched_dep_time)) +
    geom_freqpoly(
      mapping = aes(color = cancelled),
      binwidth = 1/4
    )
```

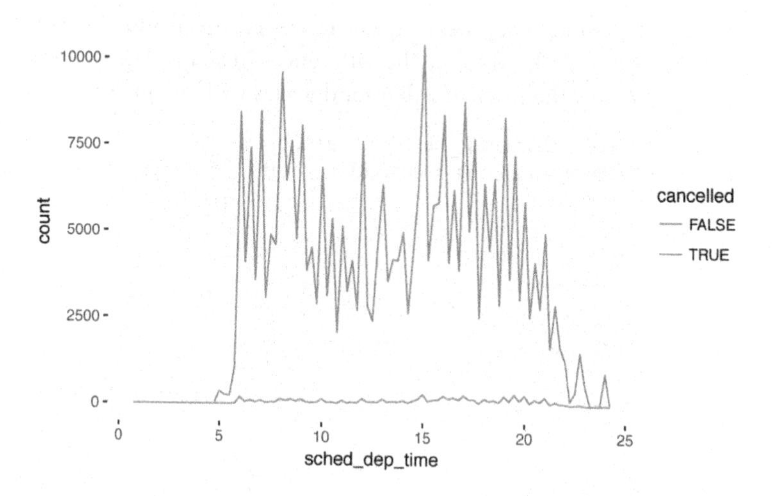

However, this plot isn't great because there are many more noncancelled flights than cancelled flights. In the next section we'll explore some techniques for improving this comparison.

Exercises

1. What happens to missing values in a histogram? What happens to missing values in a bar chart? Why is there a difference?
2. What does `na.rm = TRUE` do in `mean()` and `sum()`?

Covariation

If variation describes the behavior *within* a variable, covariation describes the behavior *between* variables. *Covariation* is the tendency for the values of two or more variables to vary together in a related way. The best way to spot covariation is to visualize the relationship between two or more variables. How you do that should again depend on the type of variables involved.

A Categorical and Continuous Variable

It's common to want to explore the distribution of a continuous variable broken down by a categorical variable, as in the previous frequency polygon. The default appearance of `geom_freqpoly()` is not that useful for that sort of comparison because the height is given by the count. That means if one of the groups is much smaller than the others, it's hard to see the differences in shape. For example, let's explore how the price of a diamond varies with its quality:

```
ggplot(data = diamonds, mapping = aes(x = price)) +
  geom_freqpoly(mapping = aes(color = cut), binwidth = 500)
```

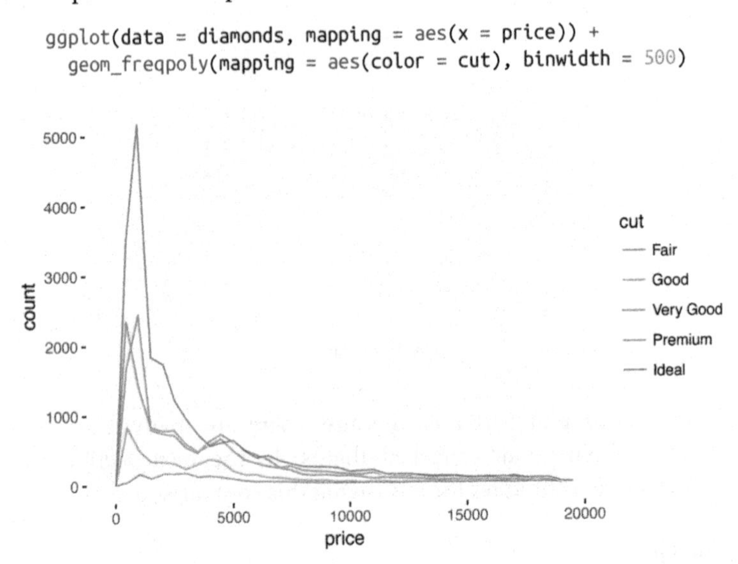

It's hard to see the difference in distribution because the overall counts differ so much:

```
ggplot(diamonds) +
  geom_bar(mapping = aes(x = cut))
```

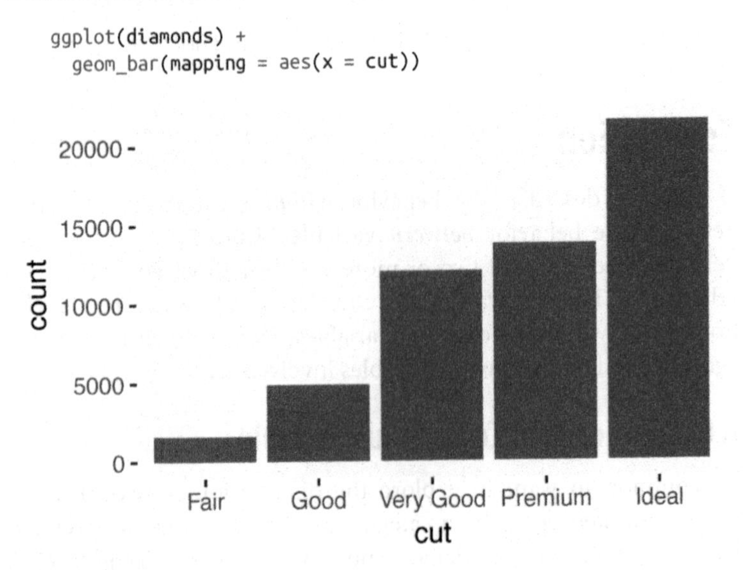

To make the comparison easier we need to swap what is displayed on the y-axis. Instead of displaying count, we'll display *density*, which is the count standardized so that the area under each frequency polygon is one:

```
ggplot(
  data = diamonds,
    mapping = aes(x = price, y = ..density..)
  ) +
   geom_freqpoly(mapping = aes(color = cut), binwidth = 500)
```

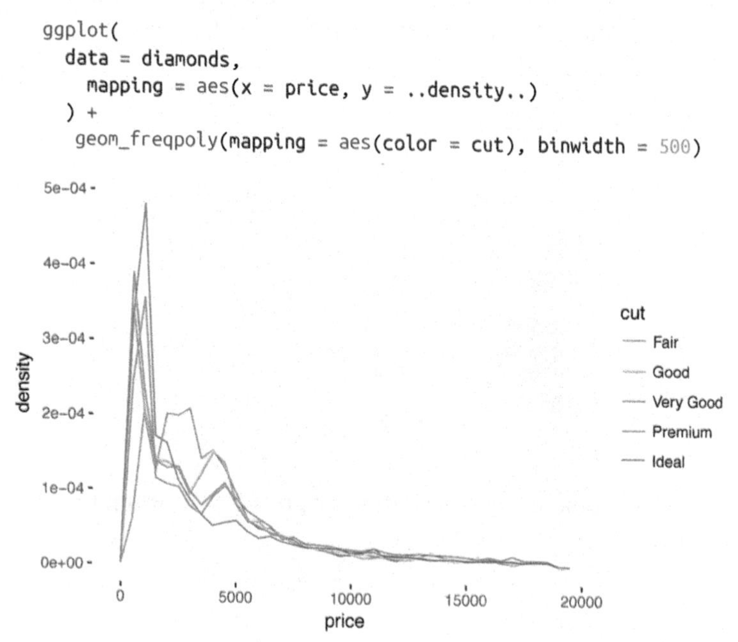

There's something rather surprising about this plot—it appears that fair diamonds (the lowest quality) have the highest average price! But maybe that's because frequency polygons are a little hard to interpret—there's a lot going on in this plot.

Another alternative to display the distribution of a continuous variable broken down by a categorical variable is the boxplot. A *boxplot* is a type of visual shorthand for a distribution of values that is popular among statisticians. Each boxplot consists of:

- A box that stretches from the 25th percentile of the distribution to the 75th percentile, a distance known as the interquartile range (IQR). In the middle of the box is a line that displays the median, i.e., 50th percentile, of the distribution. These three lines give you a sense of the spread of the distribution and whether or not the distribution is symmetric about the median or skewed to one side.

- Visual points that display observations that fall more than 1.5 times the IQR from either edge of the box. These outlying points are unusual, so they are plotted individually.
- A line (or whisker) that extends from each end of the box and goes to the farthest nonoutlier point in the distribution.

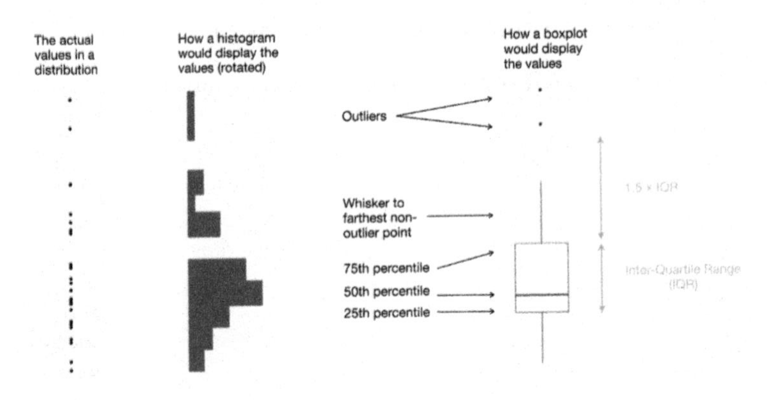

Let's take a look at the distribution of price by cut using `geom_boxplot()`:

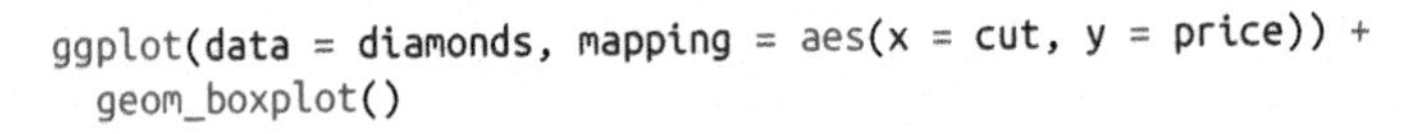

```
ggplot(data = diamonds, mapping = aes(x = cut, y = price)) +
  geom_boxplot()
```

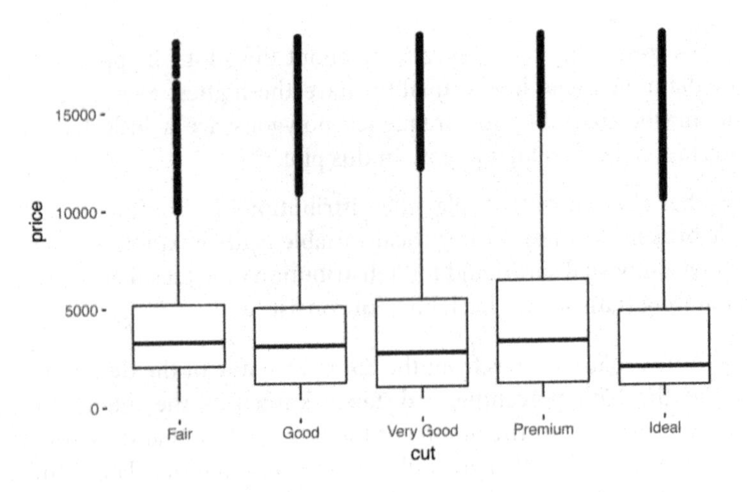

We see much less information about the distribution, but the boxplots are much more compact so we can more easily compare them (and fit more on one plot). It supports the counterintuitive finding that better quality diamonds are cheaper on average! In the exercises, you'll be challenged to figure out why.

`cut` is an ordered factor: fair is worse than good, which is worse than very good, and so on. Many categorical variables don't have such an intrinsic order, so you might want to reorder them to make a more informative display. One way to do that is with the `reorder()` function.

For example, take the `class` variable in the `mpg` dataset. You might be interested to know how highway mileage varies across classes:

```
ggplot(data = mpg, mapping = aes(x = class, y = hwy)) +
  geom_boxplot()
```

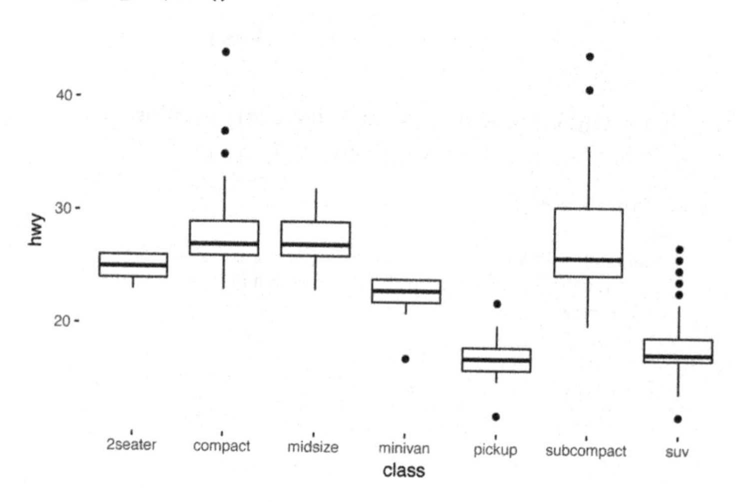

To make the trend easier to see, we can reorder `class` based on the median value of `hwy`:

```
ggplot(data = mpg) +
  geom_boxplot(
    mapping = aes(
      x = reorder(class, hwy, FUN = median),
      y = hwy
    )
  )
```

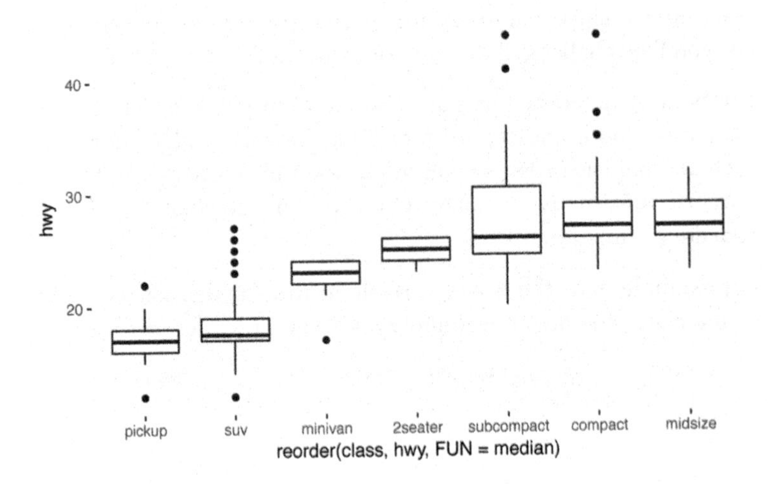

If you have long variable names, geom_boxplot() will work better if you flip it 90°. You can do that with coord_flip():

```
ggplot(data = mpg) +
  geom_boxplot(
    mapping = aes(
      x = reorder(class, hwy, FUN = median),
      y = hwy
    )
  ) +
  coord_flip()
```

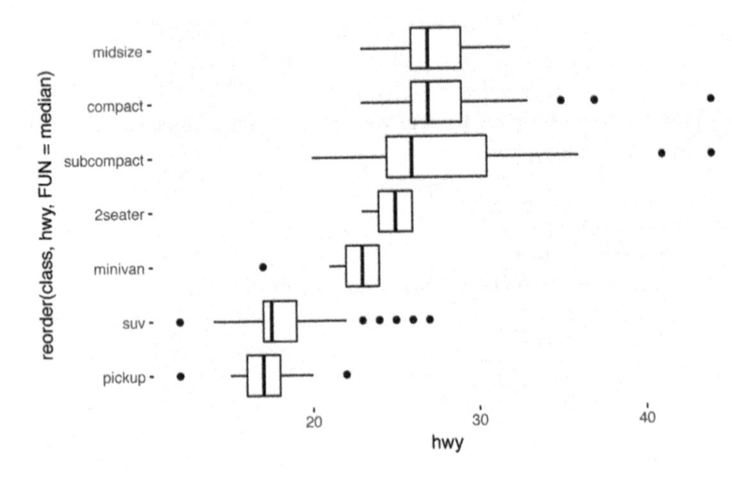

Exercises

1. Use what you've learned to improve the visualization of the departure times of cancelled versus noncancelled flights.
2. What variable in the diamonds dataset is most important for predicting the price of a diamond? How is that variable correlated with cut? Why does the combination of those two relationships lead to lower quality diamonds being more expensive?
3. Install the **ggstance** package, and create a horizontal boxplot. How does this compare to using `coord_flip()`?
4. One problem with boxplots is that they were developed in an era of much smaller datasets and tend to display a prohibitively large number of "outlying values." One approach to remedy this problem is the letter value plot. Install the **lvplot** package, and try using `geom_lv()` to display the distribution of price versus cut. What do you learn? How do you interpret the plots?
5. Compare and contrast `geom_violin()` with a faceted `geom_histogram()`, or a colored `geom_freqpoly()`. What are the pros and cons of each method?
6. If you have a small dataset, it's sometimes useful to use `geom_jitter()` to see the relationship between a continuous and categorical variable. The **ggbeeswarm** package provides a number of methods similar to `geom_jitter()`. List them and briefly describe what each one does.

Two Categorical Variables

To visualize the covariation between categorical variables, you'll need to count the number of observations for each combination. One way to do that is to rely on the built-in `geom_count()`:

```
ggplot(data = diamonds) +
  geom_count(mapping = aes(x = cut, y = color))
```

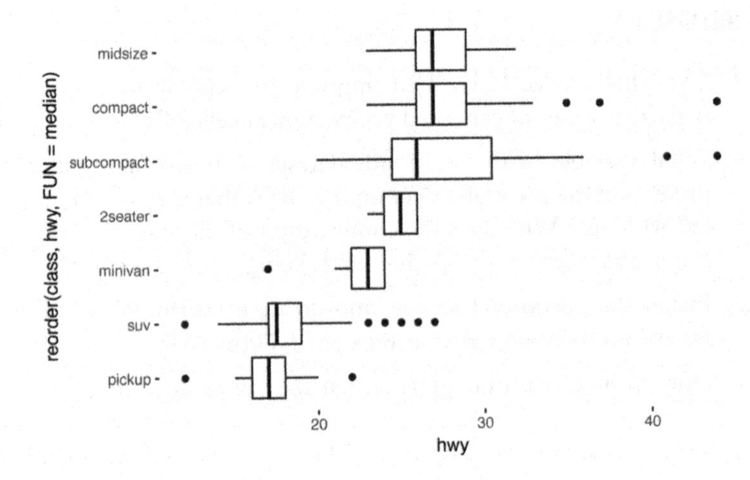

The size of each circle in the plot displays how many observations occurred at each combination of values. Covariation will appear as a strong correlation between specific x values and specific y values.

Another approach is to compute the count with **dplyr**:

```
diamonds %>%
  count(color, cut)
#> Source: local data frame [35 x 3]
#> Groups: color [?]
#>
#>    color       cut     n
#>    <ord>     <ord> <int>
#> 1      D      Fair   163
#> 2      D      Good   662
#> 3      D Very Good  1513
#> 4      D   Premium  1603
#> 5      D     Ideal  2834
#> 6      E      Fair   224
#> # ... with 29 more rows
```

Then visualize with `geom_tile()` and the fill aesthetic:

```
diamonds %>%
  count(color, cut) %>%
  ggplot(mapping = aes(x = color, y = cut)) +
    geom_tile(mapping = aes(fill = n))
```

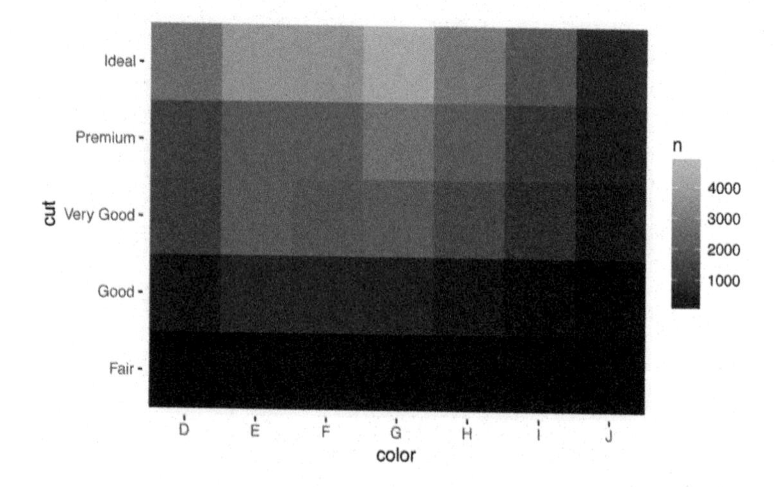

If the categorical variables are unordered, you might want to use the **seriation** package to simultaneously reorder the rows and columns in order to more clearly reveal interesting patterns. For larger plots, you might want to try the **d3heatmap** or **heatmaply** packages, which create interactive plots.

Exercises

1. How could you rescale the `count` dataset to more clearly show the distribution of cut within color, or color within cut?
2. Use `geom_tile()` together with **dplyr** to explore how average flight delays vary by destination and month of year. What makes the plot difficult to read? How could you improve it?
3. Why is it slightly better to use `aes(x = color, y = cut)` rather than `aes(x = cut, y = color)` in the previous example?

Two Continuous Variables

You've already seen one great way to visualize the covariation between two continuous variables: draw a scatterplot with `geom_point()`. You can see covariation as a pattern in the points. For example, you can see an exponential relationship between the carat size and price of a diamond:

```
ggplot(data = diamonds) +
  geom_point(mapping = aes(x = carat, y = price))
```

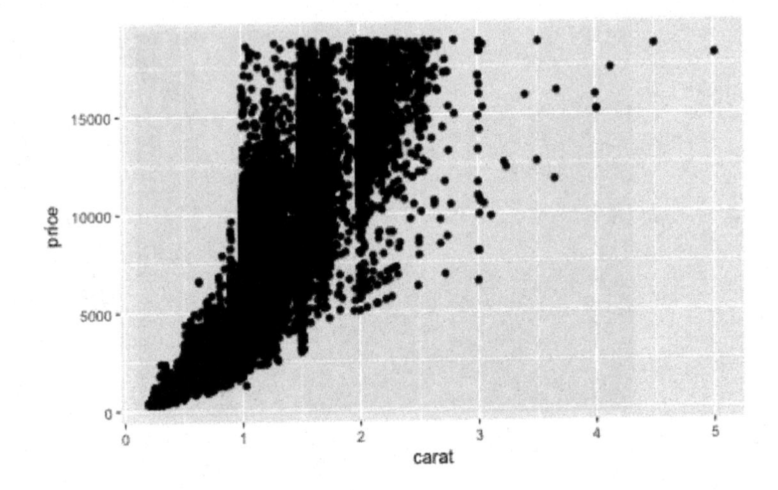

Scatterplots become less useful as the size of your dataset grows, because points begin to overplot, and pile up into areas of uniform black (as in the preceding scatterplot). You've already seen one way to fix the problem, using the `alpha` aesthetic to add transparency:

```
ggplot(data = diamonds) +
  geom_point(
    mapping = aes(x = carat, y = price),
    alpha = 1 / 100
  )
```

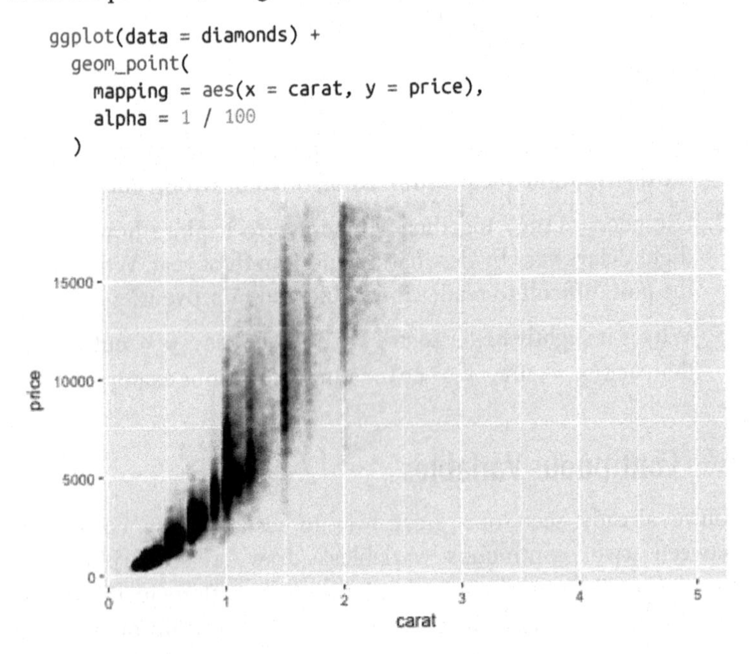

But using transparency can be challenging for very large datasets. Another solution is to use bin. Previously you used `geom_histo`

gram() and geom_freqpoly() to bin in one dimension. Now you'll learn how to use geom_bin2d() and geom_hex() to bin in two dimensions.

geom_bin2d() and geom_hex() divide the coordinate plane into 2D bins and then use a fill color to display how many points fall into each bin. geom_bin2d() creates rectangular bins. geom_hex() creates hexagonal bins. You will need to install the **hexbin** package to use geom_hex():

```
ggplot(data = smaller) +
  geom_bin2d(mapping = aes(x = carat, y = price))

# install.packages("hexbin")
ggplot(data = smaller) +
  geom_hex(mapping = aes(x = carat, y = price))
#> Loading required package: methods
```

Another option is to bin one continuous variable so it acts like a categorical variable. Then you can use one of the techniques for visualizing the combination of a categorical and a continuous variable that you learned about. For example, you could bin carat and then for each group, display a boxplot:

```
ggplot(data = smaller, mapping = aes(x = carat, y = price)) +
  geom_boxplot(mapping = aes(group = cut_width(carat, 0.1)))
```

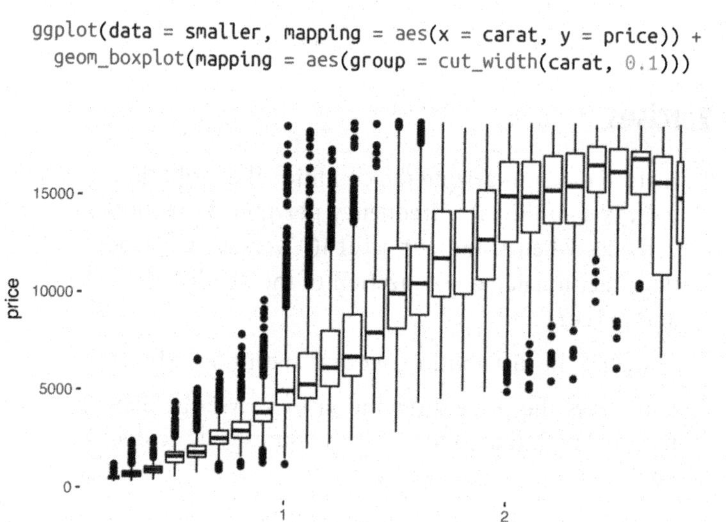

cut_width(x, width), as used here, divides x into bins of width width. By default, boxplots look roughly the same (apart from the number of outliers) regardless of how many observations there are, so it's difficult to tell that each boxplot summarizes a different number of points. One way to show that is to make the width of the boxplot proportional to the number of points with `varwidth = TRUE`.

Another approach is to display approximately the same number of points in each bin. That's the job of `cut_number()`:

```
ggplot(data = smaller, mapping = aes(x = carat, y = price)) +
  geom_boxplot(mapping = aes(group = cut_number(carat, 20)))
```

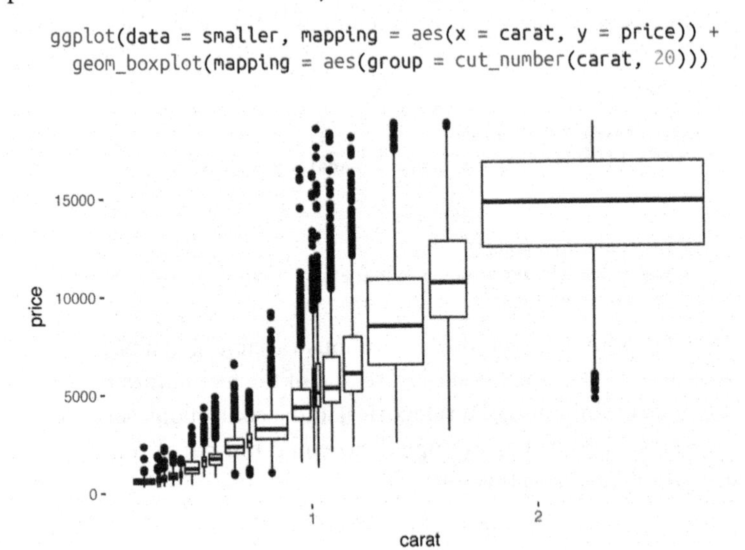

Exercises

1. Instead of summarizing the conditional distribution with a boxplot, you could use a frequency polygon. What do you need to consider when using `cut_width()` versus `cut_number()`? How does that impact a visualization of the 2D distribution of `carat` and `price`?
2. Visualize the distribution of `carat`, partitioned by `price`.
3. How does the price distribution of very large diamonds compare to small diamonds. Is it as you expect, or does it surprise you?
4. Combine two of the techniques you've learned to visualize the combined distribution of `cut`, `carat`, and `price`.
5. Two-dimensional plots reveal outliers that are not visible in one-dimensional plots. For example, some points in the following plot have an unusual combination of `x` and `y` values, which

makes the points outliers even though their x and y values appear normal when examined separately:

```
ggplot(data = diamonds) +
  geom_point(mapping = aes(x = x, y = y)) +
  coord_cartesian(xlim = c(4, 11), ylim = c(4, 11))
```

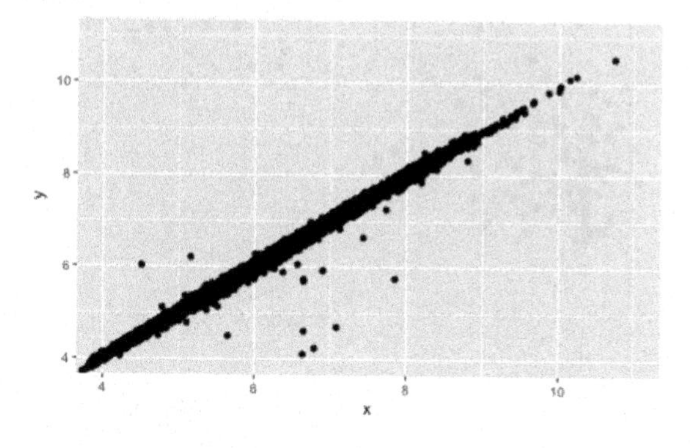

Why is a scatterplot a better display than a binned plot for this case?

Patterns and Models

Patterns in your data provide clues about relationships. If a systematic relationship exists between two variables it will appear as a pattern in the data. If you spot a pattern, ask yourself:

- Could this pattern be due to coincidence (i.e., random chance)?
- How can you describe the relationship implied by the pattern?
- How strong is the relationship implied by the pattern?
- What other variables might affect the relationship?
- Does the relationship change if you look at individual subgroups of the data?

A scatterplot of Old Faithful eruption lengths versus the wait time between eruptions shows a pattern: longer wait times are associated with longer eruptions. The scatterplot also displays the two clusters that we noticed earlier:

```
ggplot(data = faithful) +
  geom_point(mapping = aes(x = eruptions, y = waiting))
```

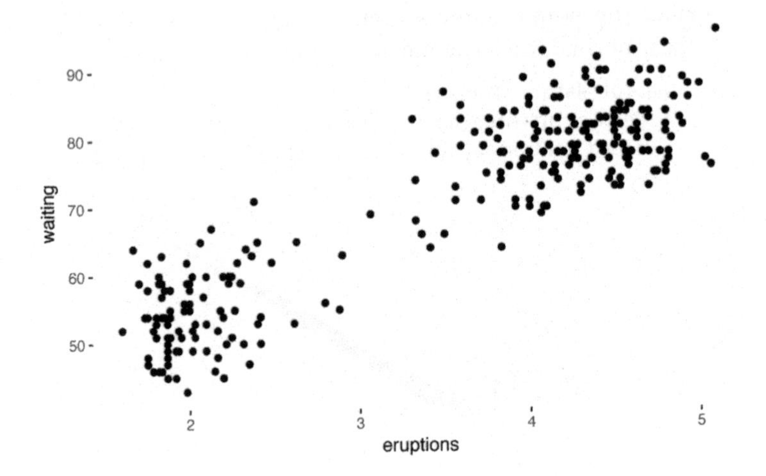

Patterns provide one of the most useful tools for data scientists because they reveal covariation. If you think of variation as a phenomenon that creates uncertainty, covariation is a phenomenon that reduces it. If two variables covary, you can use the values of one variable to make better predictions about the values of the second. If the covariation is due to a causal relationship (a special case), then you can use the value of one variable to control the value of the second.

Models are a tool for extracting patterns out of data. For example, consider the diamonds data. It's hard to understand the relationship between cut and price, because cut and carat, and carat and price, are tightly related. It's possible to use a model to remove the very strong relationship between price and carat so we can explore the subtleties that remain. The following code fits a model that predicts `price` from `carat` and then computes the residuals (the difference between the predicted value and the actual value). The residuals give us a view of the price of the diamond, once the effect of carat has been removed:

```
library(modelr)

mod <- lm(log(price) ~ log(carat), data = diamonds)

diamonds2 <- diamonds %>%
  add_residuals(mod) %>%
  mutate(resid = exp(resid))
```

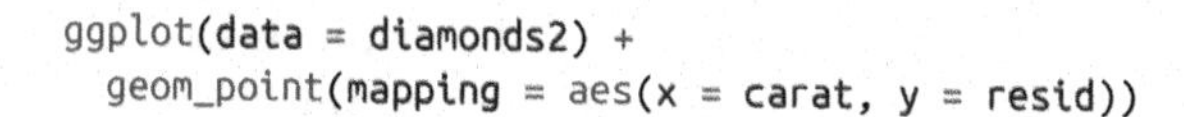

```
ggplot(data = diamonds2) +
  geom_point(mapping = aes(x = carat, y = resid))
```

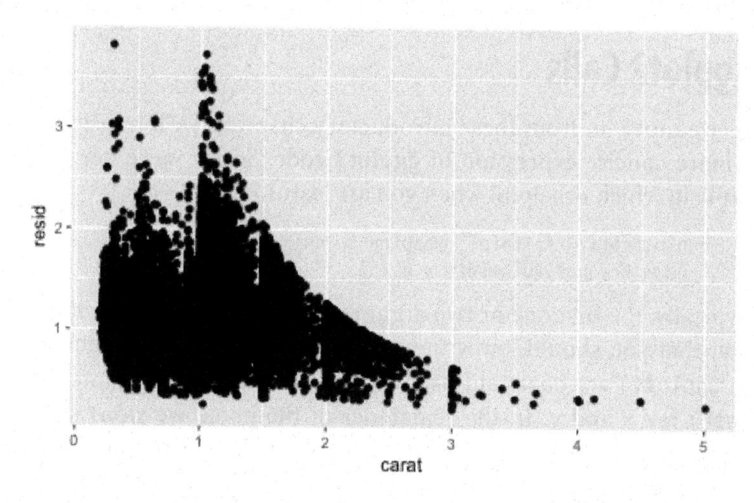

Once you've removed the strong relationship between carat and price, you can see what you expect in the relationship between cut and price—relative to their size, better quality diamonds are more expensive:

```
ggplot(data = diamonds2) +
  geom_boxplot(mapping = aes(x = cut, y = resid))
```

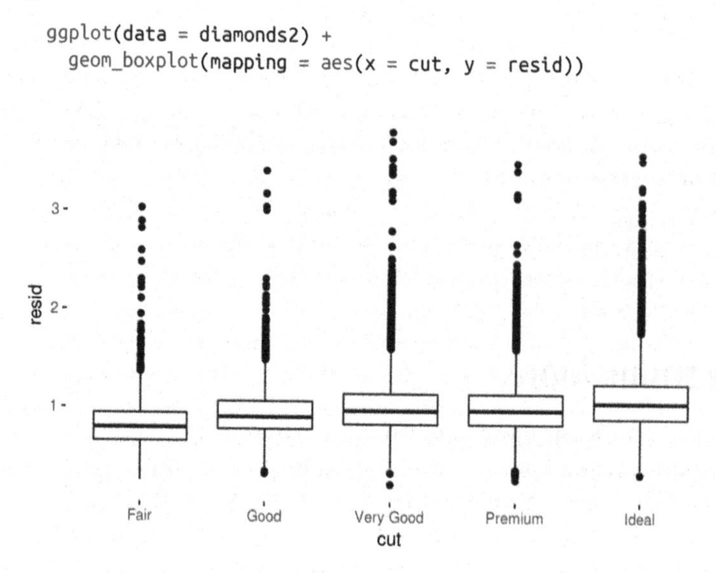

You'll learn how models, and the **modelr** package, work in the final part of the book, Part IV. We're saving modeling for later because understanding what models are and how they work is easiest once you have the tools of data wrangling and programming in hand.

ggplot2 Calls

As we move on from these introductory chapters, we'll transition to a more concise expression of **ggplot2** code. So far we've been very explicit, which is helpful when you are learning:

```
ggplot(data = faithful, mapping = aes(x = eruptions)) +
  geom_freqpoly(binwidth = 0.25)
```

Typically, the first one or two arguments to a function are so important that you should know them by heart. The first two arguments to `ggplot()` are `data` and `mapping`, and the first two arguments to `aes()` are `x` and `y`. In the remainder of the book, we won't supply those names. That saves typing, and, by reducing the amount of boilerplate, makes it easier to see what's different between plots. That's a really important programming concern that we'll come back to in Chapter 15.

Rewriting the previous plot more concisely yields:

```
ggplot(faithful, aes(eruptions)) +
  geom_freqpoly(binwidth = 0.25)
```

Sometimes we'll turn the end of a pipeline of data transformation into a plot. Watch for the transition from `%>%` to `+`. I wish this transition wasn't necessary but unfortunately **ggplot2** was created before the pipe was discovered:

```
diamonds %>%
  count(cut, clarity) %>%
  ggplot(aes(clarity, cut, fill = n)) +
    geom_tile()
```

Learning More

If you want learn more about the mechanics of **ggplot2**, I'd highly recommend grabbing a copy of the **ggplot2** book (*http://ggplot2.org/book/*). It's been recently updated, so it includes **dplyr** and **tidyr** code, and has much more space to explore all the facets of visualization. Unfortunately the book isn't generally available for free, but if you have a connection to a university you can probably get an electronic version for free through SpringerLink.

Another useful resource is the *R Graphics Cookbook* by Winston Chang. Much of the contents are available online at *http://www.cookbook-r.com/Graphs/*.

I also recommend *Graphical Data Analysis with R*, by Antony Unwin. This is a book-length treatment similar to the material covered in this chapter, but has the space to go into much greater depth.

CHAPTER 6

Workflow: Projects

One day you will need to quit R, go do something else, and return to your analysis the next day. One day you will be working on multiple analyses simultaneously that all use R and you want to keep them separate. One day you will need to bring data from the outside world into R and send numerical results and figures from R back out into the world. To handle these real-life situations, you need to make two decisions:

1. What about your analysis is "real," i.e., what will you save as your lasting record of what happened?
2. Where does your analysis "live"?

What Is Real?

As a beginning R user, it's OK to consider your environment (i.e., the objects listed in the environment pane) "real." However, in the long run, you'll be much better off if you consider your R scripts as "real."

With your R scripts (and your data files), you can re-create the environment. It's much harder to re-create your R scripts from your environment! You'll either have to retype a lot of code from memory (making mistakes all the way) or you'll have to carefully mine your R history.

To foster this behavior, I highly recommend that you instruct RStudio not to preserve your workspace between sessions:

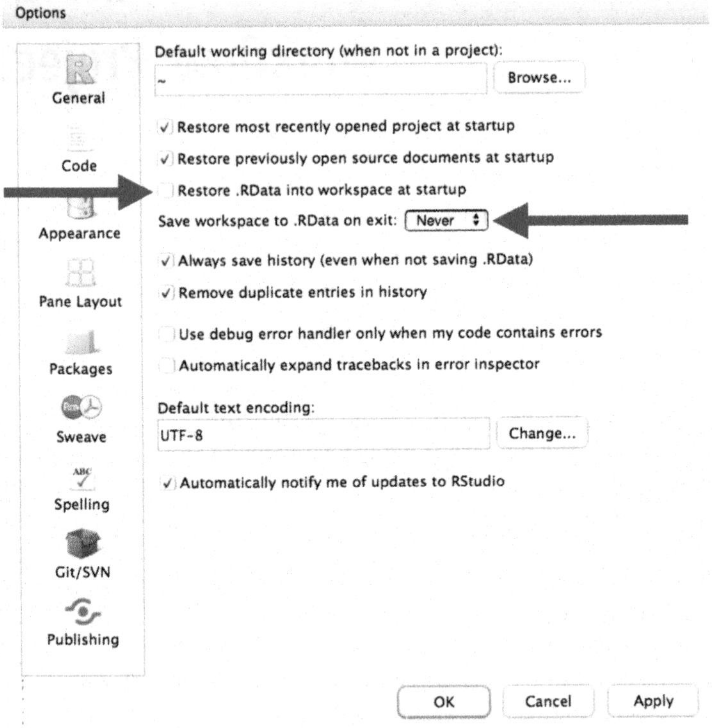

This will cause you some short-term pain, because now when you restart RStudio it will not remember the results of the code that you ran last time. But this short-term pain will save you long-term agony because it forces you to capture all important interactions in your code. There's nothing worse than discovering three months after the fact that you've only stored the results of an important calculation in your workspace, not the calculation itself in your code.

There is a great pair of keyboard shortcuts that will work together to make sure you've captured the important parts of your code in the editor:

- Press Cmd/Ctrl-Shift-F10 to restart RStudio.
- Press Cmd/Ctrl-Shift-S to rerun the current script.

I use this pattern hundreds of times a week.

Where Does Your Analysis Live?

R has a powerful notion of the *working directory*. This is where R looks for files that you ask it to load, and where it will put any files that you ask it to save. RStudio shows your current working directory at the top of the console:

And you can print this out in R code by running `getwd()`:

```
getwd()
#> [1] "/Users/hadley/Documents/r4ds/r4ds"
```

As a beginning R user, it's OK to let your home directory, documents directory, or any other weird directory on your computer be R's working directory. But you're six chapters into this book, and you're no longer a rank beginner. Very soon now you should evolve to organizing your analytical projects into directories and, when working on a project, setting R's working directory to the associated directory.

I do not recommend it, but you can also set the working directory from within R:

```
setwd("/path/to/my/CoolProject")
```

But you should never do this because there's a better way; a way that also puts you on the path to managing your R work like an expert.

Paths and Directories

Paths and directories are a little complicated because there are two basic styles of paths: Mac/Linux and Windows. There are three chief ways in which they differ:

- The most important difference is how you separate the components of the path. Mac and Linux use slashes (e.g., `plots/diamonds.pdf`) and Windows uses backslashes (e.g., `plots\diamonds.pdf`). R can work with either type (no matter what platform you're currently using), but unfortunately, backslashes mean something special to R, and to get a single backslash in the path, you need to type two backslashes! That makes life frustrating, so I recommend always using the Linux/Max style with forward slashes.

- Absolute paths (i.e., paths that point to the same place regardless of your working directory) look different. In Windows they start with a drive letter (e.g., `C:`) or two backslashes (e.g., `\\servername`) and in Mac/Linux they start with a slash "/" (e.g., `/users/hadley`). You should *never* use absolute paths in your scripts, because they hinder sharing: no one else will have exactly the same directory configuration as you.
- The last minor difference is the place that ~ points to. ~ is a convenient shortcut to your home directory. Windows doesn't really have the notion of a home directory, so it instead points to your documents directory.

RStudio Projects

R experts keep all the files associated with a project together—input data, R scripts, analytical results, figures. This is such a wise and common practice that RStudio has built-in support for this via *projects*.

Let's make a project for you to use while you're working through the rest of this book. Click File → New Project, then:

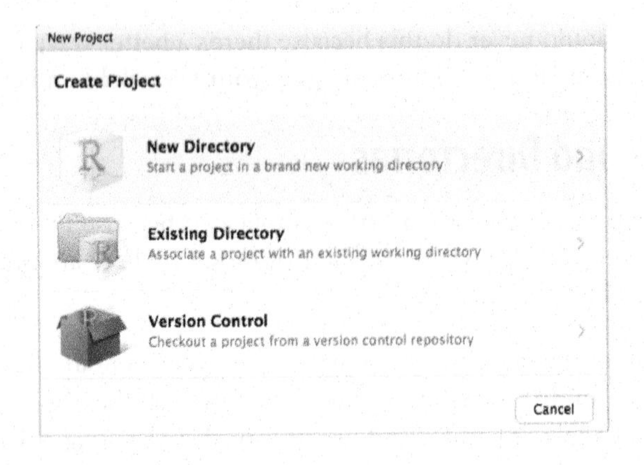

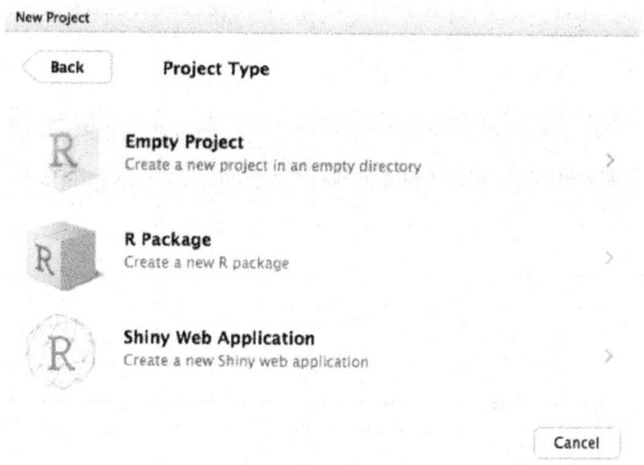

New Project

Back Project Type

Empty Project
Create a new project in an empty directory

R Package
Create a new R package

Shiny Web Application
Create a new Shiny web application

Cancel

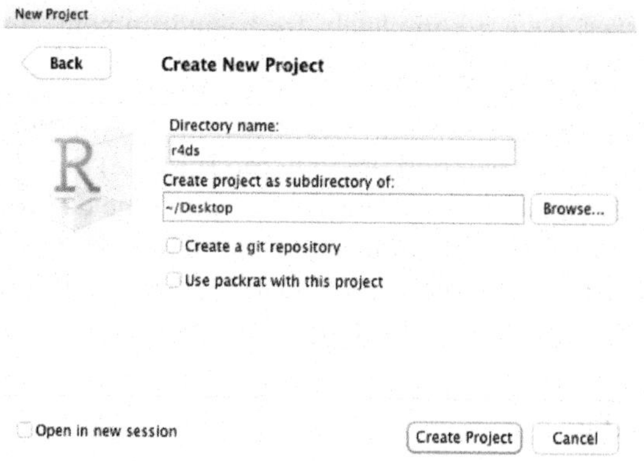

New Project

Back Create New Project

Directory name:
r4ds
Create project as subdirectory of:
~/Desktop Browse...
Create a git repository
Use packrat with this project

Open in new session Create Project Cancel

Call your project `r4ds` and think carefully about which *subdirectory* you put the project in. If you don't store it somewhere sensible, it will be hard to find it in the future!

Once this process is complete, you'll get a new RStudio project just for this book. Check that the "home" directory of your project is the current working directory:

```
getwd()
#> [1] /Users/hadley/Documents/r4ds/r4ds
```

Whenever you refer to a file with a relative path it will look for it here.

Now enter the following commands in the script editor, and save the file, calling it *diamonds.R*. Next, run the complete script, which will save a PDF and CSV file into your project directory. Don't worry about the details, you'll learn them later in the book:

```
library(tidyverse)

ggplot(diamonds, aes(carat, price)) +
  geom_hex()
ggsave("diamonds.pdf")

write_csv(diamonds, "diamonds.csv")
```

Quit RStudio. Inspect the folder associated with your project—notice the *.Rproj* file. Double-click that file to reopen the project. Notice you get back to where you left off: it's the same working directory and command history, and all the files you were working on are still open. Because you followed my instructions above, you will, however, have a completely fresh environment, guaranteeing that you're starting with a clean slate.

In your favorite OS-specific way, search your computer for *diamonds.pdf* and you will find the PDF (no surprise) but *also the script that created it* (*diamonds.r*). This is huge win! One day you will want to remake a figure or just understand where it came from. If you rigorously save figures to files *with R code* and never with the mouse or the clipboard, you will be able to reproduce old work with ease!

Summary

In summary, RStudio projects give you a solid workflow that will serve you well in the future:

- Create an RStudio project for each data analyis project.
- Keep data files there; we'll talk about loading them into R in Chapter 8.
- Keep scripts there; edit them, and run them in bits or as a whole.
- Save your outputs (plots and cleaned data) there.
- Only ever use relative paths, not absolute paths.

Everything you need is in one place, and cleanly separated from all the other projects that you are working on.

PART II
Wrangle

In this part of the book, you'll learn about data wrangling, the art of getting your data into R in a useful form for visualization and modeling. Data wrangling is very important: without it you can't work with your own data! There are three main parts to data wrangling:

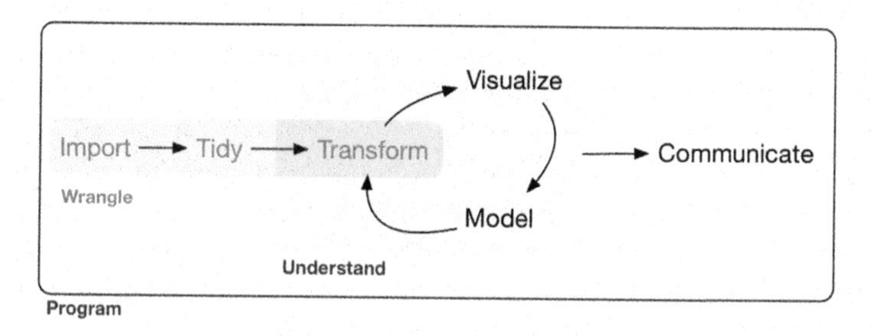

This part of the book proceeds as follows:

- In Chapter 7, you'll learn about the variant of the data frame that we use in this book: the *tibble*. You'll learn what makes them different from regular data frames, and how you can construct them "by hand."
- In Chapter 8, you'll learn how to get your data from disk and into R. We'll focus on plain-text rectangular formats, but will give you pointers to packages that help with other types of data.

- In Chapter 9, you'll learn about tidy data, a consistent way of storing your data that makes transformation, visualization, and modeling easier. You'll learn the underlying principles, and how to get your data into a tidy form.

Data wrangling also encompasses data transformation, which you've already learned a little about. Now we'll focus on new skills for three specific types of data you will frequently encounter in practice:

- Chapter 10 will give you tools for working with multiple interrelated datasets.
- Chapter 11 will introduce regular expressions, a powerful tool for manipulating strings.
- Chapter 12 will show you how R stores categorical data. They are used when a variable has a fixed set of possible values, or when you want to use a nonalphabetical ordering of a string.
- Chapter 13 will give you the key tools for working with dates and date-times.

CHAPTER 7

Tibbles with tibble

Introduction

Throughout this book we work with "tibbles" instead of R's traditional `data.frame`. Tibbles *are* data frames, but they tweak some older behaviors to make life a little easier. R is an old language, and some things that were useful 10 or 20 years ago now get in your way. It's difficult to change base R without breaking existing code, so most innovation occurs in packages. Here we will describe the *tibble* package, which provides opinionated data frames that make working in the tidyverse a little easier. In most places, I'll use the terms tibble and data frame interchangeably; when I want to draw particular attention to R's built-in data frame, I'll call them `data.frame`s.

If this chapter leaves you wanting to learn more about tibbles, you might enjoy `vignette("tibble")`.

Prerequisites

In this chapter we'll explore the **tibble** package, part of the core tidyverse.

```
library(tidyverse)
```

Creating Tibbles

Almost all of the functions that you'll use in this book produce tibbles, as tibbles are one of the unifying features of the tidyverse. Most other R packages use regular data frames, so you might want to coerce a data frame to a tibble. You can do that with `as_tibble()`:

```
as_tibble(iris)
#> # A tibble: 150 × 5
#>   Sepal.Length Sepal.Width Petal.Length Petal.Width Species
#>          <dbl>       <dbl>        <dbl>       <dbl>  <fctr>
#> 1          5.1         3.5          1.4         0.2  setosa
#> 2          4.9         3.0          1.4         0.2  setosa
#> 3          4.7         3.2          1.3         0.2  setosa
#> 4          4.6         3.1          1.5         0.2  setosa
#> 5          5.0         3.6          1.4         0.2  setosa
#> 6          5.4         3.9          1.7         0.4  setosa
#> # ... with 144 more rows
```

You can create a new tibble from individual vectors with `tibble()`. `tibble()` will automatically recycle inputs of length 1, and allows you to refer to variables that you just created, as shown here:

```
tibble(
  x = 1:5,
  y = 1,
  z = x ^ 2 + y
)
#> # A tibble: 5 × 3
#>       x     y     z
#>   <int> <dbl> <dbl>
#> 1     1     1     2
#> 2     2     1     5
#> 3     3     1    10
#> 4     4     1    17
#> 5     5     1    26
```

If you're already familiar with `data.frame()`, note that `tibble()` does much less: it never changes the type of the inputs (e.g., it never converts strings to factors!), it never changes the names of variables, and it never creates row names.

It's possible for a tibble to have column names that are not valid R variable names, aka *nonsyntactic* names. For example, they might not start with a letter, or they might contain unusual characters like a space. To refer to these variables, you need to surround them with backticks, `:

```
tb <- tibble(
  `:)` = "smile",
  ` ` = "space",
  `2000` = "number"
)
tb
```

```
#> # A tibble: 1 × 3
#>     `:)`    ` ` `2000`
#>    <chr> <chr>  <chr>
#> 1 smile space number
```

You'll also need the backticks when working with these variables in other packages, like **ggplot2**, **dplyr**, and **tidyr**.

Another way to create a tibble is with `tribble()`, short for *transposed tibble*. `tribble()` is customized for data entry in code: column headings are defined by formulas (i.e., they start with ~), and entries are separated by commas. This makes it possible to lay out small amounts of data in easy-to-read form:

```
tribble(
  ~x, ~y, ~z,
  #--|--|----
  "a", 2, 3.6,
  "b", 1, 8.5
)
#> # A tibble: 2 × 3
#>       x     y     z
#>   <chr> <dbl> <dbl>
#> 1     a     2   3.6
#> 2     b     1   8.5
```

I often add a comment (the line starting with #) to make it really clear where the header is.

Tibbles Versus data.frame

There are two main differences in the usage of a tibble versus a classic `data.frame`: printing and subsetting.

Printing

Tibbles have a refined print method that shows only the first 10 rows, and all the columns that fit on screen. This makes it much easier to work with large data. In addition to its name, each column reports its type, a nice feature borrowed from `str()`:

```
tibble(
  a = lubridate::now() + runif(1e3) * 86400,
  b = lubridate::today() + runif(1e3) * 30,
  c = 1:1e3,
  d = runif(1e3),
  e = sample(letters, 1e3, replace = TRUE)
)
```

```
#> # A tibble: 1,000 × 5
#>                     a          b     c     d     e
#>                <dttm>     <date> <int> <dbl> <chr>
#> 1 2016-10-10 17:14:14 2016-10-17     1 0.368     h
#> 2 2016-10-11 11:19:24 2016-10-22     2 0.612     n
#> 3 2016-10-11 05:43:03 2016-11-01     3 0.415     l
#> 4 2016-10-10 19:04:20 2016-10-31     4 0.212     x
#> 5 2016-10-10 15:28:37 2016-10-28     5 0.733     a
#> 6 2016-10-11 02:29:34 2016-10-24     6 0.460     v
#> # ... with 994 more rows
```

Tibbles are designed so that you don't accidentally overwhelm your console when you print large data frames. But sometimes you need more output than the default display. There are a few options that can help.

First, you can explicitly `print()` the data frame and control the number of rows (`n`) and the `width` of the display. `width = Inf` will display all columns:

```
nycflights13::flights %>%
  print(n = 10, width = Inf)
```

You can also control the default print behavior by setting options:

- `options(tibble.print_max = n, tibble.print_min = m)`: if more than `m` rows, print only `n` rows. Use `options(dplyr.print_min = Inf)` to always show all rows.
- Use `options(tibble.width = Inf)` to always print all columns, regardless of the width of the screen.

You can see a complete list of options by looking at the package help with `package?tibble`.

A final option is to use RStudio's built-in data viewer to get a scrollable view of the complete dataset. This is also often useful at the end of a long chain of manipulations:

```
nycflights13::flights %>%
  View()
```

Subsetting

So far all the tools you've learned have worked with complete data frames. If you want to pull out a single variable, you need some new tools, `$` and `[[`. `[[` can extract by name or position; `$` only extracts by name but is a little less typing:

```
df <- tibble(
  x = runif(5),
  y = rnorm(5)
)

# Extract by name
df$x
#> [1] 0.434 0.395 0.548 0.762 0.254
df[["x"]]
#> [1] 0.434 0.395 0.548 0.762 0.254

# Extract by position
df[[1]]
#> [1] 0.434 0.395 0.548 0.762 0.254
```

To use these in a pipe, you'll need to use the special placeholder `.`:

```
df %>% .$x
#> [1] 0.434 0.395 0.548 0.762 0.254
df %>% .[["x"]]
#> [1] 0.434 0.395 0.548 0.762 0.254
```

Compared to a `data.frame`, tibbles are more strict: they never do partial matching, and they will generate a warning if the column you are trying to access does not exist.

Interacting with Older Code

Some older functions don't work with tibbles. If you encounter one of these functions, use `as.data.frame()` to turn a tibble back to a `data.frame`:

```
class(as.data.frame(tb))
#> [1] "data.frame"
```

The main reason that some older functions don't work with tibbles is the `[` function. We don't use `[` much in this book because `dplyr::filter()` and `dplyr::select()` allow you to solve the same problems with clearer code (but you will learn a little about it in "Subsetting" on page 300). With base R data frames, `[` sometimes returns a data frame, and sometimes returns a vector. With tibbles, `[` always returns another tibble.

Exercises

1. How can you tell if an object is a tibble? (Hint: try printing `mtcars`, which is a regular data frame.)

2. Compare and contrast the following operations on a `data.frame` and equivalent tibble. What is different? Why might the default data frame behaviors cause you frustration?

```
df <- data.frame(abc = 1, xyz = "a")
df$x
df[, "xyz"]
df[, c("abc", "xyz")]
```

3. If you have the name of a variable stored in an object, e.g., `var <- "mpg"`, how can you extract the reference variable from a tibble?

4. Practice referring to nonsyntactic names in the following data frame by:

   a. Extracting the variable called `1`.

   b. Plotting a scatterplot of `1` versus `2`.

   c. Creating a new column called `3`, which is `2` divided by `1`.

   d. Renaming the columns to `one`, `two`, and `three`:

   ```
   annoying <- tibble(
     `1` = 1:10,
     `2` = `1` * 2 + rnorm(length(`1`))
   )
   ```

5. What does `tibble::enframe()` do? When might you use it?

6. What option controls how many additional column names are printed at the footer of a tibble?

CHAPTER 8

Data Import with readr

Introduction

Working with data provided by R packages is a great way to learn the tools of data science, but at some point you want to stop learning and start working with your own data. In this chapter, you'll learn how to read plain-text rectangular files into R. Here, we'll only scratch the surface of data import, but many of the principles will translate to other forms of data. We'll finish with a few pointers to packages that are useful for other types of data.

Prerequisites

In this chapter, you'll learn how to load flat files in R with the **readr** package, which is part of the core tidyverse.

```
library(tidyverse)
```

Getting Started

Most of **readr**'s functions are concerned with turning flat files into data frames:

- `read_csv()` reads comma-delimited files, `read_csv2()` reads semicolon-separated files (common in countries where `,` is used as the decimal place), `read_tsv()` reads tab-delimited files, and `read_delim()` reads in files with any delimiter.

- `read_fwf()` reads fixed-width files. You can specify fields either by their widths with `fwf_widths()` or their position with `fwf_positions()`. `read_table()` reads a common variation of fixed-width files where columns are separated by white space.
- `read_log()` reads Apache style log files. (But also check out **webreadr** (*https://github.com/Ironholds/webreadr*), which is built on top of `read_log()` and provides many more helpful tools.)

These functions all have similar syntax: once you've mastered one, you can use the others with ease. For the rest of this chapter we'll focus on `read_csv()`. Not only are CSV files one of the most common forms of data storage, but once you understand `read_csv()`, you can easily apply your knowledge to all the other functions in **readr**.

The first argument to `read_csv()` is the most important; it's the path to the file to read:

```
heights <- read_csv("data/heights.csv")
#> Parsed with column specification:
#> cols(
#>   earn = col_double(),
#>   height = col_double(),
#>   sex = col_character(),
#>   ed = col_integer(),
#>   age = col_integer(),
#>   race = col_character()
#> )
```

When you run `read_csv()` it prints out a column specification that gives the name and type of each column. That's an important part of **readr**, which we'll come back to in "Parsing a File" on page 137.

You can also supply an inline CSV file. This is useful for experimenting with **readr** and for creating reproducible examples to share with others:

```
read_csv("a,b,c
1,2,3
4,5,6")
#> # A tibble: 2 × 3
#>       a     b     c
#>   <int> <int> <int>
#> 1     1     2     3
#> 2     4     5     6
```

In both cases `read_csv()` uses the first line of the data for the column names, which is a very common convention. There are two cases where you might want to tweak this behavior:

- Sometimes there are a few lines of metadata at the top of the file. You can use `skip = n` to skip the first `n` lines; or use `comment = "#"` to drop all lines that start with (e.g.) `#`:

```
read_csv("The first line of metadata
  The second line of metadata
  x,y,z
  1,2,3", skip = 2)
#> # A tibble: 1 × 3
#>       x     y     z
#>   <int> <int> <int>
#> 1     1     2     3

read_csv("# A comment I want to skip
  x,y,z
  1,2,3", comment = "#")
#> # A tibble: 1 × 3
#>       x     y     z
#>   <int> <int> <int>
#> 1     1     2     3
```

- The data might not have column names. You can use `col_names = FALSE` to tell `read_csv()` not to treat the first row as headings, and instead label them sequentially from `X1` to `Xn`:

```
read_csv("1,2,3\n4,5,6", col_names = FALSE)
#> # A tibble: 2 × 3
#>      X1    X2    X3
#>   <int> <int> <int>
#> 1     1     2     3
#> 2     4     5     6
```

  (`"\n"` is a convenient shortcut for adding a new line. You'll learn more about it and other types of string escape in "String Basics" on page 195.)

  Alternatively you can pass `col_names` a character vector, which will be used as the column names:

```
read_csv("1,2,3\n4,5,6", col_names = c("x", "y", "z"))
#> # A tibble: 2 × 3
#>       x     y     z
#>   <int> <int> <int>
```

```
#> 1     1     2     3
#> 2     4     5     6
```

Another option that commonly needs tweaking is `na`. This specifies the value (or values) that are used to represent missing values in your file:

```
read_csv("a,b,c\n1,2,.", na = ".")
#> # A tibble: 1 × 3
#>       a     b     c
#>   <int> <int> <chr>
#> 1     1     2  <NA>
```

This is all you need to know to read ~75% of CSV files that you'll encounter in practice. You can also easily adapt what you've learned to read tab-separated files with `read_tsv()` and fixed-width files with `read_fwf()`. To read in more challenging files, you'll need to learn more about how **readr** parses each column, turning them into R vectors.

Compared to Base R

If you've used R before, you might wonder why we're not using `read.csv()`. There are a few good reasons to favor **readr** functions over the base equivalents:

- They are typically much faster (~10x) than their base equivalents. Long-running jobs have a progress bar, so you can see what's happening. If you're looking for raw speed, try `data.table::fread()`. It doesn't fit quite so well into the tidyverse, but it can be quite a bit faster.
- They produce tibbles, and they don't convert character vectors to factors, use row names, or munge the column names. These are common sources of frustration with the base R functions.
- They are more reproducible. Base R functions inherit some behavior from your operating system and environment variables, so import code that works on your computer might not work on someone else's.

Exercises

1. What function would you use to read a file where fields are separated with "|"?

2. Apart from `file`, `skip`, and `comment`, what other arguments do `read_csv()` and `read_tsv()` have in common?
3. What are the most important arguments to `read_fwf()`?
4. Sometimes strings in a CSV file contain commas. To prevent them from causing problems they need to be surrounded by a quoting character, like `"` or `'`. By convention, `read_csv()` assumes that the quoting character will be `"`, and if you want to change it you'll need to use `read_delim()` instead. What arguments do you need to specify to read the following text into a data frame?

    ```
    "x,y\n1,'a,b'"
    ```

5. Identify what is wrong with each of the following inline CSV files. What happens when you run the code?

    ```
    read_csv("a,b\n1,2,3\n4,5,6")
    read_csv("a,b,c\n1,2\n1,2,3,4")
    read_csv("a,b\n\"1")
    read_csv("a,b\n1,2\na,b")
    read_csv("a;b\n1;3")
    ```

Parsing a Vector

Before we get into the details of how **readr** reads files from disk, we need to take a little detour to talk about the `parse_*()` functions. These functions take a character vector and return a more specialized vector like a logical, integer, or date:

```
str(parse_logical(c("TRUE", "FALSE", "NA")))
#>  logi [1:3] TRUE FALSE NA
str(parse_integer(c("1", "2", "3")))
#>  int [1:3] 1 2 3
str(parse_date(c("2010-01-01", "1979-10-14")))
#>  Date[1:2], format: "2010-01-01" "1979-10-14"
```

These functions are useful in their own right, but are also an important building block for **readr**. Once you've learned how the individual parsers work in this section, we'll circle back and see how they fit together to parse a complete file in the next section.

Like all functions in the tidyverse, the `parse_*()` functions are uniform; the first argument is a character vector to parse, and the `na` argument specifies which strings should be treated as missing:

```
parse_integer(c("1", "231", ".", "456"), na = ".")
#> [1]   1 231  NA 456
```

If parsing fails, you'll get a warning:

```
x <- parse_integer(c("123", "345", "abc", "123.45"))
#> Warning: 2 parsing failures.
#> row col               expected actual
#>   3  -- an integer               abc
#>   4  -- no trailing characters   .45
```

And the failures will be missing in the output:

```
x
#> [1] 123 345  NA  NA
#> attr(,"problems")
#> # A tibble: 2 × 4
#>     row   col               expected actual
#>   <int> <int>                  <chr>  <chr>
#> 1     3    NA             an integer    abc
#> 2     4    NA no trailing characters    .45
```

If there are many parsing failures, you'll need to use `problems()` to get the complete set. This returns a tibble, which you can then manipulate with **dplyr**:

```
problems(x)
#> # A tibble: 2 × 4
#>     row   col               expected actual
#>   <int> <int>                  <chr>  <chr>
#> 1     3    NA             an integer    abc
#> 2     4    NA no trailing characters    .45
```

Using parsers is mostly a matter of understanding what's available and how they deal with different types of input. There are eight particularly important parsers:

- `parse_logical()` and `parse_integer()` parse logicals and integers, respectively. There's basically nothing that can go wrong with these parsers so I won't describe them here further.
- `parse_double()` is a strict numeric parser, and `parse_number()` is a flexible numeric parser. These are more complicated than you might expect because different parts of the world write numbers in different ways.
- `parse_character()` seems so simple that it shouldn't be necessary. But one complication makes it quite important: character encodings.

- `parse_factor()` creates factors, the data structure that R uses to represent categorical variables with fixed and known values.
- `parse_datetime()`, `parse_date()`, and `parse_time()` allow you to parse various date and time specifications. These are the most complicated because there are so many different ways of writing dates.

The following sections describe these parsers in more detail.

Numbers

It seems like it should be straightforward to parse a number, but three problems make it tricky:

- People write numbers differently in different parts of the world. For example, some countries use `.` in between the integer and fractional parts of a real number, while others use `,`.
- Numbers are often surrounded by other characters that provide some context, like "$1000" or "10%".
- Numbers often contain "grouping" characters to make them easier to read, like "1,000,000", and these grouping characters vary around the world.

To address the first problem, **readr** has the notion of a "locale," an object that specifies parsing options that differ from place to place. When parsing numbers, the most important option is the character you use for the decimal mark. You can override the default value of `.` by creating a new locale and setting the `decimal_mark` argument:

```
parse_double("1.23")
#> [1] 1.23
parse_double("1,23", locale = locale(decimal_mark = ","))
#> [1] 1.23
```

readr's default locale is US-centric, because generally R is US-centric (i.e., the documentation of base R is written in American English). An alternative approach would be to try and guess the defaults from your operating system. This is hard to do well, and, more importantly, makes your code fragile: even if it works on your computer, it might fail when you email it to a colleague in another country.

parse_number() addresses the second problem: it ignores non-numeric characters before and after the number. This is particularly useful for currencies and percentages, but also works to extract numbers embedded in text:

```
parse_number("$100")
#> [1] 100
parse_number("20%")
#> [1] 20
parse_number("It cost $123.45")
#> [1] 123
```

The final problem is addressed by the combination of parse_number() and the locale as parse_number() will ignore the "grouping mark":

```
# Used in America
parse_number("$123,456,789")
#> [1] 1.23e+08

# Used in many parts of Europe
parse_number(
  "123.456.789",
  locale = locale(grouping_mark = ".")
)
#> [1] 1.23e+08

# Used in Switzerland
parse_number(
  "123'456'789",
  locale = locale(grouping_mark = "'")
)
#> [1] 1.23e+08
```

Strings

It seems like parse_character() should be really simple—it could just return its input. Unfortunately life isn't so simple, as there are multiple ways to represent the same string. To understand what's going on, we need to dive into the details of how computers represent strings. In R, we can get at the underlying representation of a string using charToRaw():

```
charToRaw("Hadley")
#> [1] 48 61 64 6c 65 79
```

Each hexadecimal number represents a byte of information: 48 is H, 61 is a, and so on. The mapping from hexadecimal number to character is called the encoding, and in this case the encoding is called

ASCII. ASCII does a great job of representing English characters, because it's the *American* Standard Code for Information Interchange.

Things get more complicated for languages other than English. In the early days of computing there were many competing standards for encoding non-English characters, and to correctly interpret a string you needed to know both the values and the encoding. For example, two common encodings are Latin1 (aka ISO-8859-1, used for Western European languages) and Latin2 (aka ISO-8859-2, used for Eastern European languages). In Latin1, the byte `b1` is "±", but in Latin2, it's "ą"! Fortunately, today there is one standard that is supported almost everywhere: UTF-8. UTF-8 can encode just about every character used by humans today, as well as many extra symbols (like emoji!).

readr uses UTF-8 everywhere: it assumes your data is UTF-8 encoded when you read it, and always uses it when writing. This is a good default, but will fail for data produced by older systems that don't understand UTF-8. If this happens to you, your strings will look weird when you print them. Sometimes just one or two characters might be messed up; other times you'll get complete gibberish. For example:

```
x1 <- "El Ni\xf1o was particularly bad this year"
x2 <- "\x82\xb1\x82\xf1\x82\xc9\x82\xbf\x82\xcd"
```

To fix the problem you need to specify the encoding in `parse_character()`:

```
parse_character(x1, locale = locale(encoding = "Latin1"))
#> [1] "El Niño was particularly bad this year"
parse_character(x2, locale = locale(encoding = "Shift-JIS"))
#> [1] "こんにちは"
```

How do you find the correct encoding? If you're lucky, it'll be included somewhere in the data documentation. Unfortunately, that's rarely the case, so **readr** provides `guess_encoding()` to help you figure it out. It's not foolproof, and it works better when you have lots of text (unlike here), but it's a reasonable place to start. Expect to try a few different encodings before you find the right one:

```
guess_encoding(charToRaw(x1))
#>     encoding confidence
#> 1 ISO-8859-1       0.46
#> 2 ISO-8859-9       0.23
guess_encoding(charToRaw(x2))
```

```
#>   encoding confidence
#> 1   KOI8-R       0.42
```

The first argument to guess_encoding() can either be a path to a file, or, as in this case, a raw vector (useful if the strings are already in R).

Encodings are a rich and complex topic, and I've only scratched the surface here. If you'd like to learn more I'd recommend reading the detailed explanation at *http://kunststube.net/encoding/*.

Factors

R uses factors to represent categorical variables that have a known set of possible values. Give parse_factor() a vector of known levels to generate a warning whenever an unexpected value is present:

```
fruit <- c("apple", "banana")
parse_factor(c("apple", "banana", "bananana"), levels = fruit)
#> Warning: 1 parsing failure.
#> row col           expected   actual
#>   3  -- value in level set bananana
#> [1] apple  banana <NA>
#> attr(,"problems")
#> # A tibble: 1 × 4
#>     row   col           expected   actual
#>   <int> <int>              <chr>    <chr>
#> 1     3    NA value in level set bananana
#> Levels: apple banana
```

But if you have many problematic entries, it's often easier to leave them as character vectors and then use the tools you'll learn about in Chapter 11 and Chapter 12 to clean them up.

Dates, Date-Times, and Times

You pick between three parsers depending on whether you want a date (the number of days since 1970-01-01), a date-time (the number of seconds since midnight 1970-01-01), or a time (the number of seconds since midnight). When called without any additional arguments:

- parse_datetime() expects an ISO8601 date-time. ISO8601 is an international standard in which the components of a date are organized from biggest to smallest: year, month, day, hour, minute, second:

```
parse_datetime("2010-10-01T2010")
#> [1] "2010-10-01 20:10:00 UTC"

# If time is omitted, it will be set to midnight
parse_datetime("20101010")
#> [1] "2010-10-10 UTC"
```

This is the most important date/time standard, and if you work with dates and times frequently, I recommend reading *https://en.wikipedia.org/wiki/ISO_8601*.

- `parse_date()` expects a four-digit year, a `-` or `/`, the month, a `-` or `/`, then the day:

```
parse_date("2010-10-01")
#> [1] "2010-10-01"
```

- `parse_time()` expects the hour, `:`, minutes, optionally `:` and seconds, and an optional a.m./p.m. specifier:

```
library(hms)
parse_time("01:10 am")
#> 01:10:00
parse_time("20:10:01")
#> 20:10:01
```

Base R doesn't have a great built-in class for time data, so we use the one provided in the **hms** package.

If these defaults don't work for your data you can supply your own date-time `format`, built up of the following pieces:

Year

`%Y` (4 digits).

`%y` (2 digits; 00-69 → 2000-2069, 70-99 → 1970-1999).

Month

`%m` (2 digits).

`%b` (abbreviated name, like "Jan").

`%B` (full name, "January").

Day

`%d` (2 digits).

`%e` (optional leading space).

Time

`%H` (0-23 hour format).

`%I` (0-12, must be used with `%p`).

`%p` (a.m./p.m. indicator).

`%M` (minutes).

`%S` (integer seconds).

`%OS` (real seconds).

`%Z` (time zone [a name, e.g., `America/Chicago`]). Note: beware of abbreviations. If you're American, note that "EST" is a Canadian time zone that does not have daylight saving time. It is Eastern Standard Time! We'll come back to this in "Time Zones" on page 254.

`%z` (as offset from UTC, e.g., `+0800`).

Nondigits

`%.` (skips one nondigit character).

`%*` (skips any number of nondigits).

The best way to figure out the correct format is to create a few examples in a character vector, and test with one of the parsing functions. For example:

```
parse_date("01/02/15", "%m/%d/%y")
#> [1] "2015-01-02"
parse_date("01/02/15", "%d/%m/%y")
#> [1] "2015-02-01"
parse_date("01/02/15", "%y/%m/%d")
#> [1] "2001-02-15"
```

If you're using `%b` or `%B` with non-English month names, you'll need to set the `lang` argument to `locale()`. See the list of built-in languages in `date_names_langs()`, or if your language is not already included, create your own with `date_names()`:

```
parse_date("1 janvier 2015", "%d %B %Y", locale = locale("fr"))
#> [1] "2015-01-01"
```

Exercises

1. What are the most important arguments to `locale()`?

2. What happens if you try and set `decimal_mark` and `grouping_mark` to the same character? What happens to the default value of `grouping_mark` when you set `decimal_mark` to ","? What happens to the default value of `decimal_mark` when you set the `grouping_mark` to "."?
3. I didn't discuss the `date_format` and `time_format` options to `locale()`. What do they do? Construct an example that shows when they might be useful.
4. If you live outside the US, create a new locale object that encapsulates the settings for the types of files you read most commonly.
5. What's the difference between `read_csv()` and `read_csv2()`?
6. What are the most common encodings used in Europe? What are the most common encodings used in Asia? Do some googling to find out.
7. Generate the correct format string to parse each of the following dates and times:

```
d1 <- "January 1, 2010"
d2 <- "2015-Mar-07"
d3 <- "06-Jun-2017"
d4 <- c("August 19 (2015)", "July 1 (2015)")
d5 <- "12/30/14" # Dec 30, 2014
t1 <- "1705"
t2 <- "11:15:10.12 PM"
```

Parsing a File

Now that you've learned how to parse an individual vector, it's time to return to the beginning and explore how **readr** parses a file. There are two new things that you'll learn about in this section:

- How **readr** automatically guesses the type of each column.
- How to override the default specification.

Strategy

readr uses a heuristic to figure out the type of each column: it reads the first 1000 rows and uses some (moderately conservative) heuristics to figure out the type of each column. You can emulate this process with a character vector using `guess_parser()`, which returns **readr**'s best guess, and `parse_guess()`, which uses that guess to parse the column:

```
guess_parser("2010-10-01")
#> [1] "date"
guess_parser("15:01")
#> [1] "time"
guess_parser(c("TRUE", "FALSE"))
#> [1] "logical"
guess_parser(c("1", "5", "9"))
#> [1] "integer"
guess_parser(c("12,352,561"))
#> [1] "number"

str(parse_guess("2010-10-10"))
#>  Date[1:1], format: "2010-10-10"
```

The heuristic tries each of the following types, stopping when it finds a match:

logical
: Contains only "F", "T", "FALSE", or "TRUE".

integer
: Contains only numeric characters (and -).

double
: Contains only valid doubles (including numbers like `4.5e-5`).

number
: Contains valid doubles with the grouping mark inside.

time
: Matches the default `time_format`.

date
: Matches the default `date_format`.

date-time
: Any ISO8601 date.

If none of these rules apply, then the column will stay as a vector of strings.

Problems

These defaults don't always work for larger files. There are two basic problems:

- The first thousand rows might be a special case, and **readr** guesses a type that is not sufficiently general. For example, you might have a column of doubles that only contains integers in the first 1000 rows.
- The column might contain a lot of missing values. If the first 1000 rows contain only `NA`s, **readr** will guess that it's a character vector, whereas you probably want to parse it as something more specific.

readr contains a challenging CSV that illustrates both of these problems:

```
challenge <- read_csv(readr_example("challenge.csv"))
#> Parsed with column specification:
#> cols(
#>   x = col_integer(),
#>   y = col_character()
#> )
#> Warning: 1000 parsing failures.
#>  row col               expected             actual
#> 1001   x no trailing characters .23837975086644292
#> 1002   x no trailing characters .41167997173033655
#> 1003   x no trailing characters .7460716762579978
#> 1004   x no trailing characters .723450553836301
#> 1005   x no trailing characters .614524137461558
#> .... ... ...................... ..................
#> See problems(...) for more details.
```

(Note the use of `readr_example()`, which finds the path to one of the files included with the package.)

There are two printed outputs: the column specification generated by looking at the first 1000 rows, and the first five parsing failures. It's always a good idea to explicitly pull out the `problems()`, so you can explore them in more depth:

```
problems(challenge)
#> # A tibble: 1,000 × 4
#>     row   col               expected             actual
#>   <int> <chr>                  <chr>              <chr>
#> 1  1001     x no trailing characters .23837975086644292
#> 2  1002     x no trailing characters .41167997173033655
#> 3  1003     x no trailing characters  .7460716762579978
```

```
#> 4  1004     x no trailing characters   .723450553836301
#> 5  1005     x no trailing characters   .614524137461558
#> 6  1006     x no trailing characters   .473980569280684
#> # ... with 994 more rows
```

A good strategy is to work column by column until there are no problems remaining. Here we can see that there are a lot of parsing problems with the x column—there are trailing characters after the integer value. That suggests we need to use a double parser instead.

To fix the call, start by copying and pasting the column specification into your original call:

```
challenge <- read_csv(
  readr_example("challenge.csv"),
  col_types = cols(
    x = col_integer(),
    y = col_character()
  )
)
```

Then you can tweak the type of the x column:

```
challenge <- read_csv(
  readr_example("challenge.csv"),
  col_types = cols(
    x = col_double(),
    y = col_character()
  )
)
```

That fixes the first problem, but if we look at the last few rows, you'll see that they're dates stored in a character vector:

```
tail(challenge)
#> # A tibble: 6 × 2
#>        x          y
#>    <dbl>      <chr>
#> 1 0.805 2019-11-21
#> 2 0.164 2018-03-29
#> 3 0.472 2014-08-04
#> 4 0.718 2015-08-16
#> 5 0.270 2020-02-04
#> 6 0.608 2019-01-06
```

You can fix that by specifying that y is a date column:

```
challenge <- read_csv(
  readr_example("challenge.csv"),
  col_types = cols(
    x = col_double(),
    y = col_date()
```

```
  )
)
tail(challenge)
#> # A tibble: 6 × 2
#>       x          y
#>   <dbl>     <date>
#> 1 0.805 2019-11-21
#> 2 0.164 2018-03-29
#> 3 0.472 2014-08-04
#> 4 0.718 2015-08-16
#> 5 0.270 2020-02-04
#> 6 0.608 2019-01-06
```

Every `parse_xyz()` function has a corresponding `col_xyz()` function. You use `parse_xyz()` when the data is in a character vector in R already; you use `col_xyz()` when you want to tell **readr** how to load the data.

I highly recommend always supplying `col_types`, building up from the printout provided by **readr**. This ensures that you have a consistent and reproducible data import script. If you rely on the default guesses and your data changes, **readr** will continue to read it in. If you want to be really strict, use `stop_for_problems()`: that will throw an error and stop your script if there are any parsing problems.

Other Strategies

There are a few other general strategies to help you parse files:

- In the previous example, we just got unlucky: if we look at just one more row than the default, we can correctly parse in one shot:

  ```
  challenge2 <- read_csv(
                  readr_example("challenge.csv"),
                  guess_max = 1001
                )
  #> Parsed with column specification:
  #> cols(
  #>   x = col_double(),
  #>   y = col_date(format = "")
  #> )
  challenge2
  #> # A tibble: 2,000 × 2
  #>       x      y
  #>   <dbl> <date>
  #> 1   404   <NA>
  ```

```
#> 2  4172    <NA>
#> 3  3004    <NA>
#> 4   787    <NA>
#> 5    37    <NA>
#> 6  2332    <NA>
#> # ... with 1,994 more rows
```

- Sometimes it's easier to diagnose problems if you just read in all the columns as character vectors:

  ```
  challenge2 <- read_csv(readr_example("challenge.csv"),
    col_types = cols(.default = col_character())
  )
  ```

  This is particularly useful in conjunction with `type_convert()`, which applies the parsing heuristics to the character columns in a data frame:

  ```
  df <- tribble(
    ~x,  ~y,
    "1", "1.21",
    "2", "2.32",
    "3", "4.56"
  )
  df
  #> # A tibble: 3 × 2
  #>       x     y
  #>   <chr> <chr>
  #> 1     1  1.21
  #> 2     2  2.32
  #> 3     3  4.56

  # Note the column types
  type_convert(df)
  #> Parsed with column specification:
  #> cols(
  #>   x = col_integer(),
  #>   y = col_double()
  #> )
  #> # A tibble: 3 × 2
  #>       x     y
  #>   <int> <dbl>
  #> 1     1  1.21
  #> 2     2  2.32
  #> 3     3  4.56
  ```

- If you're reading a very large file, you might want to set `n_max` to a smallish number like 10,000 or 100,000. That will accelerate your iterations while you eliminate common problems.

- If you're having major parsing problems, sometimes it's easier to just read into a character vector of lines with `read_lines()`, or even a character vector of length 1 with `read_file()`. Then you can use the string parsing skills you'll learn later to parse more exotic formats.

Writing to a File

readr also comes with two useful functions for writing data back to disk: `write_csv()` and `write_tsv()`. Both functions increase the chances of the output file being read back in correctly by:

- Always encoding strings in UTF-8.
- Saving dates and date-times in ISO8601 format so they are easily parsed elsewhere.

If you want to export a CSV file to Excel, use `write_excel_csv()`—this writes a special character (a "byte order mark") at the start of the file, which tells Excel that you're using the UTF-8 encoding.

The most important arguments are `x` (the data frame to save) and `path` (the location to save it). You can also specify how missing values are written with `na`, and if you want to `append` to an existing file:

```
write_csv(challenge, "challenge.csv")
```

Note that the type information is lost when you save to CSV:

```
challenge
#> # A tibble: 2,000 × 2
#>       x      y
#>   <dbl> <date>
#> 1   404   <NA>
#> 2  4172   <NA>
#> 3  3004   <NA>
#> 4   787   <NA>
#> 5    37   <NA>
#> 6  2332   <NA>
#> # ... with 1,994 more rows
write_csv(challenge, "challenge-2.csv")
read_csv("challenge-2.csv")
#> Parsed with column specification:
#> cols(
#>   x = col_double(),
#>   y = col_character()
#> )
```

```
#> # A tibble: 2,000 × 2
#>        x     y
#>    <dbl> <chr>
#> 1    404  <NA>
#> 2   4172  <NA>
#> 3   3004  <NA>
#> 4    787  <NA>
#> 5     37  <NA>
#> 6   2332  <NA>
#> # ... with 1,994 more rows
```

This makes CSVs a little unreliable for caching interim results—you need to re-create the column specification every time you load in. There are two alternatives:

- `write_rds()` and `read_rds()` are uniform wrappers around the base functions `readRDS()` and `saveRDS()`. These store data in R's custom binary format called RDS:

  ```
  write_rds(challenge, "challenge.rds")
  read_rds("challenge.rds")
  #> # A tibble: 2,000 × 2
  #>        x      y
  #>    <dbl> <date>
  #> 1    404    <NA>
  #> 2   4172    <NA>
  #> 3   3004    <NA>
  #> 4    787    <NA>
  #> 5     37    <NA>
  #> 6   2332    <NA>
  #> # ... with 1,994 more rows
  ```

- The **feather** package implements a fast binary file format that can be shared across programming languages:

  ```
  library(feather)
  write_feather(challenge, "challenge.feather")
  read_feather("challenge.feather")
  #> # A tibble: 2,000 x 2
  #>        x      y
  #>    <dbl> <date>
  #> 1    404    <NA>
  #> 2   4172    <NA>
  #> 3   3004    <NA>
  #> 4    787    <NA>
  #> 5     37    <NA>
  #> 6   2332    <NA>
  #> # ... with 1,994 more rows
  ```

feather tends to be faster than RDS and is usable outside of R. RDS supports list-columns (which you'll learn about in Chapter 20); **feather** currently does not.

Other Types of Data

To get other types of data into R, we recommend starting with the tidyverse packages listed next. They're certainly not perfect, but they are a good place to start. For rectangular data:

- **haven** reads SPSS, Stata, and SAS files.
- **readxl** reads Excel files (both *.xls* and *.xlsx*).
- **DBI**, along with a database-specific backend (e.g., **RMySQL**, **RSQLite**, **RPostgreSQL**, etc.) allows you to run SQL queries against a database and return a data frame.

For hierarchical data: use **jsonlite** (by Jeroen Ooms) for JSON, and **xml2** for XML. Jenny Bryan has some excellent worked examples at *https://jennybc.github.io/purrr-tutorial/*.

For other file types, try the R data import/export manual (*https://cran.r-project.org/doc/manuals/r-release/R-data.html*) and the **rio** (*https://github.com/leeper/rio*) package.

CHAPTER 9

Tidy Data with tidyr

Introduction

> Happy families are all alike; every unhappy family is unhappy in its own way.
>
> —Leo Tolstoy

> Tidy datasets are all alike, but every messy dataset is messy in its own way.
>
> —Hadley Wickham

In this chapter, you will learn a consistent way to organize your data in R, an organization called *tidy data*. Getting your data into this format requires some up-front work, but that work pays off in the long term. Once you have tidy data and the tidy tools provided by packages in the tidyverse, you will spend much less time munging data from one representation to another, allowing you to spend more time on the analytic questions at hand.

This chapter will give you a practical introduction to tidy data and the accompanying tools in the **tidyr** package. If you'd like to learn more about the underlying theory, you might enjoy the *Tidy Data* paper (*http://www.jstatsoft.org/v59/i10/paper*) published in the *Journal of Statistical Software*.

Prerequisites

In this chapter we'll focus on **tidyr**, a package that provides a bunch of tools to help tidy up your messy datasets. **tidyr** is a member of the core tidyverse.

```
library(tidyverse)
```

Tidy Data

You can represent the same underlying data in multiple ways. The following example shows the same data organized in four different ways. Each dataset shows the same values of four variables, *country*, *year*, *population*, and *cases*, but each dataset organizes the values in a different way:

```
table1
#> # A tibble: 6 × 4
#>       country  year  cases population
#>         <chr> <int>  <int>      <int>
#> 1 Afghanistan  1999    745   19987071
#> 2 Afghanistan  2000   2666   20595360
#> 3      Brazil  1999  37737  172006362
#> 4      Brazil  2000  80488  174504898
#> 5       China  1999 212258 1272915272
#> 6       China  2000 213766 1280428583
table2
#> # A tibble: 12 × 4
#>       country  year       type      count
#>         <chr> <int>      <chr>      <int>
#> 1 Afghanistan  1999      cases        745
#> 2 Afghanistan  1999 population   19987071
#> 3 Afghanistan  2000      cases       2666
#> 4 Afghanistan  2000 population   20595360
#> 5      Brazil  1999      cases      37737
#> 6      Brazil  1999 population  172006362
#> # ... with 6 more rows
table3
#> # A tibble: 6 × 3
#>       country  year              rate
#> *       <chr> <int>             <chr>
#> 1 Afghanistan  1999      745/19987071
#> 2 Afghanistan  2000     2666/20595360
#> 3      Brazil  1999   37737/172006362
#> 4      Brazil  2000   80488/174504898
#> 5       China  1999 212258/1272915272
#> 6       China  2000 213766/1280428583

# Spread across two tibbles
```

```r
table4a  # cases
#> # A tibble: 3 × 3
#>       country `1999` `2000`
#> *       <chr>  <int>  <int>
#> 1 Afghanistan    745   2666
#> 2      Brazil  37737  80488
#> 3       China 212258 213766
table4b  # population
#> # A tibble: 3 × 3
#>       country     `1999`     `2000`
#> *       <chr>      <int>      <int>
#> 1 Afghanistan   19987071   20595360
#> 2      Brazil  172006362  174504898
#> 3       China 1272915272 1280428583
```

These are all representations of the same underlying data, but they are not equally easy to use. One dataset, the tidy dataset, will be much easier to work with inside the tidyverse.

There are three interrelated rules which make a dataset tidy:

1. Each variable must have its own column.
2. Each observation must have its own row.
3. Each value must have its own cell.

Figure 9-1 shows the rules visually.

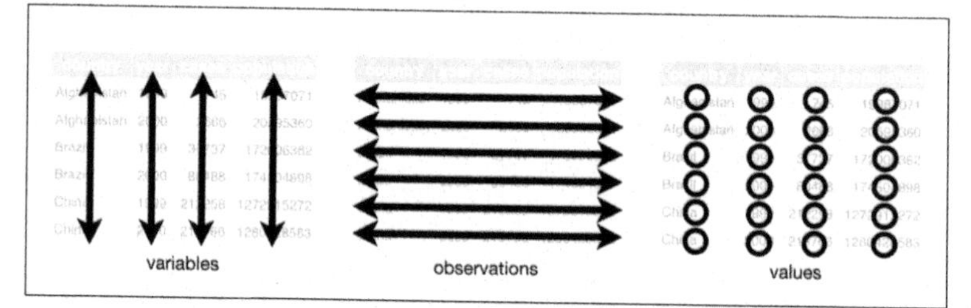

Figure 9-1. The following three rules make a dataset tidy: variables are in columns, observations are in rows, and values are in cells

These three rules are interrelated because it's impossible to only satisfy two of the three. That interrelationship leads to an even simpler set of practical instructions:

1. Put each dataset in a tibble.
2. Put each variable in a column.

In this example, only `table1` is tidy. It's the only representation where each column is a variable.

Why ensure that your data is tidy? There are two main advantages:

- There's a general advantage to picking one consistent way of storing data. If you have a consistent data structure, it's easier to learn the tools that work with it because they have an underlying uniformity.
- There's a specific advantage to placing variables in columns because it allows R's vectorized nature to shine. As you learned in "Useful Creation Functions" on page 56 and "Useful Summary Functions" on page 66, most built-in R functions work with vectors of values. That makes transforming tidy data feel particularly natural.

dplyr, **ggplot2**, and all the other packages in the tidyverse are designed to work with tidy data. Here are a couple of small examples showing how you might work with `table1`:

```
# Compute rate per 10,000
table1 %>%
  mutate(rate = cases / population * 10000)
#> # A tibble: 6 × 5
#>       country  year  cases population  rate
#>         <chr> <int>  <int>      <int> <dbl>
#> 1 Afghanistan  1999    745   19987071 0.373
#> 2 Afghanistan  2000   2666   20595360 1.294
#> 3      Brazil  1999  37737  172006362 2.194
#> 4      Brazil  2000  80488  174504898 4.612
#> 5       China  1999 212258 1272915272 1.667
#> 6       China  2000 213766 1280428583 1.669

# Compute cases per year
table1 %>%
  count(year, wt = cases)
#> # A tibble: 2 × 2
#>    year      n
#>   <int>  <int>
#> 1  1999 250740
#> 2  2000 296920

# Visualize changes over time
library(ggplot2)
ggplot(table1, aes(year, cases)) +
  geom_line(aes(group = country), color = "grey50") +
  geom_point(aes(color = country))
```

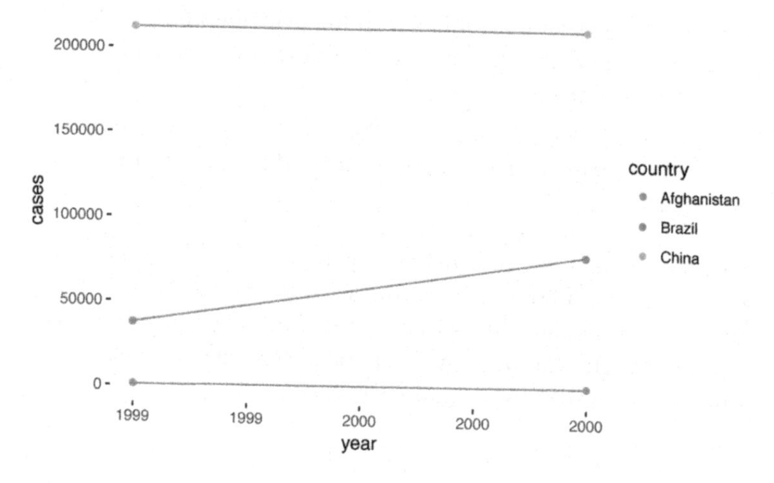

Exercises

1. Using prose, describe how the variables and observations are organized in each of the sample tables.
2. Compute the `rate` for `table2`, and `table4a` + `table4b`. You will need to perform four operations:

   a. Extract the number of TB cases per country per year.

   b. Extract the matching population per country per year.

   c. Divide cases by population, and multiply by 10,000.

   d. Store back in the appropriate place.

   Which representation is easiest to work with? Which is hardest? Why?
3. Re-create the plot showing change in cases over time using `table2` instead of `table1`. What do you need to do first?

Spreading and Gathering

The principles of tidy data seem so obvious that you might wonder if you'll ever encounter a dataset that isn't tidy. Unfortunately, however, most data that you will encounter will be untidy. There are two main reasons:

- Most people aren't familiar with the principles of tidy data, and it's hard to derive them yourself unless you spend a *lot* of time working with data.
- Data is often organized to facilitate some use other than analysis. For example, data is often organized to make entry as easy as possible.

This means for most real analyses, you'll need to do some tidying. The first step is always to figure out what the variables and observations are. Sometimes this is easy; other times you'll need to consult with the people who originally generated the data. The second step is to resolve one of two common problems:

- One variable might be spread across multiple columns.
- One observation might be scattered across multiple rows.

Typically a dataset will only suffer from one of these problems; it'll only suffer from both if you're really unlucky! To fix these problems, you'll need the two most important functions in **tidyr**: `gather()` and `spread()`.

Gathering

A common problem is a dataset where some of the column names are not names of variables, but *values* of a variable. Take `table4a`; the column names `1999` and `2000` represent values of the `year` variable, and each row represents two observations, not one:

```
table4a
#> # A tibble: 3 × 3
#>       country `1999` `2000`
#> *       <chr>  <int>  <int>
#> 1 Afghanistan    745   2666
#> 2      Brazil  37737  80488
#> 3       China 212258 213766
```

To tidy a dataset like this, we need to *gather* those columns into a new pair of variables. To describe that operation we need three parameters:

- The set of columns that represent values, not variables. In this example, those are the columns `1999` and `2000`.

- The name of the variable whose values form the column names. I call that the `key`, and here it is `year`.
- The name of the variable whose values are spread over the cells. I call that `value`, and here it's the number of `cases`.

Together those parameters generate the call to `gather()`:

```
table4a %>%
  gather(`1999`, `2000`, key = "year", value = "cases")
#> # A tibble: 6 × 3
#>       country  year  cases
#>         <chr> <chr>  <int>
#> 1 Afghanistan  1999    745
#> 2      Brazil  1999  37737
#> 3       China  1999 212258
#> 4 Afghanistan  2000   2666
#> 5      Brazil  2000  80488
#> 6       China  2000 213766
```

The columns to gather are specified with `dplyr::select()` style notation. Here there are only two columns, so we list them individually. Note that "1999" and "2000" are nonsyntactic names so we have to surround them in backticks. To refresh your memory of the other ways to select columns, see "Select Columns with select()" on page 51.

In the final result, the gathered columns are dropped, and we get new `key` and `value` columns. Otherwise, the relationships between the original variables are preserved. Visually, this is shown in Figure 9-2. We can use `gather()` to tidy `table4b` in a similar fashion. The only difference is the variable stored in the cell values:

```
table4b %>%
  gather(`1999`, `2000`, key = "year", value = "population")
#> # A tibble: 6 × 3
#>       country  year population
#>         <chr> <chr>      <int>
#> 1 Afghanistan  1999   19987071
#> 2      Brazil  1999  172006362
#> 3       China  1999 1272915272
#> 4 Afghanistan  2000   20595360
#> 5      Brazil  2000  174504898
#> 6       China  2000 1280428583
```

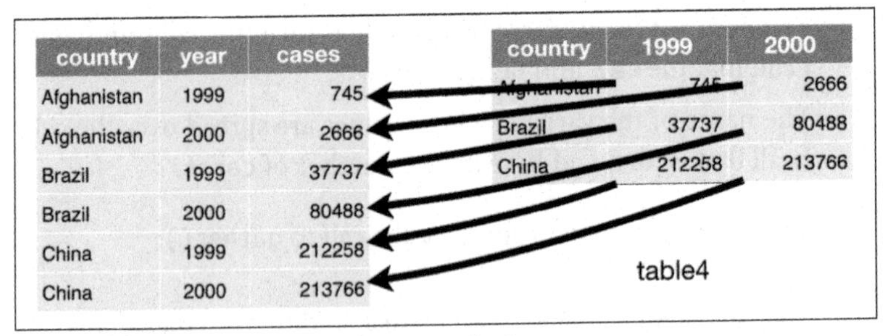

Figure 9-2. Gathering table4 into a tidy form

To combine the tidied versions of `table4a` and `table4b` into a single tibble, we need to use `dplyr::left_join()`, which you'll learn about in Chapter 10:

```
tidy4a <- table4a %>%
  gather(`1999`, `2000`, key = "year", value = "cases")
tidy4b <- table4b %>%
  gather(`1999`, `2000`, key = "year", value = "population")
left_join(tidy4a, tidy4b)
#> Joining, by = c("country", "year")
#> # A tibble: 6 × 4
#>       country  year  cases population
#>         <chr> <chr>  <int>      <int>
#> 1 Afghanistan  1999    745   19987071
#> 2      Brazil  1999  37737  172006362
#> 3       China  1999 212258 1272915272
#> 4 Afghanistan  2000   2666   20595360
#> 5      Brazil  2000  80488  174504898
#> 6       China  2000 213766 1280428583
```

Spreading

Spreading is the opposite of gathering. You use it when an observation is scattered across multiple rows. For example, take `table2`—an observation is a country in a year, but each observation is spread across two rows:

```
table2
#> # A tibble: 12 × 4
#>       country  year       type     count
#>         <chr> <int>      <chr>     <int>
#> 1 Afghanistan  1999      cases       745
#> 2 Afghanistan  1999 population  19987071
#> 3 Afghanistan  2000      cases      2666
#> 4 Afghanistan  2000 population  20595360
#> 5      Brazil  1999      cases     37737
```

```
#> 6      Brazil  1999 population 172006362
#> # ... with 6 more rows
```

To tidy this up, we first analyze the representation in a similar way to `gather()`. This time, however, we only need two parameters:

- The column that contains variable names, the `key` column. Here, it's `type`.
- The column that contains values forms multiple variables, the `value` column. Here, it's `count`.

Once we've figured that out, we can use `spread()`, as shown programmatically here, and visually in Figure 9-3:

```
spread(table2, key = type, value = count)
#> # A tibble: 6 × 4
#>       country  year  cases population
#> *       <chr> <int>  <int>      <int>
#> 1 Afghanistan  1999    745   19987071
#> 2 Afghanistan  2000   2666   20595360
#> 3      Brazil  1999  37737  172006362
#> 4      Brazil  2000  80488  174504898
#> 5       China  1999 212258 1272915272
#> 6       China  2000 213766 1280428583
```

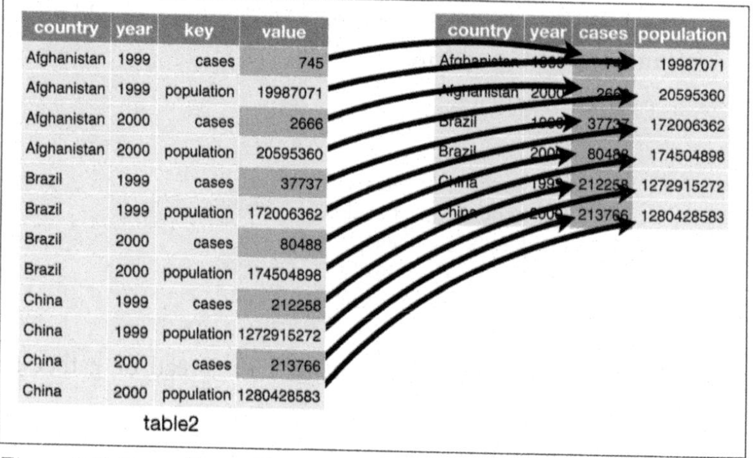

Figure 9-3. Spreading table2 makes it tidy

As you might have guessed from the common `key` and `value` arguments, `spread()` and `gather()` are complements. `gather()` makes wide tables narrower and longer; `spread()` makes long tables shorter and wider.

Exercises

1. Why are `gather()` and `spread()` not perfectly symmetrical? Carefully consider the following example:

```
stocks <- tibble(
  year   = c(2015, 2015, 2016, 2016),
  half  = c(   1,    2,     1,    2),
  return = c(1.88, 0.59, 0.92, 0.17)
)
stocks %>%
  spread(year, return) %>%
  gather("year", "return", `2015`:`2016`)
```

   (Hint: look at the variable types and think about column *names*.)

   Both `spread()` and `gather()` have a `convert` argument. What does it do?

2. Why does this code fail?

```
table4a %>%
  gather(1999, 2000, key = "year", value = "cases")
#> Error in eval(expr, envir, enclos):
#> Position must be between 0 and n
```

3. Why does spreading this tibble fail? How could you add a new column to fix the problem?

```
people <- tribble(
  ~name,             ~key,    ~value,
  #-----------------|--------|------
  "Phillip Woods",   "age",       45,
  "Phillip Woods",   "height",   186,
  "Phillip Woods",   "age",       50,
  "Jessica Cordero", "age",       37,
  "Jessica Cordero", "height",   156
)
```

4. Tidy this simple tibble. Do you need to spread or gather it? What are the variables?

```
preg <- tribble(
  ~pregnant, ~male, ~female,
  "yes",     NA,    10,
  "no",      20,    12
)
```

Separating and Pull

So far you've learned how to tidy `table2` and `table4`, but not `table3`. `table3` has a different problem: we have one column (`rate`) that contains two variables (`cases` and `population`). To fix this problem, we'll need the `separate()` function. You'll also learn about the complement of `separate()`: `unite()`, which you use if a single variable is spread across multiple columns.

Separate

`separate()` pulls apart one column into multiple columns, by splitting wherever a separator character appears. Take `table3`:

```
table3
#> # A tibble: 6 × 3
#>       country  year              rate
#> *       <chr> <int>             <chr>
#> 1 Afghanistan  1999      745/19987071
#> 2 Afghanistan  2000     2666/20595360
#> 3      Brazil  1999   37737/172006362
#> 4      Brazil  2000   80488/174504898
#> 5       China  1999 212258/1272915272
#> 6       China  2000 213766/1280428583
```

The `rate` column contains both `cases` and `population` variables, and we need to split it into two variables. `separate()` takes the name of the column to separate, and the names of the columns to separate into, as shown in Figure 9-4 and the following code:

```
table3 %>%
  separate(rate, into = c("cases", "population"))
#> # A tibble: 6 × 4
#>       country  year  cases population
#> *       <chr> <int>  <chr>      <chr>
#> 1 Afghanistan  1999    745   19987071
#> 2 Afghanistan  2000   2666   20595360
#> 3      Brazil  1999  37737  172006362
#> 4      Brazil  2000  80488  174504898
#> 5       China  1999 212258 1272915272
#> 6       China  2000 213766 1280428583
```

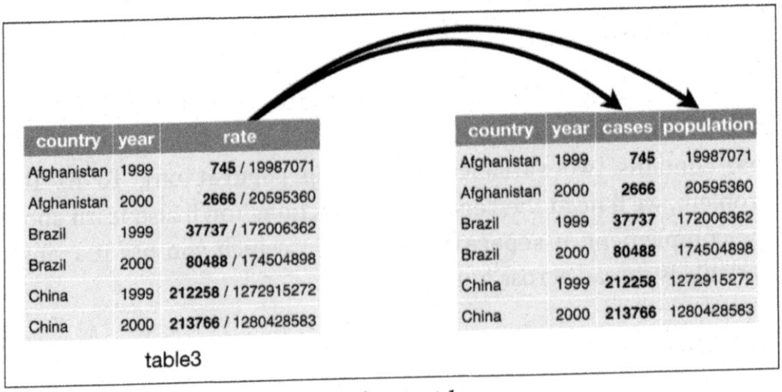

country	year	rate
Afghanistan	1999	**745** / 19987071
Afghanistan	2000	**2666** / 20595360
Brazil	1999	**37737** / 172006362
Brazil	2000	**80488** / 174504898
China	1999	**212258** / 1272915272
China	2000	**213766** / 1280428583

table3

country	year	cases	population
Afghanistan	1999	**745**	19987071
Afghanistan	2000	**2666**	20595360
Brazil	1999	**37737**	172006362
Brazil	2000	**80488**	174504898
China	1999	**212258**	1272915272
China	2000	**213766**	1280428583

Figure 9-4. Separating table3 makes it tidy

By default, `separate()` will split values wherever it sees a non-alphanumeric character (i.e., a character that isn't a number or letter). For example, in the preceding code, `separate()` split the values of `rate` at the forward slash characters. If you wish to use a specific character to separate a column, you can pass the character to the `sep` argument of `separate()`. For example, we could rewrite the preceding code as:

```
table3 %>%
  separate(rate, into = c("cases", "population"), sep = "/")
```

(Formally, `sep` is a regular expression, which you'll learn more about in Chapter 11.)

Look carefully at the column types: you'll notice that `case` and `population` are character columns. This is the default behavior in `separate()`: it leaves the type of the column as is. Here, however, it's not very useful as those really are numbers. We can ask `separate()` to try and convert to better types using `convert = TRUE`:

```
table3 %>%
  separate(
    rate,
    into = c("cases", "population"),
    convert = TRUE
  )
#> # A tibble: 6 × 4
#>       country  year  cases population
#> *       <chr> <int>  <int>      <int>
#> 1 Afghanistan  1999    745   19987071
#> 2 Afghanistan  2000   2666   20595360
#> 3      Brazil  1999  37737  172006362
#> 4      Brazil  2000  80488  174504898
```

```
#> 5       China  1999 212258 1272915272
#> 6       China  2000 213766 1280428583
```

You can also pass a vector of integers to `sep`. `separate()` will interpret the integers as positions to split at. Positive values start at 1 on the far left of the strings; negative values start at –1 on the far right of the strings. When using integers to separate strings, the length of `sep` should be one less than the number of names in `into`.

You can use this arrangement to separate the last two digits of each year. This makes this data less tidy, but is useful in other cases, as you'll see in a little bit:

```
table3 %>%
  separate(year, into = c("century", "year"), sep = 2)
#> # A tibble: 6 × 4
#>       country century  year              rate
#> *       <chr>   <chr> <chr>             <chr>
#> 1 Afghanistan      19    99      745/19987071
#> 2 Afghanistan      20    00     2666/20595360
#> 3      Brazil      19    99   37737/172006362
#> 4      Brazil      20    00   80488/174504898
#> 5       China      19    99 212258/1272915272
#> 6       China      20    00 213766/1280428583
```

Unite

`unite()` is the inverse of `separate()`: it combines multiple columns into a single column. You'll need it much less frequently than `separate()`, but it's still a useful tool to have in your back pocket.

We can use `unite()` to rejoin the *century* and *year* columns that we created in the last example. That data is saved as `tidyr::table5`. `unite()` takes a data frame, the name of the new variable to create, and a set of columns to combine, again specified in `dplyr::select()` . The result is shown in Figure 9-5 and in the following code:

```
table5 %>%
  unite(new, century, year)
#> # A tibble: 6 × 3
#>       country   new              rate
#> *       <chr> <chr>             <chr>
#> 1 Afghanistan 19_99      745/19987071
#> 2 Afghanistan 20_00     2666/20595360
#> 3      Brazil 19_99   37737/172006362
#> 4      Brazil 20_00   80488/174504898
```

```
#> 5        China 19_99 212258/1272915272
#> 6        China 20_00 213766/1280428583
```

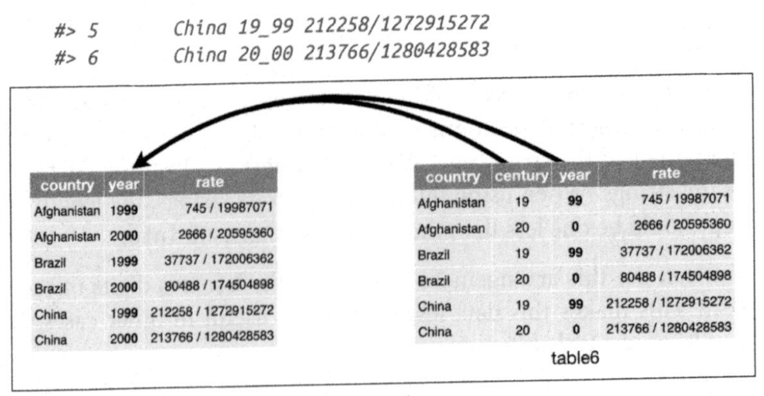

Figure 9-5. Uniting table5 makes it tidy

In this case we also need to use the `sep` argument. The default will place an underscore (_) between the values from different columns. Here we don't want any separator so we use `""`:

```
table5 %>%
  unite(new, century, year, sep = "")
#> # A tibble: 6 × 3
#>       country   new              rate
#> *       <chr> <chr>             <chr>
#> 1 Afghanistan  1999      745/19987071
#> 2 Afghanistan  2000     2666/20595360
#> 3      Brazil  1999    37737/172006362
#> 4      Brazil  2000    80488/174504898
#> 5       China  1999 212258/1272915272
#> 6       China  2000 213766/1280428583
```

Exercises

1. What do the `extra` and `fill` arguments do in `separate()`? Experiment with the various options for the following two toy datasets:

    ```
    tibble(x = c("a,b,c", "d,e,f,g", "h,i,j")) %>%
      separate(x, c("one", "two", "three"))

    tibble(x = c("a,b,c", "d,e", "f,g,i")) %>%
      separate(x, c("one", "two", "three"))
    ```

2. Both `unite()` and `separate()` have a `remove` argument. What does it do? Why would you set it to `FALSE`?

3. Compare and contrast separate() and extract(). Why are there three variations of separation (by position, by separator, and with groups), but only one unite?

Missing Values

Changing the representation of a dataset brings up an important subtlety of missing values. Surprisingly, a value can be missing in one of two possible ways:

- *Explicitly*, i.e., flagged with NA.
- *Implicitly*, i.e., simply not present in the data.

Let's illustrate this idea with a very simple dataset:

```
stocks <- tibble(
  year   = c(2015, 2015, 2015, 2015, 2016, 2016, 2016),
  qtr    = c(   1,    2,    3,    4,    2,    3,    4),
  return = c(1.88, 0.59, 0.35,   NA, 0.92, 0.17, 2.66)
)
```

There are two missing values in this dataset:

- The return for the fourth quarter of 2015 is explicitly missing, because the cell where its value should be instead contains NA.
- The return for the first quarter of 2016 is implicitly missing, because it simply does not appear in the dataset.

One way to think about the difference is with this Zen-like koan: an explicit missing value is the presence of an absence; an implicit missing value is the absence of a presence.

The way that a dataset is represented can make implicit values explicit. For example, we can make the implicit missing value explicit by putting years in the columns:

```
stocks %>%
  spread(year, return)
#> # A tibble: 4 × 3
#>     qtr `2015` `2016`
#> * <dbl>  <dbl>  <dbl>
#> 1     1   1.88     NA
#> 2     2   0.59   0.92
#> 3     3   0.35   0.17
#> 4     4     NA   2.66
```

Because these explicit missing values may not be important in other representations of the data, you can set `na.rm = TRUE` in `gather()` to turn explicit missing values implicit:

```
stocks %>%
  spread(year, return) %>%
  gather(year, return, `2015`:`2016`, na.rm = TRUE)
#> # A tibble: 6 × 3
#>     qtr  year return
#> * <dbl> <chr>  <dbl>
#> 1     1  2015   1.88
#> 2     2  2015   0.59
#> 3     3  2015   0.35
#> 4     2  2016   0.92
#> 5     3  2016   0.17
#> 6     4  2016   2.66
```

Another important tool for making missing values explicit in tidy data is `complete()`:

```
stocks %>%
  complete(year, qtr)
#> # A tibble: 8 × 3
#>    year   qtr return
#>   <dbl> <dbl>  <dbl>
#> 1  2015     1   1.88
#> 2  2015     2   0.59
#> 3  2015     3   0.35
#> 4  2015     4     NA
#> 5  2016     1     NA
#> 6  2016     2   0.92
#> # ... with 2 more rows
```

`complete()` takes a set of columns, and finds all unique combinations. It then ensures the original dataset contains all those values, filling in explicit NAs where necessary.

There's one other important tool that you should know for working with missing values. Sometimes when a data source has primarily been used for data entry, missing values indicate that the previous value should be carried forward:

```
treatment <- tribble(
  ~ person,           ~ treatment, ~response,
  "Derrick Whitmore", 1,           7,
  NA,                 2,           10,
  NA,                 3,           9,
  "Katherine Burke",  1,           4
)
```

You can fill in these missing values with `fill()`. It takes a set of columns where you want missing values to be replaced by the most recent nonmissing value (sometimes called last observation carried forward):

```
treatment %>%
  fill(person)
#> # A tibble: 4 × 3
#>             person treatment response
#>              <chr>     <dbl>    <dbl>
#> 1 Derrick Whitmore         1        7
#> 2 Derrick Whitmore         2       10
#> 3 Derrick Whitmore         3        9
#> 4  Katherine Burke         1        4
```

Exercises

1. Compare and contrast the `fill` arguments to `spread()` and `complete()`.
2. What does the direction argument to `fill()` do?

Case Study

To finish off the chapter, let's pull together everything you've learned to tackle a realistic data tidying problem. The `tidyr::who` dataset contains tuberculosis (TB) cases broken down by year, country, age, gender, and diagnosis method. The data comes from the *2014 World Health Organization Global Tuberculosis Report*, available at *http://www.who.int/tb/country/data/download/en/*.

There's a wealth of epidemiological information in this dataset, but it's challenging to work with the data in the form that it's provided:

```
who
#> # A tibble: 7,240 × 60
#>       country  iso2  iso3  year new_sp_m014 new_sp_m1524
#>         <chr> <chr> <chr> <int>       <int>        <int>
#> 1 Afghanistan    AF   AFG  1980          NA           NA
#> 2 Afghanistan    AF   AFG  1981          NA           NA
#> 3 Afghanistan    AF   AFG  1982          NA           NA
#> 4 Afghanistan    AF   AFG  1983          NA           NA
#> 5 Afghanistan    AF   AFG  1984          NA           NA
#> 6 Afghanistan    AF   AFG  1985          NA           NA
#> # ... with 7,234 more rows, and 54 more variables:
#> #   new_sp_m2534 <int>, new_sp_m3544 <int>,
#> #   new_sp_m4554 <int>, new_sp_m5564 <int>,
```

```
#> #   new_sp_m65 <int>, new_sp_f014 <int>,
#> #   new_sp_f1524 <int>, new_sp_f2534 <int>,
#> #   new_sp_f3544 <int>, new_sp_f4554 <int>,
#> #   new_sp_f5564 <int>, new_sp_f65 <int>,
#> #   new_sn_m014 <int>, new_sn_m1524 <int>,
#> #   new_sn_m2534 <int>, new_sn_m3544 <int>,
#> #   new_sn_m4554 <int>, new_sn_m5564 <int>,
#> #   new_sn_m65 <int>, new_sn_f014 <int>,
#> #   new_sn_f1524 <int>, new_sn_f2534 <int>,
#> #   new_sn_f3544 <int>, new_sn_f4554 <int>,
#> #   new_sn_f5564 <int>, new_sn_f65 <int>,
#> #   new_ep_m014 <int>, new_ep_m1524 <int>,
#> #   new_ep_m2534 <int>, new_ep_m3544 <int>,
#> #   new_ep_m4554 <int>, new_ep_m5564 <int>,
#> #   new_ep_m65 <int>, new_ep_f014 <int>,
#> #   new_ep_f1524 <int>, new_ep_f2534 <int>,
#> #   new_ep_f3544 <int>, new_ep_f4554 <int>,
#> #   new_ep_f5564 <int>, new_ep_f65 <int>,
#> #   newrel_m014 <int>, newrel_m1524 <int>,
#> #   newrel_m2534 <int>, newrel_m3544 <int>,
#> #   newrel_m4554 <int>, newrel_m5564 <int>,
#> #   newrel_m65 <int>, newrel_f014 <int>,
#> #   newrel_f1524 <int>, newrel_f2534 <int>,
#> #   newrel_f3544 <int>, newrel_f4554 <int>,
#> #   newrel_f5564 <int>, newrel_f65 <int>
```

This is a very typical real-life dataset. It contains redundant columns, odd variable codes, and many missing values. In short, `who` is messy, and we'll need multiple steps to tidy it. Like **dplyr**, **tidyr** is designed so that each function does one thing well. That means in real-life situations you'll usually need to string together multiple verbs into a pipeline.

The best place to start is almost always to gather together the columns that are not variables. Let's have a look at what we've got:

- It looks like `country`, `iso2`, and `iso3` are three variables that redundantly specify the country.
- `year` is clearly also a variable.
- We don't know what all the other columns are yet, but given the structure in the variable names (e.g., `new_sp_m014`, `new_ep_m014`, `new_ep_f014`) these are likely to be values, not variables.

So we need to gather together all the columns from `new_sp_m014` to `newrel_f65`. We don't know what those values represent yet, so we'll give them the generic name "key". We know the cells represent the count of cases, so we'll use the variable cases. There are a lot of missing values in the current representation, so for now we'll use na.rm just so we can focus on the values that are present:

```
who1 <- who %>%
  gather(
    new_sp_m014:newrel_f65, key = "key",
    value = "cases",
    na.rm = TRUE
  )
who1
#> # A tibble: 76,046 × 6
#>       country  iso2  iso3  year         key cases
#> *       <chr> <chr> <chr> <int>       <chr> <int>
#> 1 Afghanistan    AF   AFG  1997 new_sp_m014     0
#> 2 Afghanistan    AF   AFG  1998 new_sp_m014    30
#> 3 Afghanistan    AF   AFG  1999 new_sp_m014     8
#> 4 Afghanistan    AF   AFG  2000 new_sp_m014    52
#> 5 Afghanistan    AF   AFG  2001 new_sp_m014   129
#> 6 Afghanistan    AF   AFG  2002 new_sp_m014    90
#> # ... with 7.604e+04 more rows
```

We can get some hint of the structure of the values in the new key column by counting them:

```
who1 %>%
  count(key)
#> # A tibble: 56 × 2
#>             key     n
#>           <chr> <int>
#> 1   new_ep_f014  1032
#> 2  new_ep_f1524  1021
#> 3  new_ep_f2534  1021
#> 4  new_ep_f3544  1021
#> 5  new_ep_f4554  1017
#> 6  new_ep_f5564  1017
#> # ... with 50 more rows
```

You might be able to parse this out by yourself with a little thought and some experimentation, but luckily we have the data dictionary handy. It tells us:

1. The first three letters of each column denote whether the column contains new or old cases of TB. In this dataset, each column contains new cases.
2. The next two letters describe the type of TB:
    - rel stands for cases of relapse.

- ep stands for cases of extrapulmonary TB.
- sn stands for cases of pulmonary TB that could not be diagnosed by a pulmonary smear (smear negative).
- sp stands for cases of pulmonary TB that could be diagnosed be a pulmonary smear (smear positive).

3. The sixth letter gives the sex of TB patients. The dataset groups cases by males (m) and females (f).
4. The remaining numbers give the age group. The dataset groups cases into seven age groups:

   - 014 = 0–14 years old
   - 1524 = 15–24 years old
   - 2534 = 25–34 years old
   - 3544 = 35–44 years old
   - 4554 = 45–54 years old
   - 5564 = 55–64 years old
   - 65 = 65 or older

We need to make a minor fix to the format of the column names: unfortunately the names are slightly inconsistent because instead of new_rel we have newrel (it's hard to spot this here but if you don't fix it we'll get errors in subsequent steps). You'll learn about str_replace() in Chapter 11, but the basic idea is pretty simple: replace the characters "newrel" with "new_rel". This makes all variable names consistent:

```
who2 <- who1 %>%
  mutate(key = stringr::str_replace(key, "newrel", "new_rel"))
who2
#> # A tibble: 76,046 × 6
#>       country  iso2  iso3  year         key cases
#>         <chr> <chr> <chr> <int>       <chr> <int>
#> 1 Afghanistan    AF   AFG  1997 new_sp_m014     0
#> 2 Afghanistan    AF   AFG  1998 new_sp_m014    30
#> 3 Afghanistan    AF   AFG  1999 new_sp_m014     8
#> 4 Afghanistan    AF   AFG  2000 new_sp_m014    52
#> 5 Afghanistan    AF   AFG  2001 new_sp_m014   129
#> 6 Afghanistan    AF   AFG  2002 new_sp_m014    90
#> # ... with 7.604e+04 more rows
```

We can separate the values in each code with two passes of `separate()`. The first pass will split the codes at each underscore:

```
who3 <- who2 %>%
  separate(key, c("new", "type", "sexage"), sep = "_")
who3
#> # A tibble: 76,046 × 8
#>       country  iso2  iso3  year   new  type sexage cases
#> *       <chr> <chr> <chr> <int> <chr> <chr>  <chr> <int>
#> 1 Afghanistan    AF   AFG  1997   new    sp   m014     0
#> 2 Afghanistan    AF   AFG  1998   new    sp   m014    30
#> 3 Afghanistan    AF   AFG  1999   new    sp   m014     8
#> 4 Afghanistan    AF   AFG  2000   new    sp   m014    52
#> 5 Afghanistan    AF   AFG  2001   new    sp   m014   129
#> 6 Afghanistan    AF   AFG  2002   new    sp   m014    90
#> # ... with 7.604e+04 more rows
```

Then we might as well drop the `new` column because it's constant in this dataset. While we're dropping columns, let's also drop `iso2` and `iso3` since they're redundant:

```
who3 %>%
  count(new)
#> # A tibble: 1 × 2
#>     new     n
#>   <chr> <int>
#> 1   new 76046
who4 <- who3 %>%
  select(-new, -iso2, -iso3)
```

Next we'll separate `sexage` into `sex` and `age` by splitting after the first character:

```
who5 <- who4 %>%
  separate(sexage, c("sex", "age"), sep = 1)
who5
#> # A tibble: 76,046 × 6
#>       country  year  type   sex   age cases
#> *       <chr> <int> <chr> <chr> <chr> <int>
#> 1 Afghanistan  1997    sp     m   014     0
#> 2 Afghanistan  1998    sp     m   014    30
#> 3 Afghanistan  1999    sp     m   014     8
#> 4 Afghanistan  2000    sp     m   014    52
#> 5 Afghanistan  2001    sp     m   014   129
#> 6 Afghanistan  2002    sp     m   014    90
#> # ... with 7.604e+04 more rows
```

The `who` dataset is now tidy!

I've shown you the code a piece at a time, assigning each interim result to a new variable. This typically isn't how you'd work interactively. Instead, you'd gradually build up a complex pipe:

```
who %>%
  gather(code, value, new_sp_m014:newrel_f65, na.rm = TRUE) %>%
  mutate(
    code = stringr::str_replace(code, "newrel", "new_rel")
  ) %>%
  separate(code, c("new", "var", "sexage")) %>%
  select(-new, -iso2, -iso3) %>%
  separate(sexage, c("sex", "age"), sep = 1)
```

Exercises

1. In this case study I set `na.rm = TRUE` just to make it easier to check that we had the correct values. Is this reasonable? Think about how missing values are represented in this dataset. Are there implicit missing values? What's the difference between an `NA` and zero?
2. What happens if you neglect the `mutate()` step? (`mutate(key = stringr::str_replace(key, "newrel", "new_rel"))`).
3. I claimed that `iso2` and `iso3` were redundant with `country`. Confirm this claim.
4. For each country, year, and sex compute the total number of cases of TB. Make an informative visualization of the data.

Nontidy Data

Before we continue on to other topics, it's worth talking briefly about nontidy data. Earlier in the chapter, I used the pejorative term "messy" to refer to nontidy data. That's an oversimplification: there are lots of useful and well-founded data structures that are not tidy data. There are two main reasons to use other data structures:

- Alternative representations may have substantial performance or space advantages.
- Specialized fields have evolved their own conventions for storing data that may be quite different to the conventions of tidy data.

Either of these reasons means you'll need something other than a tibble (or data frame). If your data does fit naturally into a rectangular structure composed of observations and variables, I think tidy data should be your default choice. But there are good reasons to use other structures; tidy data is not the only way. If you'd like to learn more about nontidy data, I'd highly recommend this thoughtful blog post by Jeff Leek (*http://simplystatistics.org/2016/02/17/non-tidy-data/*).

CHAPTER 10

Relational Data with dplyr

Introduction

It's rare that a data analysis involves only a single table of data. Typically you have many tables of data, and you must combine them to answer the questions that you're interested in. Collectively, multiple tables of data are called *relational data* because it is the relations, not just the individual datasets, that are important.

Relations are always defined between a pair of tables. All other relations are built up from this simple idea: the relations of three or more tables are always a property of the relations between each pair. Sometimes both elements of a pair can be the same table! This is needed if, for example, you have a table of people, and each person has a reference to their parents.

To work with relational data you need verbs that work with pairs of tables. There are three families of verbs designed to work with relational data:

- *Mutating joins*, which add new variables to one data frame from matching observations in another.
- *Filtering joins*, which filter observations from one data frame based on whether or not they match an observation in the other table.
- *Set operations*, which treat observations as if they were set elements.

The most common place to find relational data is in a *relational* database management system (or RDBMS), a term that encompasses almost all modern databases. If you've used a database before, you've almost certainly used SQL. If so, you should find the concepts in this chapter familiar, although their expression in **dplyr** is a little different. Generally, **dplyr** is a little easier to use than SQL because **dplyr** is specialized to do data analysis: it makes common data analysis operations easier, at the expense of making it more difficult to do other things that aren't commonly needed for data analysis.

Prerequisites

We will explore relational data from **nycflights13** using the two-table verbs from **dplyr**.

```
library(tidyverse)
library(nycflights13)
```

nycflights13

We will use the **nycflights13** package to learn about relational data. **nycflights13** contains four tibbles that are related to the `flights` table that you used in Chapter 3:

- `airlines` lets you look up the full carrier name from its abbreviated code:

  ```
  airlines
  #> # A tibble: 16 × 2
  #>   carrier                     name
  #>     <chr>                    <chr>
  #> 1      9E        Endeavor Air Inc.
  #> 2      AA   American Airlines Inc.
  #> 3      AS     Alaska Airlines Inc.
  #> 4      B6          JetBlue Airways
  #> 5      DL     Delta Air Lines Inc.
  #> 6      EV ExpressJet Airlines Inc.
  #> # ... with 10 more rows
  ```

- `airports` gives information about each airport, identified by the `faa` airport code:

```
airports
#> # A tibble: 1,396 × 7
#>     faa                           name   lat   lon
#>   <chr>                          <chr> <dbl> <dbl>
#> 1   04G              Lansdowne Airport  41.1 -80.6
#> 2   06A  Moton Field Municipal Airport  32.5 -85.7
#> 3   06C            Schaumburg Regional  42.0 -88.1
#> 4   06N                Randall Airport  41.4 -74.4
#> 5   09J          Jekyll Island Airport  31.1 -81.4
#> 6   0A9 Elizabethton Municipal Airport  36.4 -82.2
#> # ... with 1,390 more rows, and 3 more variables:
#> #   alt <int>, tz <dbl>, dst <chr>
```

- `planes` gives information about each plane, identified by its `tailnum`:

```
planes
#> # A tibble: 3,322 × 9
#>   tailnum  year                    type
#>     <chr> <int>                   <chr>
#> 1  N10156  2004 Fixed wing multi engine
#> 2  N102UW  1998 Fixed wing multi engine
#> 3  N103US  1999 Fixed wing multi engine
#> 4  N104UW  1999 Fixed wing multi engine
#> 5  N10575  2002 Fixed wing multi engine
#> 6  N105UW  1999 Fixed wing multi engine
#> # ... with 3,316 more rows, and 6 more variables:
#> #   manufacturer <chr>, model <chr>, engines <int>,
#> #   seats <int>, speed <int>, engine <chr>
```

- `weather` gives the weather at each NYC airport for each hour:

```
weather
#> # A tibble: 26,130 × 15
#>   origin  year month   day  hour  temp  dewp humid
#>    <chr> <dbl> <dbl> <int> <int> <dbl> <dbl> <dbl>
#> 1    EWR  2013     1     1     0  37.0  21.9  54.0
#> 2    EWR  2013     1     1     1  37.0  21.9  54.0
#> 3    EWR  2013     1     1     2  37.9  21.9  52.1
#> 4    EWR  2013     1     1     3  37.9  23.0  54.5
#> 5    EWR  2013     1     1     4  37.9  24.1  57.0
#> 6    EWR  2013     1     1     6  39.0  26.1  59.4
#> # ... with 2.612e+04 more rows, and 7 more variables:
#> #   wind_dir <dbl>, wind_speed <dbl>, wind_gust <dbl>,
#> #   precip <dbl>, pressure <dbl>, visib <dbl>,
#> #   time_hour <dttm>
```

One way to show the relationships between the different tables is with a drawing:

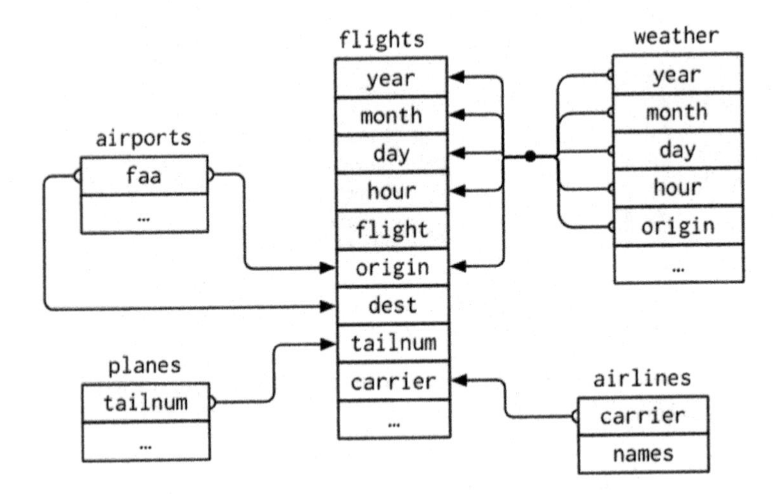

This diagram is a little overwhelming, but it's simple compared to some you'll see in the wild! The key to understanding diagrams like this is to remember each relation always concerns a pair of tables. You don't need to understand the whole thing; you just need to understand the chain of relations between the tables that you are interested in.

For **nycflights13**:

- `flights` connects to `planes` via a single variable, `tailnum`.
- `flights` connects to `airlines` through the `carrier` variable.
- `flights` connects to `airports` in two ways: via the `origin` and `dest` variables.
- `flights` connects to `weather` via `origin` (the location), and `year`, `month`, `day`, and `hour` (the time).

Exercises

1. Imagine you wanted to draw (approximately) the route each plane flies from its origin to its destination. What variables would you need? What tables would you need to combine?
2. I forgot to draw the relationship between `weather` and `airports`. What is the relationship and how should it appear in the diagram?

3. weather only contains information for the origin (NYC) airports. If it contained weather records for all airports in the USA, what additional relation would it define with flights?
4. We know that some days of the year are "special," and fewer people than usual fly on them. How might you represent that data as a data frame? What would be the primary keys of that table? How would it connect to the existing tables?

Keys

The variables used to connect each pair of tables are called *keys*. A key is a variable (or set of variables) that uniquely identifies an observation. In simple cases, a single variable is sufficient to identify an observation. For example, each plane is uniquely identified by its tailnum. In other cases, multiple variables may be needed. For example, to identify an observation in weather you need five variables: year, month, day, hour, and origin.

There are two types of keys:

- A *primary key* uniquely identifies an observation in its own table. For example, planes$tailnum is a primary key because it uniquely identifies each plane in the planes table.
- A *foreign key* uniquely identifies an observation in another table. For example, flights$tailnum is a foreign key because it appears in the flights table where it matches each flight to a unique plane.

A variable can be both a primary key *and* a foreign key. For example, origin is part of the weather primary key, and is also a foreign key for the airport table.

Once you've identified the primary keys in your tables, it's good practice to verify that they do indeed uniquely identify each observation. One way to do that is to count() the primary keys and look for entries where n is greater than one:

```
planes %>%
  count(tailnum) %>%
  filter(n > 1)
#> # A tibble: 0 × 2
#> # ... with 2 variables: tailnum <chr>, n <int>
```

```
weather %>%
  count(year, month, day, hour, origin) %>%
  filter(n > 1)
#> Source: local data frame [0 x 6]
#> Groups: year, month, day, hour [0]
#>
#> # ... with 6 variables: year <dbl>, month <dbl>, day <int>,
#> #   hour <int>, origin <chr>, n <int>
```

Sometimes a table doesn't have an explicit primary key: each row is an observation, but no combination of variables reliably identifies it. For example, what's the primary key in the `flights` table? You might think it would be the date plus the flight or tail number, but neither of those are unique:

```
flights %>%
  count(year, month, day, flight) %>%
  filter(n > 1)
#> Source: local data frame [29,768 x 5]
#> Groups: year, month, day [365]
#>
#>    year month   day flight     n
#>   <int> <int> <int>  <int> <int>
#> 1  2013     1     1      1     2
#> 2  2013     1     1      3     2
#> 3  2013     1     1      4     2
#> 4  2013     1     1     11     3
#> 5  2013     1     1     15     2
#> 6  2013     1     1     21     2
#> # ... with 2.976e+04 more rows

flights %>%
  count(year, month, day, tailnum) %>%
  filter(n > 1)
#> Source: local data frame [64,928 x 5]
#> Groups: year, month, day [365]
#>
#>    year month   day tailnum     n
#>   <int> <int> <int>   <chr> <int>
#> 1  2013     1     1  N0EGMQ     2
#> 2  2013     1     1  N11189     2
#> 3  2013     1     1  N11536     2
#> 4  2013     1     1  N11544     3
#> 5  2013     1     1  N11551     2
#> 6  2013     1     1  N12540     2
#> # ... with 6.492e+04 more rows
```

When starting to work with this data, I had naively assumed that each flight number would be only used once per day: that would make it much easier to communicate problems with a specific flight. Unfortunately that is not the case! If a table lacks a primary key, it's sometimes useful to add one with `mutate()` and `row_number()`. That makes it easier to match observations if you've done some filtering and want to check back in with the original data. This is called a *surrogate key*.

A primary key and the corresponding foreign key in another table form a *relation*. Relations are typically one-to-many. For example, each flight has one plane, but each plane has many flights. In other data, you'll occasionally see a 1-to-1 relationship. You can think of this as a special case of 1-to-many. You can model many-to-many relations with a many-to-1 relation plus a 1-to-many relation. For example, in this data there's a many-to-many relationship between airlines and airports: each airline flies to many airports; each airport hosts many airlines.

Exercises

1. Add a surrogate key to `flights`.
2. Identify the keys in the following datasets:
   a. `Lahman::Batting`
   b. `babynames::babynames`
   c. `nasaweather::atmos`
   d. `fueleconomy::vehicles`
   e. `ggplot2::diamonds`

   (You might need to install some packages and read some documentation.)
3. Draw a diagram illustrating the connections between the `Batting`, `Master`, and `Salaries` tables in the **Lahman** package. Draw another diagram that shows the relationship between `Master`, `Managers`, and `AwardsManagers`.

   How would you characterize the relationship between the `Batting`, `Pitching`, and `Fielding` tables?

Mutating Joins

The first tool we'll look at for combining a pair of tables is the *mutating join*. A mutating join allows you to combine variables from two tables. It first matches observations by their keys, then copies across variables from one table to the other.

Like `mutate()`, the join functions add variables to the right, so if you have a lot of variables already, the new variables won't get printed out. For these examples, we'll make it easier to see what's going on in the examples by creating a narrower dataset:

```
flights2 <- flights %>%
  select(year:day, hour, origin, dest, tailnum, carrier)
flights2
#> # A tibble: 336,776 × 8
#>    year month   day  hour origin  dest tailnum carrier
#>   <int> <int> <int> <dbl>  <chr> <chr>   <chr>   <chr>
#> 1  2013     1     1     5    EWR   IAH  N14228      UA
#> 2  2013     1     1     5    LGA   IAH  N24211      UA
#> 3  2013     1     1     5    JFK   MIA  N619AA      AA
#> 4  2013     1     1     5    JFK   BQN  N804JB      B6
#> 5  2013     1     1     6    LGA   ATL  N668DN      DL
#> 6  2013     1     1     5    EWR   ORD  N39463      UA
#> # ... with 3.368e+05 more rows
```

(Remember, when you're in RStudio, you can also use `View()` to avoid this problem.)

Imagine you want to add the full airline name to the `flights2` data. You can combine the `airlines` and `flights2` data frames with `left_join()`:

```
flights2 %>%
  select(-origin, -dest) %>%
  left_join(airlines, by = "carrier")
#> # A tibble: 336,776 × 7
#>    year month   day  hour tailnum carrier
#>   <int> <int> <int> <dbl>   <chr>   <chr>
#> 1  2013     1     1     5  N14228      UA
#> 2  2013     1     1     5  N24211      UA
#> 3  2013     1     1     5  N619AA      AA
#> 4  2013     1     1     5  N804JB      B6
#> 5  2013     1     1     6  N668DN      DL
#> 6  2013     1     1     5  N39463      UA
#> # ... with 3.368e+05 more rows, and 1 more variable:
#> #   name <chr>
```

The result of joining airlines to `flights2` is an additional variable: `name`. This is why I call this type of join a mutating join. In this case, you could have got to the same place using `mutate()` and R's base subsetting:

```
flights2 %>%
  select(-origin, -dest) %>%
  mutate(name = airlines$name[match(carrier, airlines$carrier)])
#> # A tibble: 336,776 × 7
#>    year month   day  hour tailnum carrier
#>   <int> <int> <int> <dbl>   <chr>   <chr>
#> 1  2013     1     1     5  N14228      UA
#> 2  2013     1     1     5  N24211      UA
#> 3  2013     1     1     5  N619AA      AA
#> 4  2013     1     1     5  N804JB      B6
#> 5  2013     1     1     6  N668DN      DL
#> 6  2013     1     1     5  N39463      UA
#> # ... with 3.368e+05 more rows, and 1 more variable:
#> #   name <chr>
```

But this is hard to generalize when you need to match multiple variables, and takes close reading to figure out the overall intent.

The following sections explain, in detail, how mutating joins work. You'll start by learning a useful visual representation of joins. We'll then use that to explain the four mutating join functions: the inner join, and the three outer joins. When working with real data, keys don't always uniquely identify observations, so next we'll talk about what happens when there isn't a unique match. Finally, you'll learn how to tell **dplyr** which variables are the keys for a given join.

Understanding Joins

To help you learn how joins work, I'm going to use a visual representation:

x	
1	x1
2	x2
3	x3

y	
1	y1
2	y2
4	y3

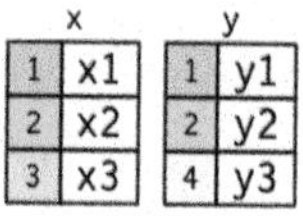

```
x <- tribble(
  ~key, ~val_x,
     1, "x1",
     2, "x2",
     3, "x3"
)
```

```
y <- tribble(
  ~key, ~val_y,
     1, "y1",
     2, "y2",
     4, "y3"
)
```

The colored column represents the "key" variable: these are used to match the rows between the tables. The gray column represents the "value" column that is carried along for the ride. In these examples I'll show a single key variable and single value variable, but the idea generalizes in a straightforward way to multiple keys and multiple values.

A join is a way of connecting each row in x to zero, one, or more rows in y. The following diagram shows each potential match as an intersection of a pair of lines:

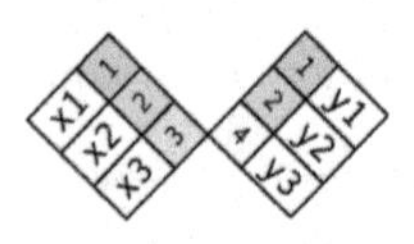

(If you look closely, you might notice that we've switched the order of the key and value columns in x. This is to emphasize that joins match based on the key; the value is just carried along for the ride.)

In an actual join, matches will be indicated with dots. The number of dots = the number of matches = the number of rows in the output.

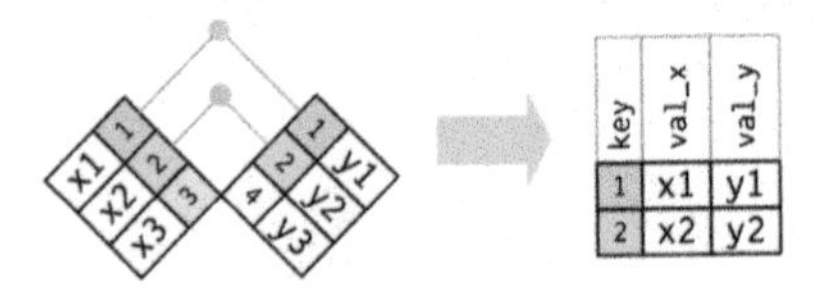

Inner Join

The simplest type of join is the *inner join*. An inner join matches pairs of observations whenever their keys are equal:

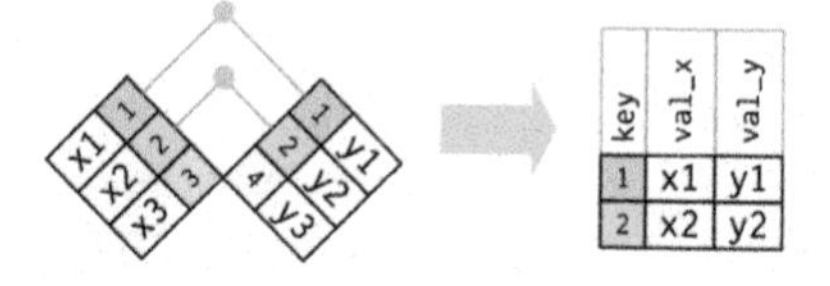

(To be precise, this is an inner *equijoin* because the keys are matched using the equality operator. Since most joins are equijoins we usually drop that specification.)

The output of an inner join is a new data frame that contains the key, the x values, and the y values. We use by to tell **dplyr** which variable is the key:

```
x %>%
  inner_join(y, by = "key")
#> # A tibble: 2 × 3
#>     key val_x val_y
#>   <dbl> <chr> <chr>
#> 1     1    x1    y1
#> 2     2    x2    y2
```

The most important property of an inner join is that unmatched rows are not included in the result. This means that generally inner joins are usually not appropriate for use in analysis because it's too easy to lose observations.

Outer Joins

An inner join keeps observations that appear in both tables. An *outer join* keeps observations that appear in at least one of the tables. There are three types of outer joins:

- A *left join* keeps all observations in x.
- A *right join* keeps all observations in y.
- A *full join* keeps all observations in x and y.

These joins work by adding an additional "virtual" observation to each table. This observation has a key that always matches (if no other key matches), and a value filled with NA.

Graphically, that looks like:

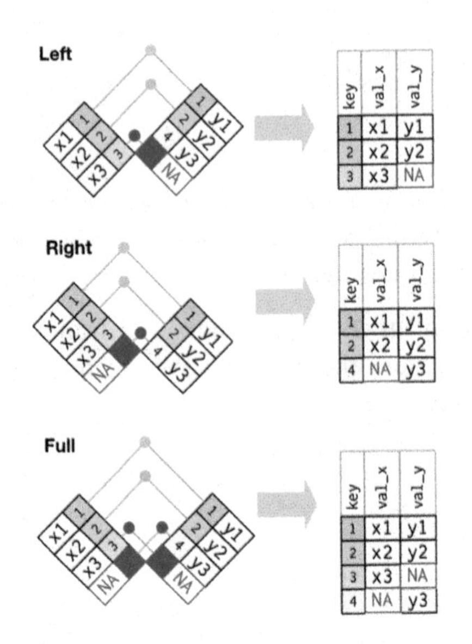

The most commonly used join is the left join: you use this whenever you look up additional data from another table, because it preserves the original observations even when there isn't a match. The left join should be your default join: use it unless you have a strong reason to prefer one of the others.

Another way to depict the different types of joins is with a Venn diagram:

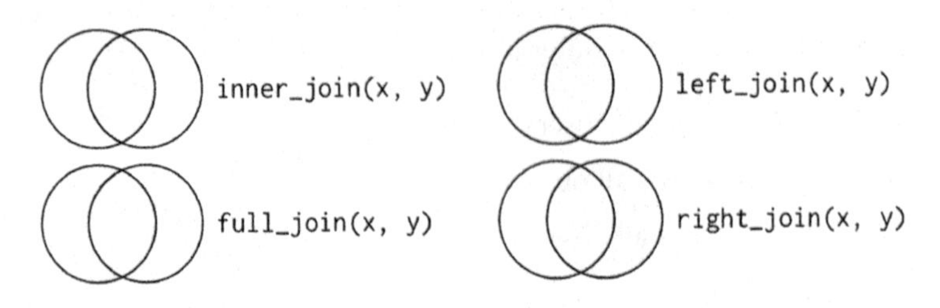

However, this is not a great representation. It might jog your memory about which join preserves the observations in which table, but it suffers from a major limitation: a Venn diagram can't show what happens when keys don't uniquely identify an observation.

Duplicate Keys

So far all the diagrams have assumed that the keys are unique. But that's not always the case. This section explains what happens when the keys are not unique. There are two possibilities:

- One table has duplicate keys. This is useful when you want to add in additional information as there is typically a one-to-many relationship:

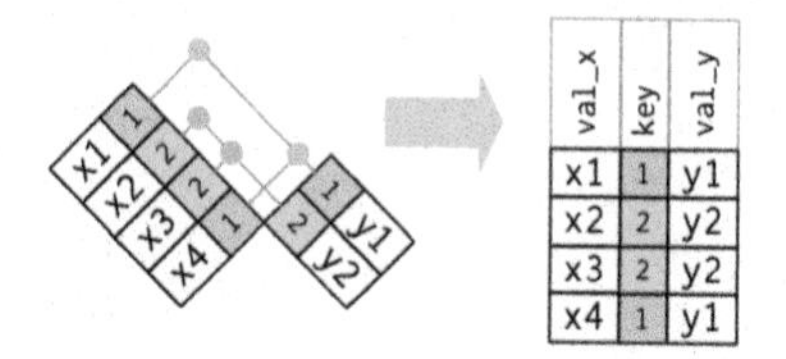

  Note that I've put the key column in a slightly different position in the output. This reflects that the key is a primary key in y and a foreign key in x:

```
x <- tribble(
  ~key, ~val_x,
     1, "x1",
     2, "x2",
     2, "x3",
     1, "x4"
)
y <- tribble(
  ~key, ~val_y,
     1, "y1",
     2, "y2"
)
left_join(x, y, by = "key")
#> # A tibble: 4 × 3
#>     key val_x val_y
#>   <dbl> <chr> <chr>
#> 1     1    x1    y1
#> 2     2    x2    y2
#> 3     2    x3    y2
#> 4     1    x4    y1
```

- Both tables have duplicate keys. This is usually an error because in neither table do the keys uniquely identify an observation. When you join duplicated keys, you get all possible combinations, the Cartesian product:

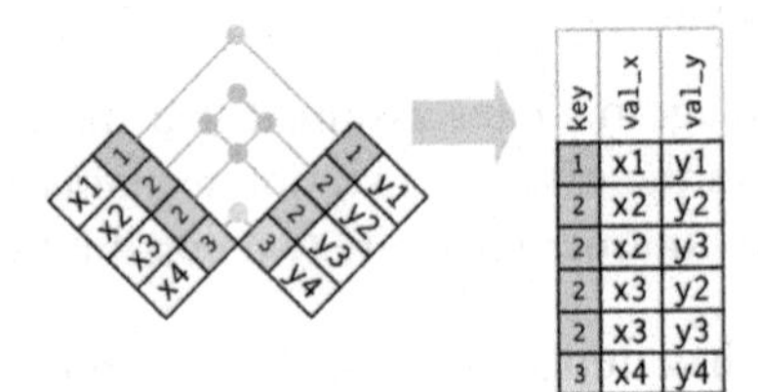


```
x <- tribble(
  ~key, ~val_x,
     1, "x1",
     2, "x2",
     2, "x3",
     3, "x4"
)
y <- tribble(
  ~key, ~val_y,
     1, "y1",
     2, "y2",
     2, "y3",
     3, "y4"
)
left_join(x, y, by = "key")
#> # A tibble: 6 × 3
#>     key val_x val_y
#>   <dbl> <chr> <chr>
#> 1     1    x1    y1
#> 2     2    x2    y2
#> 3     2    x2    y3
#> 4     2    x3    y2
#> 5     2    x3    y3
#> 6     3    x4    y4
```

Defining the Key Columns

So far, the pairs of tables have always been joined by a single variable, and that variable has the same name in both tables. That constraint was encoded by `by = "key"`. You can use other values for `by` to connect the tables in other ways:

- The default, `by = NULL`, uses all variables that appear in both tables, the so-called *natural* join. For example, the flights and weather tables match on their common variables: `year`, `month`, `day`, `hour`, and `origin`:

```
flights2 %>%
  left_join(weather)
#> Joining, by = c("year", "month", "day", "hour",
#>   "origin")
#> # A tibble: 336,776 × 18
#>    year month   day  hour origin  dest tailnum
#>   <dbl> <dbl> <int> <dbl>  <chr> <chr>   <chr>
#> 1  2013     1     1     5    EWR   IAH  N14228
#> 2  2013     1     1     5    LGA   IAH  N24211
#> 3  2013     1     1     5    JFK   MIA  N619AA
#> 4  2013     1     1     5    JFK   BQN  N804JB
#> 5  2013     1     1     6    LGA   ATL  N668DN
#> 6  2013     1     1     5    EWR   ORD  N39463
#> # ... with 3.368e+05 more rows, and 11 more variables:
#> #   carrier <chr>, temp <dbl>, dewp <dbl>,
#> #   humid <dbl>, wind_dir <dbl>, wind_speed <dbl>,
#> #   wind_gust <dbl>, precip <dbl>, pressure <dbl>,
#> #   visib <dbl>, time_hour <dttm>
```

- A character vector, `by = "x"`. This is like a natural join, but uses only some of the common variables. For example, `flights` and `planes` have `year` variables, but they mean different things so we only want to join by `tailnum`:

```
flights2 %>%
  left_join(planes, by = "tailnum")
#> # A tibble: 336,776 × 16
#>   year.x month   day  hour origin  dest tailnum
#>    <int> <int> <int> <dbl>  <chr> <chr>   <chr>
#> 1   2013     1     1     5    EWR   IAH  N14228
#> 2   2013     1     1     5    LGA   IAH  N24211
#> 3   2013     1     1     5    JFK   MIA  N619AA
#> 4   2013     1     1     5    JFK   BQN  N804JB
#> 5   2013     1     1     6    LGA   ATL  N668DN
#> 6   2013     1     1     5    EWR   ORD  N39463
#> # ... with 3.368e+05 more rows, and 9 more variables:
#> #   carrier <chr>, year.y <int>, type <chr>,
#> #   manufacturer <chr>, model <chr>, engines <int>,
#> #   seats <int>, speed <int>, engine <chr>
```

  Note that the `year` variables (which appear in both input data frames, but are not constrained to be equal) are disambiguated in the output with a suffix.

- A named character vector: `by = c("a" = "b")`. This will match variable `a` in table `x` to variable `b` in table `y`. The variables from `x` will be used in the output.

For example, if we want to draw a map we need to combine the flights data with the airports data, which contains the location (`lat` and `long`) of each airport. Each flight has an origin and destination `airport`, so we need to specify which one we want to join to:

```
flights2 %>%
  left_join(airports, c("dest" = "faa"))
#> # A tibble: 336,776 × 14
#>    year month   day  hour origin  dest tailnum
#>   <int> <int> <int> <dbl>  <chr> <chr>   <chr>
#> 1  2013     1     1     5    EWR   IAH  N14228
#> 2  2013     1     1     5    LGA   IAH  N24211
#> 3  2013     1     1     5    JFK   MIA  N619AA
#> 4  2013     1     1     5    JFK   BQN  N804JB
#> 5  2013     1     1     6    LGA   ATL  N668DN
#> 6  2013     1     1     5    EWR   ORD  N39463
#> # ... with 3.368e+05 more rows, and 7 more variables:
#> #   carrier <chr>, name <chr>, lat <dbl>, lon <dbl>,
#> #   alt <int>, tz <dbl>, dst <chr>

flights2 %>%
  left_join(airports, c("origin" = "faa"))
#> # A tibble: 336,776 × 14
#>    year month   day  hour origin  dest tailnum
#>   <int> <int> <int> <dbl>  <chr> <chr>   <chr>
#> 1  2013     1     1     5    EWR   IAH  N14228
#> 2  2013     1     1     5    LGA   IAH  N24211
#> 3  2013     1     1     5    JFK   MIA  N619AA
#> 4  2013     1     1     5    JFK   BQN  N804JB
#> 5  2013     1     1     6    LGA   ATL  N668DN
#> 6  2013     1     1     5    EWR   ORD  N39463
#> # ... with 3.368e+05 more rows, and 7 more variables:
#> #   carrier <chr>, name <chr>, lat <dbl>, lon <dbl>,
#> #   alt <int>, tz <dbl>, dst <chr>
```

Exercises

1. Compute the average delay by destination, then join on the `airports` data frame so you can show the spatial distribution of delays. Here's an easy way to draw a map of the United States:

```
airports %>%
  semi_join(flights, c("faa" = "dest")) %>%
  ggplot(aes(lon, lat)) +
    borders("state") +
```

```
    geom_point() +
    coord_quickmap()
```

(Don't worry if you don't understand what `semi_join()` does—you'll learn about it next.)

You might want to use the `size` or `color` of the points to display the average delay for each airport.

2. Add the location of the origin *and* destination (i.e., the `lat` and `lon`) to `flights`.
3. Is there a relationship between the age of a plane and its delays?
4. What weather conditions make it more likely to see a delay?
5. What happened on June 13, 2013? Display the spatial pattern of delays, and then use Google to cross-reference with the weather.

Other Implementations

`base::merge()` can perform all four types of mutating join:

dplyr	merge
`inner_join(x, y)`	`merge(x, y)`
`left_join(x, y)`	`merge(x, y, all.x = TRUE)`
`right_join(x, y)`	`merge(x, y, all.y = TRUE),`
`full_join(x, y)`	`merge(x, y, all.x = TRUE, all.y = TRUE)`

The advantages of the specific **dplyr** verbs is that they more clearly convey the intent of your code: the difference between the joins is really important but concealed in the arguments of `merge()`. **dplyr**'s joins are considerably faster and don't mess with the order of the rows.

SQL is the inspiration for **dplyr**'s conventions, so the translation is straightforward:

dplyr	SQL
`inner_join(x, y, by = "z")`	`SELECT * FROM x INNER JOIN y USING (z)`
`left_join(x, y, by = "z")`	`SELECT * FROM x LEFT OUTER JOIN y USING (z)`

dplyr	SQL
`right_join(x, y, by = "z")`	`SELECT * FROM x RIGHT OUTER JOIN y USING (z)`
`full_join(x, y, by = "z")`	`SELECT * FROM x FULL OUTER JOIN y USING (z)`

Note that "`INNER`" and "`OUTER`" are optional, and often omitted.

Joining different variables between the tables, e.g., `inner_join(x, y, by = c("a" = "b"))`, uses a slightly different syntax in SQL: `SELECT * FROM x INNER JOIN y ON x.a = y.b`. As this syntax suggests, SQL supports a wider range of join types than **dplyr** because you can connect the tables using constraints other than equality (sometimes called non-equijoins).

Filtering Joins

Filtering joins match observations in the same way as mutating joins, but affect the observations, not the variables. There are two types:

- `semi_join(x, y)` *keeps* all observations in `x` that have a match in `y`.
- `anti_join(x, y)` *drops* all observations in `x` that have a match in `y`.

Semi-joins are useful for matching filtered summary tables back to the original rows. For example, imagine you've found the top-10 most popular destinations:

```
top_dest <- flights %>%
  count(dest, sort = TRUE) %>%
  head(10)
top_dest
#> # A tibble: 10 × 2
#>    dest     n
#>   <chr> <int>
#> 1   ORD 17283
#> 2   ATL 17215
#> 3   LAX 16174
#> 4   BOS 15508
#> 5   MCO 14082
#> 6   CLT 14064
#> # ... with 4 more rows
```

Now you want to find each flight that went to one of those destinations. You could construct a filter yourself:

```
flights %>%
  filter(dest %in% top_dest$dest)
#> # A tibble: 141,145 × 19
#>    year month   day dep_time sched_dep_time dep_delay
#>   <int> <int> <int>    <int>          <int>     <dbl>
#> 1  2013     1     1      542            540         2
#> 2  2013     1     1      554            600        -6
#> 3  2013     1     1      554            558        -4
#> 4  2013     1     1      555            600        -5
#> 5  2013     1     1      557            600        -3
#> 6  2013     1     1      558            600        -2
#> # ... with 1.411e+05 more rows, and 12 more variables:
#> #   arr_time <int>, sched_arr_time <int>, arr_delay <dbl>,
#> #   carrier <chr>, flight <int>, tailnum <chr>, origin <chr>,
#> #   dest <chr>, air_time <dbl>, distance <dbl>, hour <dbl>,
#> #   minute <dbl>, time_hour <dttm>
```

But it's difficult to extend that approach to multiple variables. For example, imagine that you'd found the 10 days with the highest average delays. How would you construct the filter statement that used `year`, `month`, and `day` to match it back to `flights`?

Instead you can use a semi-join, which connects the two tables like a mutating join, but instead of adding new columns, only keeps the rows in `x` that have a match in `y`:

```
flights %>%
  semi_join(top_dest)
#> Joining, by = "dest"
#> # A tibble: 141,145 × 19
#>    year month   day dep_time sched_dep_time dep_delay
#>   <int> <int> <int>    <int>          <int>     <dbl>
#> 1  2013     1     1      554            558        -4
#> 2  2013     1     1      558            600        -2
#> 3  2013     1     1      608            600         8
#> 4  2013     1     1      629            630        -1
#> 5  2013     1     1      656            700        -4
#> 6  2013     1     1      709            700         9
#> # ... with 1.411e+05 more rows, and 13 more variables:
#> #   arr_time <int>, sched_arr_time <int>, arr_delay <dbl>,
#> #   carrier <chr>, flight <int>, tailnum <chr>, origin <chr>,
#> #   dest <chr>, air_time <dbl>, distance <dbl>, hour <dbl>,
#> #   minute <dbl>, time_hour <dttm>
```

Graphically, a semi-join looks like this:

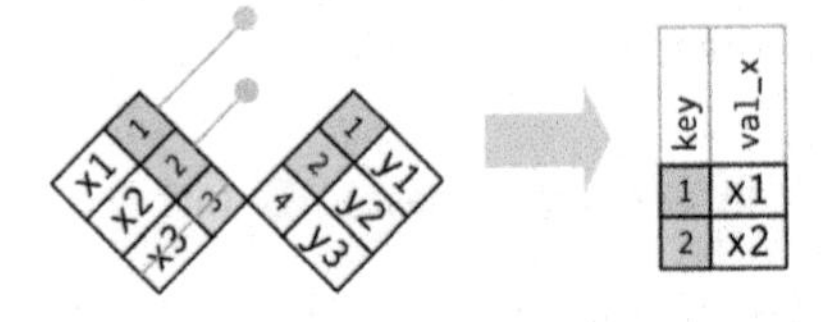

Only the existence of a match is important; it doesn't matter which observation is matched. This means that filtering joins never duplicate rows like mutating joins do:

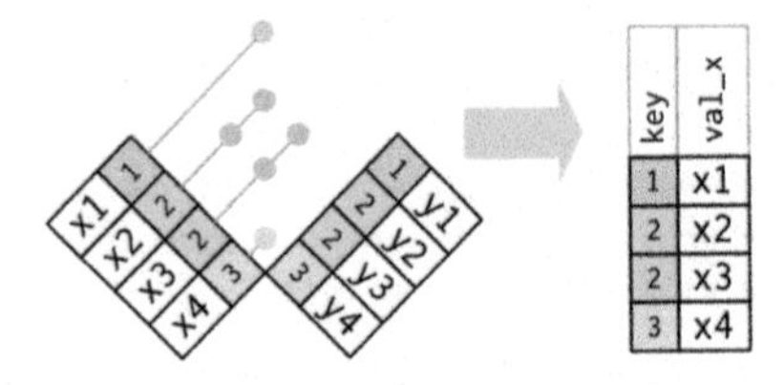

The inverse of a semi-join is an anti-join. An anti-join keeps the rows that *don't* have a match:

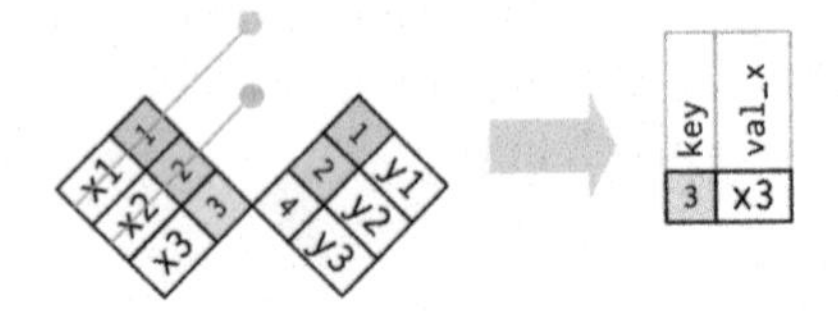

Anti-joins are useful for diagnosing join mismatches. For example, when connecting `flights` and `planes`, you might be interested to know that there are many `flights` that don't have a match in `planes`:

```
flights %>%
  anti_join(planes, by = "tailnum") %>%
  count(tailnum, sort = TRUE)
#> # A tibble: 722 × 2
#>   tailnum     n
#>     <chr> <int>
#> 1    <NA>  2512
#> 2  N725MQ   575
#> 3  N722MQ   513
#> 4  N723MQ   507
#> 5  N713MQ   483
```

```
#> 6 N735MQ    396
#> # ... with 716 more rows
```

Exercises

1. What does it mean for a flight to have a missing `tailnum`? What do the tail numbers that don't have a matching record in `planes` have in common? (Hint: one variable explains ~90% of the problems.)
2. Filter flights to only show flights with planes that have flown at least 100 flights.
3. Combine `fueleconomy::vehicles` and `fueleconomy::common` to find only the records for the most common models.
4. Find the 48 hours (over the course of the whole year) that have the worst delays. Cross-reference it with the `weather` data. Can you see any patterns?
5. What does `anti_join(flights, airports, by = c("dest" = "faa"))` tell you? What does `anti_join(airports, flights, by = c("faa" = "dest"))` tell you?
6. You might expect that there's an implicit relationship between plane and airline, because each plane is flown by a single airline. Confirm or reject this hypothesis using the tools you've learned in the preceding section.

Join Problems

The data you've been working with in this chapter has been cleaned up so that you'll have as few problems as possible. Your own data is unlikely to be so nice, so there are a few things that you should do with your own data to make your joins go smoothly:

1. Start by identifying the variables that form the primary key in each table. You should usually do this based on your understanding of the data, not empirically by looking for a combination of variables that give a unique identifier. If you just look for variables without thinking about what they mean, you might get (un)lucky and find a combination that's unique in your current data but the relationship might not be true in general.

For example, the altitude and longitude uniquely identify each airport, but they are not good identifiers!

```
airports %>% count(alt, lon) %>% filter(n > 1)
#> Source: local data frame [0 x 3]
#> Groups: alt [0]
#>
#> # ... with 3 variables: alt <int>, lon <dbl>, n <int>
```

2. Check that none of the variables in the primary key are missing. If a value is missing then it can't identify an observation!
3. Check that your foreign keys match primary keys in another table. The best way to do this is with an `anti_join()`. It's common for keys not to match because of data entry errors. Fixing these is often a lot of work.

   If you do have missing keys, you'll need to be thoughtful about your use of inner versus outer joins, carefully considering whether or not you want to drop rows that don't have a match.

Be aware that simply checking the number of rows before and after the join is not sufficient to ensure that your join has gone smoothly. If you have an inner join with duplicate keys in both tables, you might get unlucky as the number of dropped rows might exactly equal the number of duplicated rows!

Set Operations

The final type of two-table verb are the set operations. Generally, I use these the least frequently, but they are occasionally useful when you want to break a single complex filter into simpler pieces. All these operations work with a complete row, comparing the values of every variable. These expect the `x` and `y` inputs to have the same variables, and treat the observations like sets:

`intersect(x, y)`
: Return only observations in both `x` and `y`.

`union(x, y)`
: Return unique observations in `x` and `y`.

`setdiff(x, y)`
: Return observations in `x`, but not in `y`.

Given this simple data:

```
df1 <- tribble(
  ~x, ~y,
   1,  1,
   2,  1
)
df2 <- tribble(
  ~x, ~y,
   1,  1,
   1,  2
)
```

The four possibilities are:

```
intersect(df1, df2)
#> # A tibble: 1 × 2
#>       x     y
#>   <dbl> <dbl>
#> 1     1     1

# Note that we get 3 rows, not 4
union(df1, df2)
#> # A tibble: 3 × 2
#>       x     y
#>   <dbl> <dbl>
#> 1     1     2
#> 2     2     1
#> 3     1     1

setdiff(df1, df2)
#> # A tibble: 1 × 2
#>       x     y
#>   <dbl> <dbl>
#> 1     2     1

setdiff(df2, df1)
#> # A tibble: 1 × 2
#>       x     y
#>   <dbl> <dbl>
#> 1     1     2
```

CHAPTER 11

Strings with stringr

Introduction

This chapter introduces you to string manipulation in R. You'll learn the basics of how strings work and how to create them by hand, but the focus of this chapter will be on regular expressions, or *regexps* for short. Regular expressions are useful because strings usually contain unstructured or semi-structured data, and regexps are a concise language for describing patterns in strings. When you first look at a regexp, you'll think a cat walked across your keyboard, but as your understanding improves they will soon start to make sense.

Prerequisites

This chapter will focus on the **stringr** package for string manipulation. **stringr** is not part of the core tidyverse because you don't always have textual data, so we need to load it explicitly.

```
library(tidyverse)
library(stringr)
```

String Basics

You can create strings with either single quotes or double quotes. Unlike other languages, there is no difference in behavior. I recommend always using ", unless you want to create a string that contains multiple ":

```
string1 <- "This is a string"
string2 <- 'To put a "quote" inside a string, use single quotes'
```

If you forget to close a quote, you'll see +, the continuation character:

```
> "This is a string without a closing quote
+
+
+ HELP I'M STUCK
```

If this happens to you, press Esc and try again!

To include a literal single or double quote in a string you can use \ to "escape" it:

```
double_quote <- "\"" # or '"'
single_quote <- '\'' # or "'"
```

That means if you want to include a literal backslash, you'll need to double it up: "\\".

Beware that the printed representation of a string is not the same as string itself, because the printed representation shows the escapes. To see the raw contents of the string, use `writeLines()`:

```
x <- c("\"", "\\")
x
#> [1] "\"" "\\"
writeLines(x)
#> "
#> \
```

There are a handful of other special characters. The most common are `"\n"`, newline, and `"\t"`, tab, but you can see the complete list by requesting help on `?'"'`, or `?"'"`. You'll also sometimes see strings like `"\u00b5"`, which is a way of writing non-English characters that works on all platforms:

```
x <- "\u00b5"
x
#> [1] "µ"
```

Multiple strings are often stored in a character vector, which you can create with `c()`:

```
c("one", "two", "three")
#> [1] "one"   "two"   "three"
```

String Length

Base R contains many functions to work with strings but we'll avoid them because they can be inconsistent, which makes them hard to remember. Instead we'll use functions from **stringr**. These have more intuitive names, and all start with `str_`. For example, `str_length()` tells you the number of characters in a string:

```
str_length(c("a", "R for data science", NA))
#> [1]  1 18 NA
```

The common `str_` prefix is particularly useful if you use RStudio, because typing `str_` will trigger autocomplete, allowing you to see all **stringr** functions:

```
> str_c              {stringr}   str_c(..., sep = "", collapse = NULL)
> str_conv           {stringr}   To understand how str_c works, you need to imagine that you are
> str_count          {stringr}   building up a matrix of strings. Each input argument forms a
> str_detect         {stringr}   column, and is expanded to the length of the longest argument,
> str_dup            {stringr}   using the usual recyling rules. The sep string is inserted between
> str_extract        {stringr}   each column. If collapse is NULL each row is collapsed into a single
> str_extract_all    {stringr}   string. If non-NULL that string is inserted at the end of each row,
                                 and the entire matrix collapsed to a single string.
                                 Press F1 for additional help
> str_
```

Combining Strings

To combine two or more strings, use `str_c()`:

```
str_c("x", "y")
#> [1] "xy"
str_c("x", "y", "z")
#> [1] "xyz"
```

Use the `sep` argument to control how they're separated:

```
str_c("x", "y", sep = ", ")
#> [1] "x, y"
```

Like most other functions in R, missing values are contagious. If you want them to print as `"NA"`, use `str_replace_na()`:

```
x <- c("abc", NA)
str_c("|-", x, "-|")
#> [1] "|-abc-|" NA
str_c("|-", str_replace_na(x), "-|")
#> [1] "|-abc-|" "|-NA-|"
```

As shown in the preceding code, `str_c()` is vectorized, and it automatically recycles shorter vectors to the same length as the longest:

```
str_c("prefix-", c("a", "b", "c"), "-suffix")
#> [1] "prefix-a-suffix" "prefix-b-suffix" "prefix-c-suffix"
```

Objects of length 0 are silently dropped. This is particularly useful in conjunction with `if`:

```
name <- "Hadley"
time_of_day <- "morning"
birthday <- FALSE

str_c(
  "Good ", time_of_day, " ", name,
  if (birthday) " and HAPPY BIRTHDAY",
  "."
)
#> [1] "Good morning Hadley."
```

To collapse a vector of strings into a single string, use `collapse`:

```
str_c(c("x", "y", "z"), collapse = ", ")
#> [1] "x, y, z"
```

Subsetting Strings

You can extract parts of a string using `str_sub()`. As well as the string, `str_sub()` takes `start` and `end` arguments that give the (inclusive) position of the substring:

```
x <- c("Apple", "Banana", "Pear")
str_sub(x, 1, 3)
#> [1] "App" "Ban" "Pea"

# negative numbers count backwards from end
str_sub(x, -3, -1)
#> [1] "ple" "ana" "ear"
```

Note that `str_sub()` won't fail if the string is too short; it will just return as much as possible:

```
str_sub("a", 1, 5)
#> [1] "a"
```

You can also use the assignment form of `str_sub()` to modify strings:

```
str_sub(x, 1, 1) <- str_to_lower(str_sub(x, 1, 1))
x
#> [1] "apple"  "banana" "pear"
```

Locales

Earlier I used `str_to_lower()` to change the text to lowercase. You can also use `str_to_upper()` or `str_to_title()`. However, changing case is more complicated than it might at first appear because different languages have different rules for changing case. You can pick which set of rules to use by specifying a locale:

```
# Turkish has two i's: with and without a dot, and it
# has a different rule for capitalizing them:
str_to_upper(c("i", "ı"))
#> [1] "I" "I"
str_to_upper(c("i", "ı"), locale = "tr")
#> [1] "İ" "I"
```

The locale is specified as an ISO 639 language code, which is a two- or three-letter abbreviation. If you don't already know the code for your language, Wikipedia (*http://bit.ly/ISO639-1*) has a good list. If you leave the locale blank, it will use the current locale, as provided by your operating system.

Another important operation that's affected by the locale is sorting. The base R `order()` and `sort()` functions sort strings using the current locale. If you want robust behavior across different computers, you may want to use `str_sort()` and `str_order()`, which take an additional `locale` argument:

```
x <- c("apple", "eggplant", "banana")

str_sort(x, locale = "en")  # English
#> [1] "apple"    "banana"   "eggplant"

str_sort(x, locale = "haw") # Hawaiian
#> [1] "apple"    "eggplant" "banana"
```

Exercises

1. In code that doesn't use **stringr**, you'll often see `paste()` and `paste0()`. What's the difference between the two functions? What **stringr** function are they equivalent to? How do the functions differ in their handling of `NA`?

2. In your own words, describe the difference between the `sep` and `collapse` arguments to `str_c()`.

3. Use `str_length()` and `str_sub()` to extract the middle character from a string. What will you do if the string has an even number of characters?
4. What does `str_wrap()` do? When might you want to use it?
5. What does `str_trim()` do? What's the opposite of `str_trim()`?
6. Write a function that turns (e.g.) a vector `c("a", "b", "c")` into the string `a, b, and c`. Think carefully about what it should do if given a vector of length 0, 1, or 2.

Matching Patterns with Regular Expressions

Regexps are a very terse language that allow you to describe patterns in strings. They take a little while to get your head around, but once you understand them, you'll find them extremely useful.

To learn regular expressions, we'll use `str_view()` and `str_view_all()`. These functions take a character vector and a regular expression, and show you how they match. We'll start with very simple regular expressions and then gradually get more and more complicated. Once you've mastered pattern matching, you'll learn how to apply those ideas with various **stringr** functions.

Basic Matches

The simplest patterns match exact strings:

```
x <- c("apple", "banana", "pear")
str_view(x, "an")
```

```
apple
banana
pear
```

The next step up in complexity is `.`, which matches any character (except a newline):

```
str_view(x, ".a.")
```

```
apple
banana
pear
```

But if "." matches any character, how do you match the character "."? You need to use an "escape" to tell the regular expression you want to match it exactly, not use its special behavior. Like strings, regexps use the backslash, \, to escape special behavior. So to match an ., you need the regexp \.. Unfortunately this creates a problem. We use strings to represent regular expressions, and \ is also used as an escape symbol in strings. So to create the regular expression \. we need the string "\\.":

```
# To create the regular expression, we need \\
dot <- "\\."

# But the expression itself only contains one:
writeLines(dot)
#> \.

# And this tells R to look for an explicit .
str_view(c("abc", "a.c", "bef"), "a\\.c")
```

```
abc
a.c
bef
```

If \ is used as an escape character in regular expressions, how do you match a literal \? Well you need to escape it, creating the regular expression \\. To create that regular expression, you need to use a string, which also needs to escape \. That means to match a literal \ you need to write "\\\\"—you need four backslashes to match one!

```
x <- "a\\b"
writeLines(x)
#> a\b

str_view(x, "\\\\")
```

```
a\b
```

In this book, I'll write regular expressions as \. and strings that represent the regular expression as "\\.".

Exercises

1. Explain why each of these strings don't match a \: "\", "\\", "\\\".

2. How would you match the sequence `"'\`?
3. What patterns will the regular expression `\..\..\..` match? How would you represent it as a string?

Anchors

By default, regular expressions will match any part of a string. It's often useful to *anchor* the regular expression so that it matches from the start or end of the string. You can use:

- `^` to match the start of the string.
- `$` to match the end of the string.

```
x <- c("apple", "banana", "pear")
str_view(x, "^a")
```

```
apple
banana
pear
```

```
str_view(x, "a$")
```

```
apple
banana
pear
```

To remember which is which, try this mnemonic that I learned from Evan Misshula (*http://bit.ly/EvanMisshula*): if you begin with power (`^`), you end up with money (`$`).

To force a regular expression to only match a complete string, anchor it with both `^` and `$`:

```
x <- c("apple pie", "apple", "apple cake")
str_view(x, "apple")
```

```
apple pie
apple
apple cake
```

```
str_view(x, "^apple$")
```

```
apple pie
apple
apple cake
```

You can also match the boundary between words with `\b`. I don't often use this in R, but I will sometimes use it when I'm doing a search in RStudio when I want to find the name of a function that's a component of other functions. For example, I'll search for `\bsum\b` to avoid matching `summarize`, `summary`, `rowsum`, and so on.

Exercises

1. How would you match the literal string `"$^$"`?
2. Given the corpus of common words in `stringr::words`, create regular expressions that find all words that:
   a. Start with "y".
   b. End with "x".
   c. Are exactly three letters long. (Don't cheat by using `str_length()`!)
   d. Have seven letters or more.

   Since this list is long, you might want to use the `match` argument to `str_view()` to show only the matching or non-matching words.

Character Classes and Alternatives

There are a number of special patterns that match more than one character. You've already seen `.`, which matches any character apart from a newline. There are four other useful tools:

- `\d` matches any digit.
- `\s` matches any whitespace (e.g., space, tab, newline).
- `[abc]` matches a, b, or c.
- `[^abc]` matches anything except a, b, or c.

Remember, to create a regular expression containing `\d` or `\s`, you'll need to escape the `\` for the string, so you'll type `"\\d"` or `"\\s"`.

You can use *alternation* to pick between one or more alternative patterns. For example, `abc|d..f` will match either "abc", or "deaf". Note that the precedence for `|` is low, so that `abc|xyz` matches `abc` or `xyz` not `abcyz` or `abxyz`. Like with mathematical expressions, if precedence ever gets confusing, use parentheses to make it clear what you want:

```
str_view(c("grey", "gray"), "gr(e|a)y")
```

grey

gray

Exercises

1. Create regular expressions to find all words that:
   a. Start with a vowel.
   b. Only contain consonants. (Hint: think about matching "not"-vowels.)
   c. End with `ed`, but not with `eed`.
   d. End with `ing` or `ize`.
2. Empirically verify the rule "i before e except after c."
3. Is "q" always followed by a "u"?
4. Write a regular expression that matches a word if it's probably written in British English, not American English.
5. Create a regular expression that will match telephone numbers as commonly written in your country.

Repetition

The next step up in power involves controlling how many times a pattern matches:

- `?`: 0 or 1
- `+`: 1 or more
- `*`: 0 or more

```
x <- "1888 is the longest year in Roman numerals: MDCCCLXXXVIII"
str_view(x, "CC?")
```

```
1888 is the longest year in Roman numerals: MDCCCLXXXVIII
```

```
str_view(x, "CC+")
```

```
1888 is the longest year in Roman numerals: MDCCCLXXXVIII
```

```
str_view(x, 'C[LX]+')
```

```
1888 is the longest year in Roman numerals: MDCCCLXXXVIII
```

Note that the precedence of these operators is high, so you can write `colou?r` to match either American or British spellings. That means most uses will need parentheses, like `bana(na)+`.

You can also specify the number of matches precisely:

- `{n}`: exactly n
- `{n,}`: n or more
- `{,m}`: at most m
- `{n,m}`: between n and m

```
str_view(x, "C{2}")
```

```
1888 is the longest year in Roman numerals: MDCCCLXXXVIII
```

```
str_view(x, "C{2,}")
```

```
1888 is the longest year in Roman numerals: MDCCCLXXXVIII
```

```
str_view(x, "C{2,3}")
```

```
1888 is the longest year in Roman numerals: MDCCCLXXXVIII
```

By default these matches are "greedy": they will match the longest string possible. You can make them "lazy," matching the shortest string possible, by putting a ? after them. This is an advanced feature of regular expressions, but it's useful to know that it exists:

```
str_view(x, 'C{2,3}?')
```

```
1888 is the longest year in Roman numerals: MDCCCLXXXVIII
```

```
str_view(x, 'C[LX]+?')
```

```
1888 is the longest year in Roman numerals: MDCCCLXXXVIII
```

Exercises

1. Describe the equivalents of ?, +, and * in `{m,n}` form.
2. Describe in words what these regular expressions match (read carefully to see if I'm using a regular expression or a string that defines a regular expression):
   a. `^.*$`
   b. `"\\{.+\\}"`
   c. `\d{4}-\d{2}-\d{2}`
   d. `"\\\\{4}"`
3. Create regular expressions to find all words that:
   a. Start with three consonants.
   b. Have three or more vowels in a row.
   c. Have two or more vowel-consonant pairs in a row.
4. Solve the beginner regexp crosswords at *https://regexcrossword.com/challenges/beginner*.

Grouping and Backreferences

Earlier, you learned about parentheses as a way to disambiguate complex expressions. They also define "groups" that you can refer to with *backreferences*, like `\1`, `\2`, etc. For example, the following regular expression finds all fruits that have a repeated pair of letters:

```
str_view(fruit, "(..)\\1", match = TRUE)
```

```
banana
coconut
cucumber
jujube
papaya
salal berry
```

(Shortly, you'll also see how they're useful in conjunction with `str_match()`.)

Exercises

1. Describe, in words, what these expressions will match:
   a. `(.)\1\1`
   b. `"(.)(.)\\2\\1"`
   c. `(..)\1`
   d. `"(.).\\1.\\1"`
   e. `"(.)(.)(.).*\\3\\2\\1"`
2. Construct regular expressions to match words that:
   a. Start and end with the same character.
   b. Contain a repeated pair of letters (e.g., "church" contains "ch" repeated twice).
   c. Contain one letter repeated in at least three places (e.g., "eleven" contains three "e"s).

Tools

Now that you've learned the basics of regular expressions, it's time to learn how to apply them to real problems. In this section you'll learn a wide array of **stringr** functions that let you:

- Determine which strings match a pattern.
- Find the positions of matches.
- Extract the content of matches.
- Replace matches with new values.
- Split a string based on a match.

A word of caution before we continue: because regular expressions are so powerful, it's easy to try and solve every problem with a single regular expression. In the words of Jamie Zawinski:

> Some people, when confronted with a problem, think "I know, I'll use regular expressions." Now they have two problems.

As a cautionary tale, check out this regular expression that checks if an email address is valid:

```
(?:(?:\r\n)?[ \t])*(?:(?:(?:[^()<>@,;:\\".\[\] \000-\031]+(?:(?:(?:\r\n)?[ \t])+|\Z|(?=[\["()<>@,;:\\".\
[\]]))|"(?:[^\"\r\\]|\\.|(?:(?:\r\n)?[ \t]))*"(?:(?:\r\n)?[ \t])*)(?:\.(?:(?:\r\n)?[ \t])*(?:[^()<>@,;:\
\".\[\] \000-\031]+(?:(?:(?:\r\n)?[ \t])+|\Z|(?=[\["()<>@,;:\\".\[\]]))|"(?:[^\"\r\\]|\\.|(?:(?:\r\n)?
[ \t]))*"(?:(?:\r\n)?[ \t])*))*@(?:(?:\r\n)?[ \t])*(?:[^()<>@,;:\\".\[\] \000-\031]+(?:(?:(?:\r\n)?[ \t])
+|\Z|(?=[\["()<>@,;:\\".\[\]]))|\[([^\[\]\r\\]|\\.)*\](?:(?:\r\n)?[ \t])*)(?:\.(?:(?:\r\n)?[ \t])*(?:
[^()<>@,;:\\".\[\] \000-\031]+(?:(?:(?:\r\n)?[ \t])+|\Z|(?=[\["()<>@,;:\\".\[\]]))|\[([^\[\]\r\\]|\\.)*\]
(?:(?:\r\n)?[ \t])*))*|(?:[^()<>@,;:\\".\[\] \000-\031]+(?:(?:(?:\r\n)?[ \t])+|\Z|(?=[\["()<>@,;:\\".\
[\]]))|"(?:[^\"\r\\]|\\.|(?:(?:\r\n)?[ \t]))*"(?:(?:\r\n)?[ \t])*)*\<(?:(?:\r\n)?[ \t])*(?:@(?:[^()<>@,;:
\\".\[\] \000-\031]+(?:(?:(?:\r\n)?[ \t])+|\Z|(?=[\["()<>@,;:\\".\[\]]))|\[([^\[\]\r\\]|\\.)*\](?:(?:
\r\n)?[ \t])*)(?:\.(?:(?:\r\n)?[ \t])*(?:[^()<>@,;:\\".\[\] \000-\031]+(?:(?:(?:\r\n)?[ \t])+|\Z|(?=[\
["()<>@,;:\\".\[\]]))|\[([^\[\]\r\\]|\\.)*\](?:(?:\r\n)?[ \t])*))*(?:,@(?:(?:\r\n)?[ \t])*(?:[^()<>@,;:\
\".\[\] \000-\031]+(?:(?:(?:\r\n)?[ \t])+|\Z|(?=[\["()<>@,;:\\".\[\]]))|\[([^\[\]\r\\]|\\.)*\](?:(?:
\r\n)?[ \t])*)(?:\.(?:(?:\r\n)?[ \t])*(?:[^()<>@,;:\\".\[\] \000-\031]+(?:(?:(?:\r\n)?[ \t])+|\Z|(?=[\
["()<>@,;:\\".\[\]]))|\[([^\[\]\r\\]|\\.)*\](?:(?:\r\n)?[ \t])*))*)*:(?:(?:\r\n)?[ \t])*)?(?:[^()<>@,;:\
\".\[\] \000-\031]+(?:(?:(?:\r\n)?[ \t])+|\Z|(?=[\["()<>@,;:\\".\[\]]))|"(?:[^\"\r\\]|\\.|(?:(?:\r\n)?
[ \t]))*"(?:(?:\r\n)?[ \t])*)(?:\.(?:(?:\r\n)?[ \t])*(?:[^()<>@,;:\\".\[\] \000-\031]+(?:(?:(?:\r\n)?
[ \t])+|\Z|(?=[\["()<>@,;:\\".\[\]]))|"(?:[^\"\r\\]|\\.|(?:(?:\r\n)?[ \t]))*"(?:(?:\r\n)?[ \t])*))*@(?:
(?:\r\n)?[ \t])*(?:[^()<>@,;:\\".\[\] \000-\031]+(?:(?:(?:\r\n)?[ \t])+|\Z|(?=[\["()<>@,;:\\".\[\]]))|\
[([^\[\]\r\\]|\\.)*\](?:(?:\r\n)?[ \t])*)(?:\.(?:(?:\r\n)?[ \t])*(?:[^()<>@,;:\\".\[\] \000-\031]+(?:(?:
(?:\r\n)?[ \t])+|\Z|(?=[\["()<>@,;:\\".\[\]]))|\[([^\[\]\r\\]|\\.)*\](?:(?:\r\n)?[ \t])*))*\>(?:(?:\r\n)?
[ \t])*)|(?:[^()<>@,;:\\".\[\] \000-\031]+(?:(?:(?:\r\n)?[ \t])+|\Z|(?=[\["()<>@,;:\\".\[\]]))|"(?:
[^\"\r\\]|\\.|(?:(?:\r\n)?[ \t]))*"(?:(?:\r\n)?[ \t])*)*:(?:(?:\r\n)?[ \t])*(?:(?:(?:[^()<>@,;:\\".\[\]
\000-\031]+(?:(?:(?:\r\n)?[ \t])+|\Z|(?=[\["()<>@,;:\\".\[\]]))|"(?:[^\"\r\\]|\\.|(?:(?:\r\n)?
[ \t]))*"(?:(?:\r\n)?[ \t])*)(?:\.(?:(?:\r\n)?[ \t])*(?:[^()<>@,;:\\".\[\] \000-\031]+(?:(?:(?:\r\n)?
[ \t])+|\Z|(?=[\["()<>@,;:\\".\[\]]))|"(?:[^\"\r\\]|\\.|(?:(?:\r\n)?[ \t]))*"(?:(?:\r\n)?[ \t])*))*@(?:
(?:\r\n)?[ \t])*(?:[^()<>@,;:\\".\[\] \000-\031]+(?:(?:(?:\r\n)?[ \t])+|\Z|(?=[\["()<>@,;:\\".\[\]]))|\
[([^\[\]\r\\]|\\.)*\](?:(?:\r\n)?[ \t])*)(?:\.(?:(?:\r\n)?[ \t])*(?:[^()<>@,;:\\".\[\] \000-\031]+(?:(?:
(?:\r\n)?[ \t])+|\Z|(?=[\["()<>@,;:\\".\[\]]))|\[([^\[\]\r\\]|\\.)*\](?:(?:\r\n)?[ \t])*))*|(?:[^()<>@,;:
\\".\[\] \000-\031]+(?:(?:(?:\r\n)?[ \t])+|\Z|(?=[\["()<>@,;:\\".\[\]]))|"(?:[^\"\r\\]|\\.|(?:(?:\r\n)?
[ \t]))*"(?:(?:\r\n)?[ \t])*)*\<(?:(?:\r\n)?[ \t])*(?:@(?:[^()<>@,;:\\".\[\] \000-\031]+(?:(?:(?:\r\n)?
[ \t])+|\Z|(?=[\["()<>@,;:\\".\[\]]))|\[([^\[\]\r\\]|\\.)*\](?:(?:\r\n)?[ \t])*)(?:\.(?:(?:\r\n)?
[ \t])*(?:[^()<>@,;:\\".\[\] \000-\031]+(?:(?:(?:\r\n)?[ \t])+|\Z|(?=[\["()<>@,;:\\".\[\]]))|\[([^\[\]\r\
\]|\\.)*\](?:(?:\r\n)?[ \t])*))*(?:,@(?:(?:\r\n)?[ \t])*(?:[^()<>@,;:\\".\[\] \000-\031]+(?:(?:(?:\r\n)?
[ \t])+|\Z|(?=[\["()<>@,;:\\".\[\]]))|\[([^\[\]\r\\]|\\.)*\](?:(?:\r\n)?[ \t])*)(?:\.(?:(?:\r\n)?
[ \t])*(?:[^()<>@,;:\\".\[\] \000-\031]+(?:(?:(?:\r\n)?[ \t])+|\Z|(?=[\["()<>@,;:\\".\[\]]))|\[([^\[\]\r\
\]|\\.)*\](?:(?:\r\n)?[ \t])*))*)*:(?:(?:\r\n)?[ \t])*)?(?:[^()<>@,;:\\".\[\] \000-\031]+(?:(?:(?:\r\n)?
[ \t])+|\Z|(?=[\["()<>@,;:\\".\[\]]))|"(?:[^\"\r\\]|\\.|(?:(?:\r\n)?[ \t]))*"(?:(?:\r\n)?[ \t])*)(?:\.(?:
(?:\r\n)?[ \t])*(?:[^()<>@,;:\\".\[\] \000-\031]+(?:(?:(?:\r\n)?[ \t])+|\Z|(?=[\["()<>@,;:\\".\
[\]]))|"(?:[^\"\r\\]|\\.|(?:(?:\r\n)?[ \t]))*"(?:(?:\r\n)?[ \t])*))*@(?:(?:\r\n)?[ \t])*(?:[^()<>@,;:\\".
\[\] \000-\031]+(?:(?:(?:\r\n)?[ \t])+|\Z|(?=[\["()<>@,;:\\".\[\]]))|\[([^\[\]\r\\]|\\.)*\](?:(?:\r\n)?
[ \t])*)(?:\.(?:(?:\r\n)?[ \t])*(?:[^()<>@,;:\\".\[\] \000-\031]+(?:(?:(?:\r\n)?[ \t])+|\Z|(?=[\
["()<>@,;:\\".\[\]]))|\[([^\[\]\r\\]|\\.)*\](?:(?:\r\n)?[ \t])*))*\>(?:(?:\r\n)?[ \t])*)(?:,\s*(?:(?:
[^()<>@,;:\\".\[\] \000-\031]+(?:(?:(?:\r\n)?[ \t])+|\Z|(?=[\["()<>@,;:\\".\[\]]))|"(?:[^\"\r\\]|\\.|(?:
(?:\r\n)?[ \t]))*"(?:(?:\r\n)?[ \t])*)(?:\.(?:(?:\r\n)?[ \t])*(?:[^()<>@,;:\\".\[\] \000-\031]+(?:(?:(?:
\r\n)?[ \t])+|\Z|(?=[\["()<>@,;:\\".\[\]]))|"(?:[^\"\r\\]|\\.|(?:(?:\r\n)?[ \t]))*"(?:(?:\r\n)?
[ \t])*))*@(?:(?:\r\n)?[ \t])*(?:[^()<>@,;:\\".\[\] \000-\031]+(?:(?:(?:\r\n)?[ \t])+|\Z|(?=[\["()<>@,;:\
\".\[\]]))|\[([^\[\]\r\\]|\\.)*\](?:(?:\r\n)?[ \t])*)(?:\.(?:(?:\r\n)?[ \t])*(?:[^()<>@,;:\\".\[\] \000-
\031]+(?:(?:(?:\r\n)?[ \t])+|\Z|(?=[\["()<>@,;:\\".\[\]]))|\[([^\[\]\r\\]|\\.)*\](?:(?:\r\n)?[ \t])*))*|
(?:[^()<>@,;:\\".\[\] \000-\031]+(?:(?:(?:\r\n)?[ \t])+|\Z|(?=[\["()<>@,;:\\".\[\]]))|"(?:[^\"\r\\]|\\.|
(?:(?:\r\n)?[ \t]))*"(?:(?:\r\n)?[ \t])*)*\<(?:(?:\r\n)?[ \t])*(?:@(?:[^()<>@,;:\\".\[\] \000-\031]+(?:
(?:(?:\r\n)?[ \t])+|\Z|(?=[\["()<>@,;:\\".\[\]]))|\[([^\[\]\r\\]|\\.)*\](?:(?:\r\n)?[ \t])*)(?:\.(?:(?:
\r\n)?[ \t])*(?:[^()<>@,;:\\".\[\] \000-\031]+(?:(?:(?:\r\n)?[ \t])+|\Z|(?=[\["()<>@,;:\\".\[\]]))|\[([^\
[\]\r\\]|\\.)*\](?:(?:\r\n)?[ \t])*))*(?:,@(?:(?:\r\n)?[ \t])*(?:[^()<>@,;:\\".\[\] \000-\031]+(?:(?:(?:
\r\n)?[ \t])+|\Z|(?=[\["()<>@,;:\\".\[\]]))|\[([^\[\]\r\\]|\\.)*\](?:(?:\r\n)?[ \t])*)(?:\.(?:(?:\r\n)?
[ \t])*(?:[^()<>@,;:\\".\[\] \000-\031]+(?:(?:(?:\r\n)?[ \t])+|\Z|(?=[\["()<>@,;:\\".\[\]]))|\[([^\[\]\r\
\]|\\.)*\](?:(?:\r\n)?[ \t])*))*)*:(?:(?:\r\n)?[ \t])*)?(?:[^()<>@,;:\\".\[\] \000-\031]+(?:(?:(?:\r\n)?
[ \t])+|\Z|(?=[\["()<>@,;:\\".\[\]]))|"(?:[^\"\r\\]|\\.|(?:(?:\r\n)?[ \t]))*"(?:(?:\r\n)?[ \t])*)(?:\.(?:
(?:\r\n)?[ \t])*(?:[^()<>@,;:\\".\[\] \000-\031]+(?:(?:(?:\r\n)?[ \t])+|\Z|(?=[\["()<>@,;:\\".\
[\]]))|"(?:[^\"\r\\]|\\.|(?:(?:\r\n)?[ \t]))*"(?:(?:\r\n)?[ \t])*))*@(?:(?:\r\n)?[ \t])*(?:[^()<>@,;:\\".
\[\] \000-\031]+(?:(?:(?:\r\n)?[ \t])+|\Z|(?=[\["()<>@,;:\\".\[\]]))|\[([^\[\]\r\\]|\\.)*\](?:(?:\r\n)?
[ \t])*)(?:\.(?:(?:\r\n)?[ \t])*(?:[^()<>@,;:\\".\[\] \000-\031]+(?:(?:(?:\r\n)?[ \t])+|\Z|(?=[\
["()<>@,;:\\".\[\]]))|\[([^\[\]\r\\]|\\.)*\](?:(?:\r\n)?[ \t])*))*\>(?:(?:\r\n)?[ \t])*))*)?;\s*)
```

This is a somewhat pathological example (because email addresses are actually suprisingly complex), but is used in real code. See the stackoverflow discussion (*http://stackoverflow.com/a/201378*) for more details.

Don't forget that you're in a programming language and you have other tools at your disposal. Instead of creating one complex regular expression, it's often easier to create a series of simpler regexps. If you get stuck trying to create a single regexp that solves your problem, take a step back and think if you could break the problem down into smaller pieces, solving each challenge before moving on to the next one.

Detect Matches

To determine if a character vector matches a pattern, use `str_detect()`. It returns a logical vector the same length as the input:

```
x <- c("apple", "banana", "pear")
str_detect(x, "e")
#> [1]  TRUE FALSE  TRUE
```

Remember that when you use a logical vector in a numeric context, `FALSE` becomes 0 and `TRUE` becomes 1. That makes `sum()` and `mean()` useful if you want to answer questions about matches across a larger vector:

```
# How many common words start with t?
sum(str_detect(words, "^t"))
#> [1] 65
# What proportion of common words end with a vowel?
mean(str_detect(words, "[aeiou]$"))
#> [1] 0.277
```

When you have complex logical conditions (e.g., match a or b but not c unless d) it's often easier to combine multiple `str_detect()` calls with logical operators, rather than trying to create a single regular expression. For example, here are two ways to find all words that don't contain any vowels:

```
# Find all words containing at least one vowel, and negate
no_vowels_1 <- !str_detect(words, "[aeiou]")
# Find all words consisting only of consonants (non-vowels)
no_vowels_2 <- str_detect(words, "^[^aeiou]+$")
identical(no_vowels_1, no_vowels_2)
#> [1] TRUE
```

The results are identical, but I think the first approach is significantly easier to understand. If your regular expression gets overly complicated, try breaking it up into smaller pieces, giving each piece a name, and then combining the pieces with logical operations.

A common use of `str_detect()` is to select the elements that match a pattern. You can do this with logical subsetting, or the convenient `str_subset()` wrapper:

```
words[str_detect(words, "x$")]
#> [1] "box" "sex" "six" "tax"
str_subset(words, "x$")
#> [1] "box" "sex" "six" "tax"
```

Typically, however, your strings will be one column of a data frame, and you'll want to use `filter` instead:

```
df <- tibble(
  word = words,
  i = seq_along(word)
)
df %>%
  filter(str_detect(words, "x$"))
#> # A tibble: 4 × 2
#>    word     i
#>   <chr> <int>
#> 1   box   108
#> 2   sex   747
#> 3   six   772
#> 4   tax   841
```

A variation on `str_detect()` is `str_count()`: rather than a simple yes or no, it tells you how many matches there are in a string:

```
x <- c("apple", "banana", "pear")
str_count(x, "a")
#> [1] 1 3 1

# On average, how many vowels per word?
mean(str_count(words, "[aeiou]"))
#> [1] 1.99
```

It's natural to use `str_count()` with `mutate()`:

```
df %>%
  mutate(
    vowels = str_count(word, "[aeiou]"),
    consonants = str_count(word, "[^aeiou]")
  )
#> # A tibble: 980 × 4
#>         word     i vowels consonants
#>        <chr> <int>  <int>      <int>
#> 1          a     1      1          0
#> 2       able     2      2          2
#> 3      about     3      3          2
#> 4   absolute     4      4          4
#> 5     accept     5      2          4
#> 6    account     6      3          4
#> # ... with 974 more rows
```

abababa

Note that matches never overlap. For example, in "abababa", how many times will the pattern "aba" match? Regular expressions say two, not three:

```
str_count("abababa", "aba")
#> [1] 2
str_view_all("abababa", "aba")
```

abababa

Note the use of `str_view_all()`. As you'll shortly learn, many **stringr** functions come in pairs: one function works with a single match, and the other works with all matches. The second function will have the suffix `_all`.

Exercises

1. For each of the following challenges, try solving it by using both a single regular expression, and a combination of multiple `str_detect()` calls:

   a. Find all words that start or end with x.

   b. Find all words that start with a vowel and end with a consonant.

   c. Are there any words that contain at least one of each different vowel?

   d. What word has the highest number of vowels? What word has the highest proportion of vowels? (Hint: what is the denominator?)

Extract Matches

To extract the actual text of a match, use `str_extract()`. To show that off, we're going to need a more complicated example. I'm going to use the Harvard sentences (*http://bit.ly/Harvardsentences*), which were designed to test VOIP systems, but are also useful for practicing regexes. These are provided in `stringr::sentences`:

```
length(sentences)
#> [1] 720
head(sentences)
#> [1] "The birch canoe slid on the smooth planks."
#> [2] "Glue the sheet to the dark blue background."
```

```
#> [3] "It's easy to tell the depth of a well."
#> [4] "These days a chicken leg is a rare dish."
#> [5] "Rice is often served in round bowls."
#> [6] "The juice of lemons makes fine punch."
```

Imagine we want to find all sentences that contain a color. We first create a vector of color names, and then turn it into a single regular expression:

```
colors <- c(
  "red", "orange", "yellow", "green", "blue", "purple"
)
color_match <- str_c(colors, collapse = "|")
color_match
#> [1] "red|orange|yellow|green|blue|purple"
```

Now we can select the sentences that contain a color, and then extract the color to figure out which one it is:

```
has_color <- str_subset(sentences, color_match)
matches <- str_extract(has_color, color_match)
head(matches)
#> [1] "blue" "blue" "red"  "red"  "red"  "blue"
```

Note that `str_extract()` only extracts the first match. We can see that most easily by first selecting all the sentences that have more than one match:

```
more <- sentences[str_count(sentences, color_match) > 1]
str_view_all(more, color_match)
```

```
It is hard to erase blue or red ink.
The green light in the brown box flickered.
The sky in the west is tinged with orange red.
```

```
str_extract(more, color_match)
#> [1] "blue"   "green"  "orange"
```

```
It is hard to erase blue or red ink.
The green light in the brown box flickered.
The sky in the west is tinged with orange red.
```

This is a common pattern for **stringr** functions, because working with a single match allows you to use much simpler data structures. To get all matches, use `str_extract_all()`. It returns a list:

```
str_extract_all(more, color_match)
#> [[1]]
#> [1] "blue" "red"
```

```
#>
#> [[2]]
#> [1] "green" "red"
#>
#> [[3]]
#> [1] "orange" "red"
```

You'll learn more about lists in "Recursive Vectors (Lists)" on page 302 and Chapter 17.

If you use `simplify = TRUE`, `str_extract_all()` will return a matrix with short matches expanded to the same length as the longest:

```
str_extract_all(more, color_match, simplify = TRUE)
#>      [,1]     [,2]
#> [1,] "blue"   "red"
#> [2,] "green"  "red"
#> [3,] "orange" "red"

x <- c("a", "a b", "a b c")
str_extract_all(x, "[a-z]", simplify = TRUE)
#>      [,1] [,2] [,3]
#> [1,] "a"  ""   ""
#> [2,] "a"  "b"  ""
#> [3,] "a"  "b"  "c"
```

Exercises

1. In the previous example, you might have noticed that the regular expression matched "flickered," which is not a color. Modify the regex to fix the problem.
2. From the Harvard sentences data, extract:
   a. The first word from each sentence.
   b. All words ending in `ing`.
   c. All plurals.

Grouped Matches

Earlier in this chapter we talked about the use of parentheses for clarifying precedence and for backreferences when matching. You can also use parentheses to extract parts of a complex match. For example, imagine we want to extract nouns from the sentences. As a heuristic, we'll look for any word that comes after "a" or "the". Defining a "word" in a regular expression is a little tricky, so here I use a simple approximation—a sequence of at least one character that isn't a space:

```
noun <- "(a|the) ([^ ]+)"

has_noun <- sentences %>%
  str_subset(noun) %>%
  head(10)
has_noun %>%
  str_extract(noun)
#>  [1] "the smooth" "the sheet"  "the depth"  "a chicken"
#>  [5] "the parked" "the sun"    "the huge"   "the ball"
#>  [9] "the woman"  "a helps"
```

`str_extract()` gives us the complete match; `str_match()` gives each individual component. Instead of a character vector, it returns a matrix, with one column for the complete match followed by one column for each group:

```
has_noun %>%
  str_match(noun)
#>       [,1]         [,2]  [,3]
#>  [1,] "the smooth" "the" "smooth"
#>  [2,] "the sheet"  "the" "sheet"
#>  [3,] "the depth"  "the" "depth"
#>  [4,] "a chicken"  "a"   "chicken"
#>  [5,] "the parked" "the" "parked"
#>  [6,] "the sun"    "the" "sun"
#>  [7,] "the huge"   "the" "huge"
#>  [8,] "the ball"   "the" "ball"
#>  [9,] "the woman"  "the" "woman"
#> [10,] "a helps"    "a"   "helps"
```

(Unsurprisingly, our heuristic for detecting nouns is poor, and also picks up adjectives like smooth and parked.)

If your data is in a tibble, it's often easier to use `tidyr::extract()`. It works like `str_match()` but requires you to name the matches, which are then placed in new columns:

```
tibble(sentence = sentences) %>%
  tidyr::extract(
    sentence, c("article", "noun"), "(a|the) ([^ ]+)",
    remove = FALSE
  )
#> # A tibble: 720 × 3
#>                                        sentence article    noun
#> *                                         <chr>   <chr>   <chr>
#> 1  The birch canoe slid on the smooth planks.     the  smooth
#> 2 Glue the sheet to the dark blue background.     the   sheet
#> 3        It's easy to tell the depth of a well.     the   depth
```

```
#> 4    These days a chicken leg is a rare dish.       a chicken
#> 5         Rice is often served in round bowls.    <NA>     <NA>
#> 6       The juice of lemons makes fine punch.     <NA>     <NA>
#> # ... with 714 more rows
```

Like `str_extract()`, if you want all matches for each string, you'll need `str_match_all()`.

Exercises

1. Find all words that come after a "number" like "one", "two", "three", etc. Pull out both the number and the word.
2. Find all contractions. Separate out the pieces before and after the apostrophe.

Replacing Matches

`str_replace()` and `str_replace_all()` allow you to replace matches with new strings. The simplest use is to replace a pattern with a fixed string:

```
x <- c("apple", "pear", "banana")
str_replace(x, "[aeiou]", "-")
#> [1] "-pple"  "p-ar"   "b-nana"
str_replace_all(x, "[aeiou]", "-")
#> [1] "-ppl-"  "p--r"   "b-n-n-"
```

With `str_replace_all()` you can perform multiple replacements by supplying a named vector:

```
x <- c("1 house", "2 cars", "3 people")
str_replace_all(x, c("1" = "one", "2" = "two", "3" = "three"))
#> [1] "one house"    "two cars"     "three people"
```

Instead of replacing with a fixed string you can use backreferences to insert components of the match. In the following code, I flip the order of the second and third words:

```
sentences %>%
  str_replace("([^ ]+) ([^ ]+) ([^ ]+)", "\\1 \\3 \\2") %>%
  head(5)
#> [1] "The canoe birch slid on the smooth planks."
#> [2] "Glue sheet the to the dark blue background."
#> [3] "It's to easy tell the depth of a well."
#> [4] "These a days chicken leg is a rare dish."
#> [5] "Rice often is served in round bowls."
```

Exercises

1. Replace all forward slashes in a string with backslashes.
2. Implement a simple version of `str_to_lower()` using `replace_all()`.
3. Switch the first and last letters in `words`. Which of those strings are still words?

Splitting

Use `str_split()` to split a string up into pieces. For example, we could split sentences into words:

```
sentences %>%
  head(5) %>%
  str_split(" ")
#> [[1]]
#> [1] "The"     "birch"   "canoe"   "slid"    "on"      "the"
#> [7] "smooth"  "planks."
#>
#> [[2]]
#> [1] "Glue"        "the"         "sheet"       "to"
#> [5] "the"         "dark"        "blue"        "background."
#>
#> [[3]]
#> [1] "It's"  "easy"  "to"    "tell"  "the"   "depth" "of"
#> [8] "a"     "well."
#>
#> [[4]]
#> [1] "These"   "days"    "a"       "chicken" "leg"     "is"
#> [7] "a"       "rare"    "dish."
#>
#> [[5]]
#> [1] "Rice"   "is"     "often"  "served" "in"     "round"
#> [7] "bowls."
```

Because each component might contain a different number of pieces, this returns a list. If you're working with a length-1 vector, the easiest thing is to just extract the first element of the list:

```
"a|b|c|d" %>%
  str_split("\\|") %>%
  .[[1]]
#> [1] "a" "b" "c" "d"
```

Otherwise, like the other **stringr** functions that return a list, you can use `simplify = TRUE` to return a matrix:

```
sentences %>%
  head(5) %>%
  str_split(" ", simplify = TRUE)
#>      [,1]    [,2]    [,3]    [,4]      [,5]  [,6]    [,7]
#> [1,] "The"   "birch" "canoe" "slid"    "on"  "the"   "smooth"
#> [2,] "Glue"  "the"   "sheet" "to"      "the" "dark"  "blue"
#> [3,] "It's"  "easy"  "to"    "tell"    "the" "depth" "of"
#> [4,] "These" "days"  "a"     "chicken" "leg" "is"    "a"
#> [5,] "Rice"  "is"    "often" "served"  "in"  "round" "bowls."
#>      [,8]          [,9]
#> [1,] "planks."     ""
#> [2,] "background." ""
#> [3,] "a"           "well."
#> [4,] "rare"        "dish."
#> [5,] ""            ""
```

You can also request a maximum number of pieces:

```
fields <- c("Name: Hadley", "Country: NZ", "Age: 35")
fields %>% str_split(": ", n = 2, simplify = TRUE)
#>      [,1]      [,2]
#> [1,] "Name"    "Hadley"
#> [2,] "Country" "NZ"
#> [3,] "Age"     "35"
```

Instead of splitting up strings by patterns, you can also split up by character, line, sentence, and word `boundary()`s:

```
x <- "This is a sentence.  This is another sentence."
str_view_all(x, boundary("word"))
```

```
This is a sentence. This is another sentence.
```

```
str_split(x, " ")[[1]]
#> [1] "This"      "is"        "a"         "sentence." ""
#> [6] ""          "This"
#> [7] "is"        "another"   "sentence."
str_split(x, boundary("word"))[[1]]
#> [1] "This"     "is"       "a"        "sentence" "This"
#> [6] "is"
#> [7] "another"  "sentence"
```

Exercises

1. Split up a string like `"apples, pears, and bananas"` into individual components.
2. Why is it better to split up by `boundary("word")` than `" "`?

3. What does splitting with an empty string ("") do? Experiment, and then read the documentation.

Find Matches

`str_locate()` and `str_locate_all()` give you the starting and ending positions of each match. These are particularly useful when none of the other functions does exactly what you want. You can use `str_locate()` to find the matching pattern, and `str_sub()` to extract and/or modify them.

Other Types of Pattern

When you use a pattern that's a string, it's automatically wrapped into a call to `regex()`:

```
# The regular call:
str_view(fruit, "nana")
# Is shorthand for
str_view(fruit, regex("nana"))
```

You can use the other arguments of `regex()` to control details of the match:

- `ignore_case = TRUE` allows characters to match either their uppercase or lowercase forms. This always uses the current locale:

  ```
  bananas <- c("banana", "Banana", "BANANA")
  str_view(bananas, "banana")
  ```

  ```
  banana
  Banana
  BANANA
  ```

  ```
  str_view(bananas, regex("banana", ignore_case = TRUE))
  ```

- `multiline = TRUE` allows ^ and $ to match the start and end of each line rather than the start and end of the complete string:

  ```
  x <- "Line 1\nLine 2\nLine 3"
  str_extract_all(x, "^Line")[[1]]
  #> [1] "Line"
  str_extract_all(x, regex("^Line", multiline = TRUE))[[1]]
  #> [1] "Line" "Line" "Line"
  ```

- `comments = TRUE` allows you to use comments and white space to make complex regular expressions more understandable. Spaces are ignored, as is everything after #. To match a literal space, you'll need to escape it: `"\\ "`.

```
phone <- regex("
  \\(?     # optional opening parens
  (\\d{3}) # area code
  [)- ]?   # optional closing parens, dash, or space
  (\\d{3}) # another three numbers
  [ -]?    # optional space or dash
  (\\d{3}) # three more numbers
  ", comments = TRUE)

str_match("514-791-8141", phone)
#>      [,1]          [,2]  [,3]  [,4]
#> [1,] "514-791-814" "514" "791" "814"
```

- `dotall = TRUE` allows `.` to match everything, including `\n`.

There are three other functions you can use instead of `regex()`:

- `fixed()` matches exactly the specified sequence of bytes. It ignores all special regular expressions and operates at a very low level. This allows you to avoid complex escaping and can be much faster than regular expressions. The following microbenchmark shows that it's about 3x faster for a simple example:

```
microbenchmark::microbenchmark(
  fixed = str_detect(sentences, fixed("the")),
  regex = str_detect(sentences, "the"),
  times = 20
)
#> Unit: microseconds
#>   expr min  lq mean median  uq max neval cld
#>  fixed 116 117  136    120 125 389    20  a
#>  regex 333 337  346    338 342 467    20   b
```

Beware using `fixed()` with non-English data. It is problematic because there are often multiple ways of representing the same character. For example, there are two ways to define "á": either as a single character or as an "a" plus an accent:

```
a1 <- "\u00e1"
a2 <- "a\u0301"
c(a1, a2)
#> [1] "á" "á"
```

```
a1 == a2
#> [1] FALSE
```

They render identically, but because they're defined differently, fixed() doesn't find a match. Instead, you can use coll(), defined next, to respect human character comparison rules:

```
str_detect(a1, fixed(a2))
#> [1] FALSE
str_detect(a1, coll(a2))
#> [1] TRUE
```

- coll() compares strings using standard *coll*ation rules. This is useful for doing case-insensitive matching. Note that coll() takes a locale parameter that controls which rules are used for comparing characters. Unfortunately different parts of the world use different rules!

```
# That means you also need to be aware of the difference
# when doing case-insensitive matches:
i <- c("I", "İ", "i", "ı")
i
#> [1] "I" "İ" "i" "ı"

str_subset(i, coll("i", ignore_case = TRUE))
#> [1] "I" "i"
str_subset(
  i,
  coll("i", ignore_case = TRUE, locale = "tr")
)
#> [1] "İ" "i"
```

Both fixed() and regex() have ignore_case arguments, but they do not allow you to pick the locale: they always use the default locale. You can see what that is with the following code (more on **stringi** later):

```
stringi::stri_locale_info()
#> $Language
#> [1] "en"
#>
#> $Country
#> [1] "US"
#>
#> $Variant
#> [1] ""
#>
#> $Name
#> [1] "en_US"
```

The downside of `coll()` is speed; because the rules for recognizing which characters are the same are complicated, `coll()` is relatively slow compared to `regex()` and `fixed()`.

- As you saw with `str_split()`, you can use `boundary()` to match boundaries. You can also use it with the other functions:

```
x <- "This is a sentence."
str_view_all(x, boundary("word"))
```

```
This is a sentence.
```

```
str_extract_all(x, boundary("word"))
#> [[1]]
#> [1] "This"     "is"       "a"        "sentence"
```

Exercises

1. How would you find all strings containing \ with `regex()` versus with `fixed()`?
2. What are the five most common words in `sentences`?

Other Uses of Regular Expressions

There are two useful functions in base R that also use regular expressions:

- `apropos()` searches all objects available from the global environment. This is useful if you can't quite remember the name of the function:

```
apropos("replace")
#> [1] "%+replace%"  "replace"         "replace_na"
#> [4] "str_replace" "str_replace_all" "str_replace_na"
#> [7] "theme_replace"
```

- `dir()` lists all the files in a directory. The `pattern` argument takes a regular expression and only returns filenames that match the pattern. For example, you can find all the R Markdown files in the current directory with:

```
head(dir(pattern = "\\.Rmd$"))
#> [1] "communicate-plots.Rmd" "communicate.Rmd"
```

```
#> [3] "datetimes.Rmd" "EDA.Rmd"
#> [5] "explore.Rmd" "factors.Rmd"
```

(If you're more comfortable with "globs" like `*.Rmd`, you can convert them to regular expressions with `glob2rx()`).

stringi

stringr is built on top of the **stringi** package. **stringr** is useful when you're learning because it exposes a minimal set of functions, which have been carefully picked to handle the most common string manipulation functions. **stringi**, on the other hand, is designed to be comprehensive. It contains almost every function you might ever need: **stringi** has 234 functions to **stringr**'s 42.

If you find yourself struggling to do something in **stringr**, it's worth taking a look at **stringi**. The packages work very similarly, so you should be able to translate your **stringr** knowledge in a natural way. The main difference is the prefix: `str_` versus `stri_`.

Exercises

1. Find the **stringi** functions that:
   a. Count the number of words.
   b. Find duplicated strings.
   c. Generate random text.
2. How do you control the language that `stri_sort()` uses for sorting?

CHAPTER 12

Factors with forcats

Introduction

In R, factors are used to work with categorical variables, variables that have a fixed and known set of possible values. They are also useful when you want to display character vectors in a non-alphabetical order.

Historically, factors were much easier to work with than characters. As a result, many of the functions in base R automatically convert characters to factors. This means that factors often crop up in places where they're not actually helpful. Fortunately, you don't need to worry about that in the tidyverse, and can focus on situations where factors are genuinely useful.

For more historical context on factors, I recommend *stringsAsFactors: An unauthorized biography* (*http://bit.ly/stringsfactorsbio*) by Roger Peng, and *stringsAsFactors = <sigh>* (*http://bit.ly/stringsfactorsigh*) by Thomas Lumley.

Prerequisites

To work with factors, we'll use the **forcats** package, which provides tools for dealing with *cat*egorical variables (and it's an anagram of factors!). It provides a wide range of helpers for working with factors. **forcats** is not part of the core tidyverse, so we need to load it explicitly.

```
library(tidyverse)
library(forcats)
```

Creating Factors

Imagine that you have a variable that records month:

```
x1 <- c("Dec", "Apr", "Jan", "Mar")
```

Using a string to record this variable has two problems:

1. There are only twelve possible months, and there's nothing saving you from typos:

   ```
   x2 <- c("Dec", "Apr", "Jam", "Mar")
   ```

2. It doesn't sort in a useful way:

   ```
   sort(x1)
   #> [1] "Apr" "Dec" "Jan" "Mar"
   ```

You can fix both of these problems with a factor. To create a factor you must start by creating a list of the valid *levels*:

```
month_levels <- c(
  "Jan", "Feb", "Mar", "Apr", "May", "Jun",
  "Jul", "Aug", "Sep", "Oct", "Nov", "Dec"
)
```

Now you can create a factor:

```
y1 <- factor(x1, levels = month_levels)
y1
#> [1] Dec Apr Jan Mar
#> Levels: Jan Feb Mar Apr May Jun Jul Aug Sep Oct Nov Dec
sort(y1)
#> [1] Jan Mar Apr Dec
#> Levels: Jan Feb Mar Apr May Jun Jul Aug Sep Oct Nov Dec
```

And any values not in the set will be silently converted to NA:

```
y2 <- factor(x2, levels = month_levels)
y2
#> [1] Dec  Apr  <NA> Mar
#> Levels: Jan Feb Mar Apr May Jun Jul Aug Sep Oct Nov Dec
```

If you want a want an error, you can use `readr::parse_factor()`:

```
y2 <- parse_factor(x2, levels = month_levels)
#> Warning: 1 parsing failure.
#> row col           expected actual
#>   3  -- value in level set    Jam
```

If you omit the levels, they'll be taken from the data in alphabetical order:

```
factor(x1)
#> [1] Dec Apr Jan Mar
#> Levels: Apr Dec Jan Mar
```

Sometimes you'd prefer that the order of the levels match the order of the first appearance in the data. You can do that when creating the factor by setting levels to `unique(x)`, or after the fact, with `fct_inorder()`:

```
f1 <- factor(x1, levels = unique(x1))
f1
#> [1] Dec Apr Jan Mar
#> Levels: Dec Apr Jan Mar

f2 <- x1 %>% factor() %>% fct_inorder()
f2
#> [1] Dec Apr Jan Mar
#> Levels: Dec Apr Jan Mar
```

If you ever need to access the set of valid levels directly, you can do so with `levels()`:

```
levels(f2)
#> [1] "Dec" "Apr" "Jan" "Mar"
```

General Social Survey

For the rest of this chapter, we're going to focus on `forcats::gss_cat`. It's a sample of data from the General Social Survey (*http://gss.norc.org*), which is a long-running US survey conducted by the independent research organization NORC at the University of Chicago. The survey has thousands of questions, so in `gss_cat` I've selected a handful that will illustrate some common challenges you'll encounter when working with factors:

```
gss_cat
#> # A tibble: 21,483 × 9
#>     year       marital   age   race         rincome
#>    <int>        <fctr> <int> <fctr>          <fctr>
#> 1  2000 Never married    26  White  $8000 to 9999
#> 2  2000      Divorced    48  White  $8000 to 9999
#> 3  2000       Widowed    67  White Not applicable
#> 4  2000 Never married    39  White Not applicable
#> 5  2000      Divorced    25  White Not applicable
#> 6  2000       Married    25  White $20000 - 24999
#> # ... with 2.148e+04 more rows, and 4 more variables:
#> #   partyid <fctr>, relig <fctr>, denom <fctr>, tvhours <int>
```

(Remember, since this dataset is provided by a package, you can get more information about the variables with ?gss_cat.)

When factors are stored in a tibble, you can't see their levels so easily. One way to see them is with count():

```
gss_cat %>%
  count(race)
#> # A tibble: 3 × 2
#>     race     n
#>   <fctr> <int>
#> 1  Other  1959
#> 2  Black  3129
#> 3  White 16395
```

Or with a bar chart:

```
ggplot(gss_cat, aes(race)) +
  geom_bar()
```

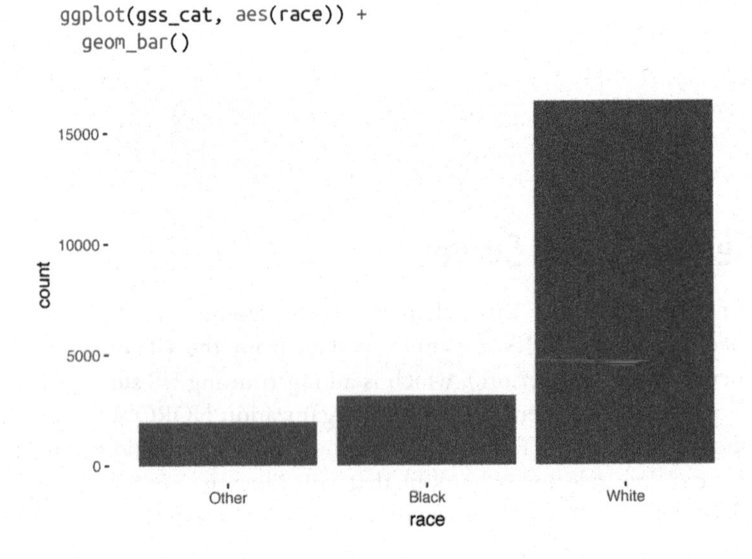

By default, **ggplot2** will drop levels that don't have any values. You can force them to display with:

```
ggplot(gss_cat, aes(race)) +
  geom_bar() +
  scale_x_discrete(drop = FALSE)
```

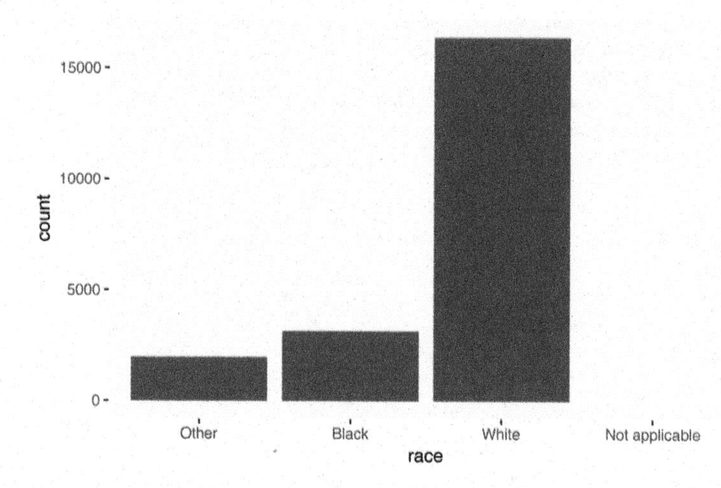

These levels represent valid values that simply did not occur in this dataset. Unfortunately, **dplyr** doesn't yet have a `drop` option, but it will in the future.

When working with factors, the two most common operations are changing the order of the levels, and changing the values of the levels. Those operations are described in the following sections.

Exercises

1. Explore the distribution of `rincome` (reported income). What makes the default bar chart hard to understand? How could you improve the plot?
2. What is the most common `relig` in this survey? What's the most common `partyid`?
3. Which `relig` does `denom` (denomination) apply to? How can you find out with a table? How can you find out with a visualization?

Modifying Factor Order

It's often useful to change the order of the factor levels in a visualization. For example, imagine you want to explore the average number of hours spent watching TV per day across religions:

```
relig <- gss_cat %>%
  group_by(relig) %>%
  summarize(
    age = mean(age, na.rm = TRUE),
    tvhours = mean(tvhours, na.rm = TRUE),
    n = n()
  )

ggplot(relig, aes(tvhours, relig)) + geom_point()
```

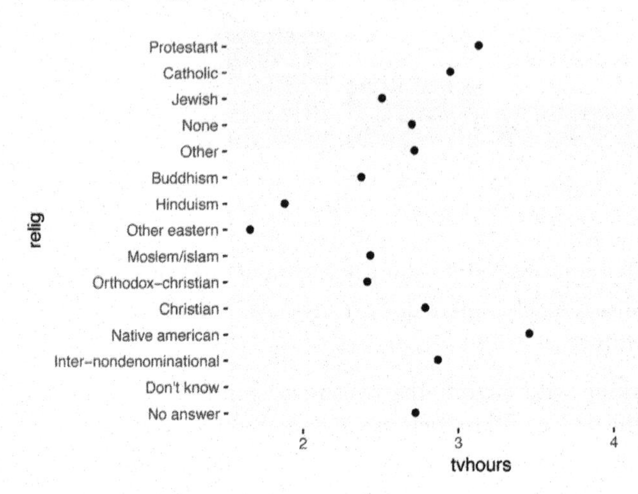

It is difficult to interpret this plot because there's no overall pattern. We can improve it by reordering the levels of `relig` using `fct_reorder()`. `fct_reorder()` takes three arguments:

- `f`, the factor whose levels you want to modify.
- `x`, a numeric vector that you want to use to reorder the levels.
- Optionally, `fun`, a function that's used if there are multiple values of `x` for each value of `f`. The default value is `median`.

```
ggplot(relig, aes(tvhours, fct_reorder(relig, tvhours))) +
  geom_point()
```

Reordering religion makes it much easier to see that people in the "Don't know" category watch much more TV, and Hinduism and other Eastern religions watch much less.

As you start making more complicated transformations, I'd recommend moving them out of `aes()` and into a separate `mutate()` step. For example, you could rewrite the preceding plot as:

```
relig %>%
  mutate(relig = fct_reorder(relig, tvhours)) %>%
  ggplot(aes(tvhours, relig)) +
    geom_point()
```

What if we create a similar plot looking at how average age varies across reported income level?

```
rincome <- gss_cat %>%
  group_by(rincome) %>%
  summarize(
    age = mean(age, na.rm = TRUE),
    tvhours = mean(tvhours, na.rm = TRUE),
    n = n()
  )

ggplot(
  rincome,
  aes(age, fct_reorder(rincome, age))
) + geom_point()
```

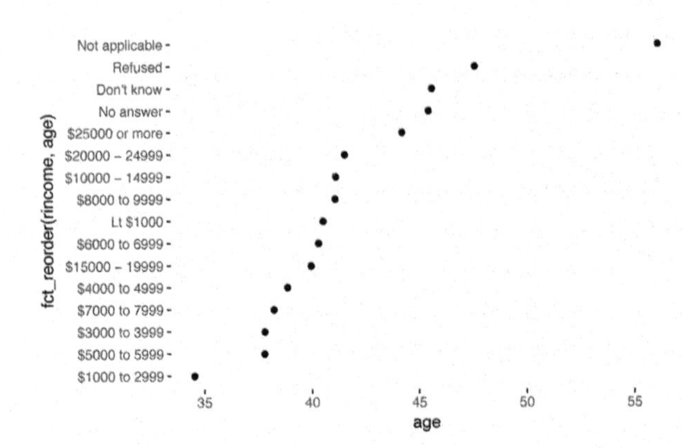

Here, arbitrarily reordering the levels isn't a good idea! That's because `rincome` already has a principled order that we shouldn't mess with. Reserve `fct_reorder()` for factors whose levels are arbitrarily ordered.

However, it does make sense to pull "Not applicable" to the front with the other special levels. You can use `fct_relevel()`. It takes a factor, `f`, and then any number of levels that you want to move to the front of the line:

```
ggplot(
 rincome,
 aes(age, fct_relevel(rincome, "Not applicable"))
) +
  geom_point()
```

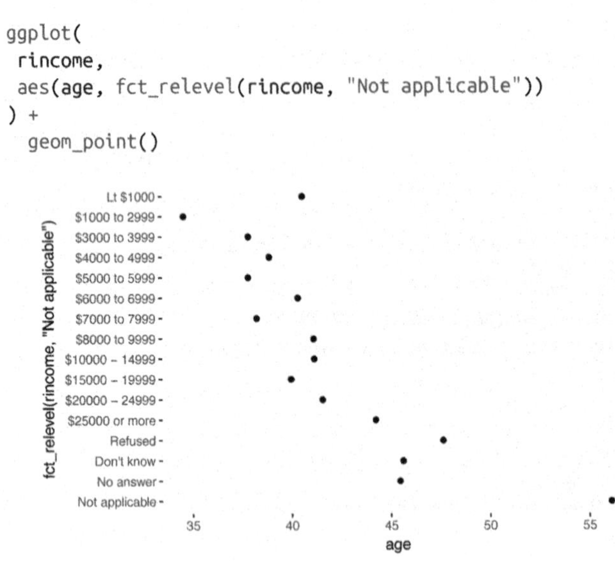

Why do you think the average age for "Not applicable" is so high?

Another type of reordering is useful when you are coloring the lines on a plot. `fct_reorder2()` reorders the factor by the `y` values associated with the largest `x` values. This makes the plot easier to read because the line colors line up with the legend:

```
by_age <- gss_cat %>%
  filter(!is.na(age)) %>%
  group_by(age, marital) %>%
  count() %>%
  mutate(prop = n / sum(n))

ggplot(by_age, aes(age, prop, color = marital)) +
  geom_line(na.rm = TRUE)

ggplot(
  by_age,
  aes(age, prop, color = fct_reorder2(marital, age, prop))
) +
  geom_line() +
  labs(color = "marital")
```

Finally, for bar plots, you can use `fct_infreq()` to order levels in increasing frequency: this is the simplest type of reordering because it doesn't need any extra variables. You may want to combine with `fct_rev()`:

```
gss_cat %>%
  mutate(marital = marital %>% fct_infreq() %>% fct_rev()) %>%
  ggplot(aes(marital)) +
    geom_bar()
```

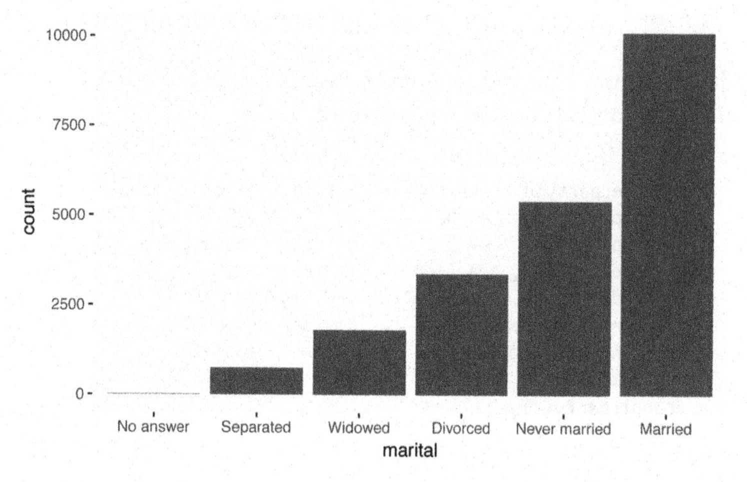

Exercises

1. There are some suspiciously high numbers in `tvhours`. Is the mean a good summary?
2. For each factor in `gss_cat` identify whether the order of the levels is arbitrary or principled.
3. Why did moving "Not applicable" to the front of the levels move it to the bottom of the plot?

Modifying Factor Levels

More powerful than changing the orders of the levels is changing their values. This allows you to clarify labels for publication, and collapse levels for high-level displays. The most general and powerful tool is `fct_recode()`. It allows you to recode, or change, the value of each level. For example, take `gss_cat$partyid`:

```
gss_cat %>% count(partyid)
#> # A tibble: 10 × 2
#>                partyid     n
#>                 <fctr> <int>
#> 1            No answer   154
#> 2           Don't know     1
#> 3          Other party   393
#> 4   Strong republican  2314
#> 5 Not str republican  3032
#> 6         Ind,near rep  1791
#> # ... with 4 more rows
```

The levels are terse and inconsistent. Let's tweak them to be longer and use a parallel construction:

```
gss_cat %>%
  mutate(partyid = fct_recode(partyid,
    "Republican, strong"    = "Strong republican",
    "Republican, weak"      = "Not str republican",
    "Independent, near rep" = "Ind,near rep",
    "Independent, near dem" = "Ind,near dem",
    "Democrat, weak"        = "Not str democrat",
    "Democrat, strong"      = "Strong democrat"
  )) %>%
  count(partyid)
#> # A tibble: 10 × 2
#>                  partyid     n
#>                   <fctr> <int>
#> 1              No answer   154
```

```
#> 2              Don't know     1
#> 3             Other party   393
#> 4     Republican, strong  2314
#> 5       Republican, weak  3032
#> 6 Independent, near rep  1791
#> # ... with 4 more rows
```

fct_recode() will leave levels that aren't explicitly mentioned as is, and will warn you if you accidentally refer to a level that doesn't exist.

To combine groups, you can assign multiple old levels to the same new level:

```
gss_cat %>%
  mutate(partyid = fct_recode(partyid,
    "Republican, strong"    = "Strong republican",
    "Republican, weak"      = "Not str republican",
    "Independent, near rep" = "Ind,near rep",
    "Independent, near dem" = "Ind,near dem",
    "Democrat, weak"        = "Not str democrat",
    "Democrat, strong"      = "Strong democrat",
    "Other"                 = "No answer",
    "Other"                 = "Don't know",
    "Other"                 = "Other party"
  )) %>%
  count(partyid)
#> # A tibble: 8 × 2
#>                 partyid     n
#>                  <fctr> <int>
#> 1                 Other   548
#> 2    Republican, strong  2314
#> 3      Republican, weak  3032
#> 4 Independent, near rep  1791
#> 5           Independent  4119
#> 6 Independent, near dem  2499
#> # ... with 2 more rows
```

You must use this technique with care: if you group together categories that are truly different you will end up with misleading results.

If you want to collapse a lot of levels, fct_collapse() is a useful variant of fct_recode(). For each new variable, you can provide a vector of old levels:

```
gss_cat %>%
  mutate(partyid = fct_collapse(partyid,
    other = c("No answer", "Don't know", "Other party"),
    rep = c("Strong republican", "Not str republican"),
    ind = c("Ind,near rep", "Independent", "Ind,near dem"),
```

```
    dem = c("Not str democrat", "Strong democrat")
  )) %>%
  count(partyid)
#> # A tibble: 4 × 2
#>   partyid     n
#>     <fctr> <int>
#> 1    other   548
#> 2      rep  5346
#> 3      ind  8409
#> 4      dem  7180
```

Sometimes you just want to lump together all the small groups to make a plot or table simpler. That's the job of `fct_lump()`:

```
gss_cat %>%
  mutate(relig = fct_lump(relig)) %>%
  count(relig)
#> # A tibble: 2 × 2
#>        relig     n
#>       <fctr> <int>
#> 1 Protestant 10846
#> 2      Other 10637
```

The default behavior is to progressively lump together the smallest groups, ensuring that the aggregate is still the smallest group. In this case it's not very helpful: it is true that the majority of Americans in this survey are Protestant, but we've probably overcollapsed.

Instead, we can use the `n` parameter to specify how many groups (excluding other) we want to keep:

```
gss_cat %>%
  mutate(relig = fct_lump(relig, n = 10)) %>%
  count(relig, sort = TRUE) %>%
  print(n = Inf)
#> # A tibble: 10 × 2
#>                      relig     n
#>                     <fctr> <int>
#> 1               Protestant 10846
#> 2                 Catholic  5124
#> 3                     None  3523
#> 4                Christian   689
#> 5                    Other   458
#> 6                   Jewish   388
#> 7                 Buddhism   147
#> 8  Inter-nondenominational   109
#> 9             Moslem/islam   104
#> 10      Orthodox-christian    95
```

Exercises

1. How have the proportions of people identifying as Democrat, Republican, and Independent changed over time?
2. How could you collapse `rincome` into a small set of categories?

CHAPTER 13

Dates and Times with lubridate

Introduction

This chapter will show you how to work with dates and times in R. At first glance, dates and times seem simple. You use them all the time in your regular life, and they don't seem to cause much confusion. However, the more you learn about dates and times, the more complicated they seem to get. To warm up, try these three seemingly simple questions:

- Does every year have 365 days?
- Does every day have 24 hours?
- Does every minute have 60 seconds?

I'm sure you know that not every year has 365 days, but do you know the full rule for determining if a year is a leap year? (It has three parts.) You might have remembered that many parts of the world use daylight saving time (DST), so that some days have 23 hours, and others have 25. You might not have known that some minutes have 61 seconds because every now and then leap seconds are added because the Earth's rotation is gradually slowing down.

Dates and times are hard because they have to reconcile two physical phenomena (the rotation of the Earth and its orbit around the sun) with a whole raft of geopolitical phenomena including months, time zones, and DST. This chapter won't teach you every last detail about dates and times, but it will give you a solid grounding of practical skills that will help you with common data analysis challenges.

Prerequisites

This chapter will focus on the **lubridate** package, which makes it easier to work with dates and times in R. **lubridate** is not part of core tidyverse because you only need it when you're working with dates/times. We will also need **nycflights13** for practice data.

```
library(tidyverse)

library(lubridate)
library(nycflights13)
```

Creating Date/Times

There are three types of date/time data that refer to an instant in time:

- A *date*. Tibbles print this as `<date>`.
- A *time* within a day. Tibbles print this as `<time>`.
- A *date-time* is a date plus a time: it uniquely identifies an instant in time (typically to the nearest second). Tibbles print this as `<dttm>`. Elsewhere in R these are called POSIXct, but I don't think that's a very useful name.

In this chapter we are only going to focus on dates and date-times as R doesn't have a native class for storing times. If you need one, you can use the **hms** package.

You should always use the simplest possible data type that works for your needs. That means if you can use a date instead of a date-time, you should. Date-times are substantially more complicated because of the need to handle time zones, which we'll come back to at the end of the chapter.

To get the current date or date-time you can use `today()` or `now()`:

```
today()
#> [1] "2016-10-10"
now()
#> [1] "2016-10-10 15:19:39 PDT"
```

Otherwise, there are three ways you're likely to create a date/time:

- From a string.
- From individual date-time components.
- From an existing date/time object.

They work as follows.

From Strings

Date/time data often comes as strings. You've seen one approach to parsing strings into date-times in "Dates, Date-Times, and Times" on page 134. Another approach is to use the helpers provided by **lubridate**. They automatically work out the format once you specify the order of the component. To use them, identify the order in which year, month, and day appear in your dates, then arrange "y", "m", and "d" in the same order. That gives you the name of the **lubridate** function that will parse your date. For example:

```
ymd("2017-01-31")
#> [1] "2017-01-31"
mdy("January 31st, 2017")
#> [1] "2017-01-31"
dmy("31-Jan-2017")
#> [1] "2017-01-31"
```

These functions also take unquoted numbers. This is the most concise way to create a single date/time object, as you might need when filtering date/time data. `ymd()` is short and unambiguous:

```
ymd(20170131)
#> [1] "2017-01-31"
```

`ymd()` and friends create dates. To create a date-time, add an underscore and one or more of "h", "m", and "s" to the name of the parsing function:

```
ymd_hms("2017-01-31 20:11:59")
#> [1] "2017-01-31 20:11:59 UTC"
mdy_hm("01/31/2017 08:01")
#> [1] "2017-01-31 08:01:00 UTC"
```

You can also force the creation of a date-time from a date by supplying a time zone:

```
ymd(20170131, tz = "UTC")
#> [1] "2017-01-31 UTC"
```

From Individual Components

Instead of a single string, sometimes you'll have the individual components of the date-time spread across multiple columns. This is what we have in the flights data:

```
flights %>%
  select(year, month, day, hour, minute)
#> # A tibble: 336,776 × 5
#>    year month   day  hour minute
#>   <int> <int> <int> <dbl>  <dbl>
#> 1  2013     1     1     5     15
#> 2  2013     1     1     5     29
#> 3  2013     1     1     5     40
#> 4  2013     1     1     5     45
#> 5  2013     1     1     6      0
#> 6  2013     1     1     5     58
#> # ... with 3.368e+05 more rows
```

To create a date/time from this sort of input, use `make_date()` for dates, or `make_datetime()` for date-times:

```
flights %>%
  select(year, month, day, hour, minute) %>%
  mutate(
    departure = make_datetime(year, month, day, hour, minute)
  )
#> # A tibble: 336,776 × 6
#>    year month   day  hour minute           departure
#>   <int> <int> <int> <dbl>  <dbl>              <dttm>
#> 1  2013     1     1     5     15 2013-01-01 05:15:00
#> 2  2013     1     1     5     29 2013-01-01 05:29:00
#> 3  2013     1     1     5     40 2013-01-01 05:40:00
#> 4  2013     1     1     5     45 2013-01-01 05:45:00
#> 5  2013     1     1     6      0 2013-01-01 06:00:00
#> 6  2013     1     1     5     58 2013-01-01 05:58:00
#> # ... with 3.368e+05 more rows
```

Let's do the same thing for each of the four time columns in `flights`. The times are represented in a slightly odd format, so we use modulus arithmetic to pull out the hour and minute components. Once I've created the date-time variables, I focus in on the variables we'll explore in the rest of the chapter:

```
make_datetime_100 <- function(year, month, day, time) {
  make_datetime(year, month, day, time %/% 100, time %% 100)
}

flights_dt <- flights %>%
  filter(!is.na(dep_time), !is.na(arr_time)) %>%
```

```
  mutate(
    dep_time = make_datetime_100(year, month, day, dep_time),
    arr_time = make_datetime_100(year, month, day, arr_time),
    sched_dep_time = make_datetime_100(
      year, month, day, sched_dep_time
    ),
    sched_arr_time = make_datetime_100(
      year, month, day, sched_arr_time
    )
  ) %>%
  select(origin, dest, ends_with("delay"), ends_with("time"))

flights_dt
#> # A tibble: 328,063 × 9
#>   origin  dest dep_delay arr_delay            dep_time
#>    <chr> <chr>     <dbl>     <dbl>              <dttm>
#> 1    EWR   IAH         2        11 2013-01-01 05:17:00
#> 2    LGA   IAH         4        20 2013-01-01 05:33:00
#> 3    JFK   MIA         2        33 2013-01-01 05:42:00
#> 4    JFK   BQN        -1       -18 2013-01-01 05:44:00
#> 5    LGA   ATL        -6       -25 2013-01-01 05:54:00
#> 6    EWR   ORD        -4        12 2013-01-01 05:54:00
#> # ... with 3.281e+05 more rows, and 4 more variables:
#> #   sched_dep_time <dttm>, arr_time <dttm>,
#> #   sched_arr_time <dttm>, air_time <dbl>
```

With this data, I can visualize the distribution of departure times across the year:

```
flights_dt %>%
  ggplot(aes(dep_time)) +
  geom_freqpoly(binwidth = 86400) # 86400 seconds = 1 day
```

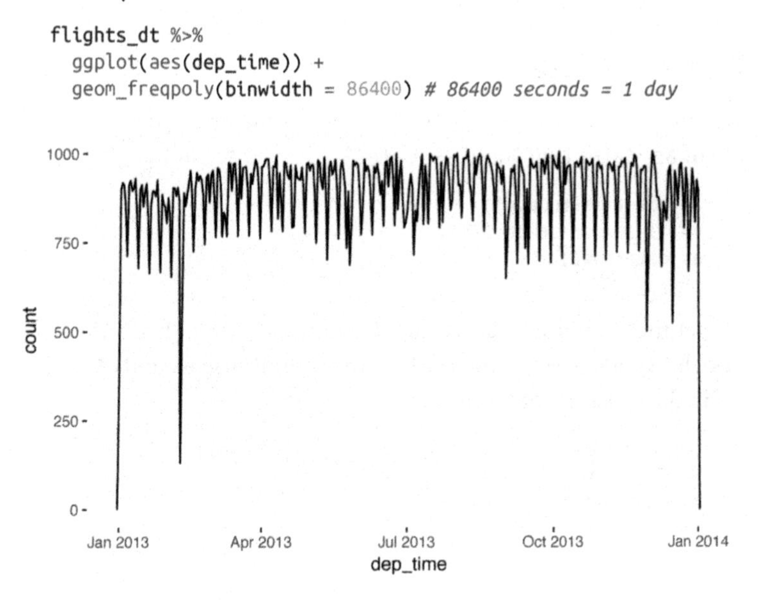

Or within a single day:

```
flights_dt %>%
  filter(dep_time < ymd(20130102)) %>%
  ggplot(aes(dep_time)) +
  geom_freqpoly(binwidth = 600) # 600 s = 10 minutes
```

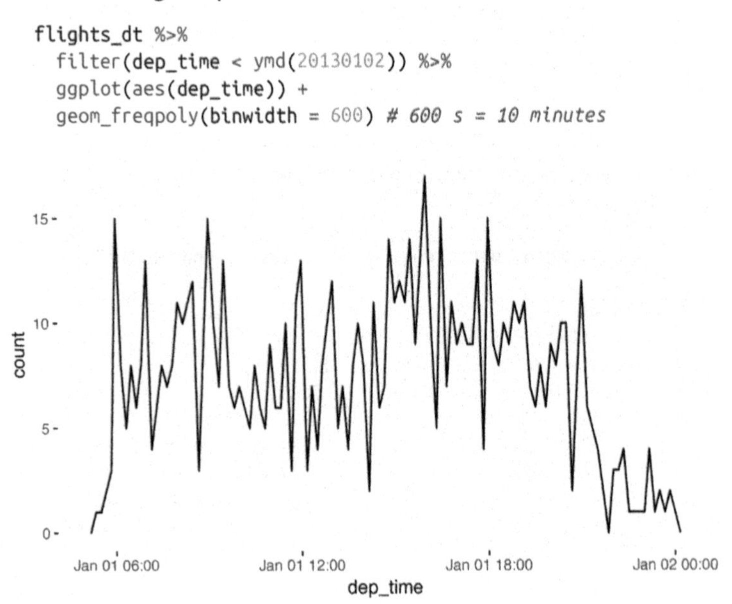

Note that when you use date-times in a numeric context (like in a histogram), 1 means 1 second, so a binwidth of 86400 means one day. For dates, 1 means 1 day.

From Other Types

You may want to switch between a date-time and a date. That's he job of `as_datetime()` and `as_date()`:

```
as_datetime(today())
#> [1] "2016-10-10 UTC"
as_date(now())
#> [1] "2016-10-10"
```

Sometimes you'll get date/times as numeric offsets from the "Unix Epoch," 1970-01-01. If the offset is in seconds, use `as_datetime()`; if it's in days, use `as_date()`:

```
as_datetime(60 * 60 * 10)
#> [1] "1970-01-01 10:00:00 UTC"
as_date(365 * 10 + 2)
#> [1] "1980-01-01"
```

Exercises

1. What happens if you parse a string that contains invalid dates?

   ```
   ymd(c("2010-10-10", "bananas"))
   ```

2. What does the `tzone` argument to `today()` do? Why is it important?

3. Use the appropriate **lubridate** function to parse each of the following dates:

   ```
   d1 <- "January 1, 2010"
   d2 <- "2015-Mar-07"
   d3 <- "06-Jun-2017"
   d4 <- c("August 19 (2015)", "July 1 (2015)")
   d5 <- "12/30/14" # Dec 30, 2014
   ```

Date-Time Components

Now that you know how to get date-time data into R's date-time data structures, let's explore what you can do with them. This section will focus on the accessor functions that let you get and set individual components. The next section will look at how arithmetic works with date-times.

Getting Components

You can pull out individual parts of the date with the accessor functions `year()`, `month()`, `mday()` (day of the month), `yday()` (day of the year), `wday()` (day of the week), `hour()`, `minute()`, and `second()`:

```
datetime <- ymd_hms("2016-07-08 12:34:56")

year(datetime)
#> [1] 2016
month(datetime)
#> [1] 7
mday(datetime)
#> [1] 8

yday(datetime)
#> [1] 190
wday(datetime)
#> [1] 6
```

For `month()` and `wday()` you can set `label = TRUE` to return the abbreviated name of the month or day of the week. Set `abbr = FALSE` to return the full name:

```
month(datetime, label = TRUE)
#> [1] Jul
#> 12 Levels: Jan < Feb < Mar < Apr < May < Jun < ... < Dec
wday(datetime, label = TRUE, abbr = FALSE)
#> [1] Friday
#> 7 Levels: Sunday < Monday < Tuesday < ... < Saturday
```

We can use `wday()` to see that more flights depart during the week than on the weekend:

```
flights_dt %>%
  mutate(wday = wday(dep_time, label = TRUE)) %>%
  ggplot(aes(x = wday)) +
    geom_bar()
```

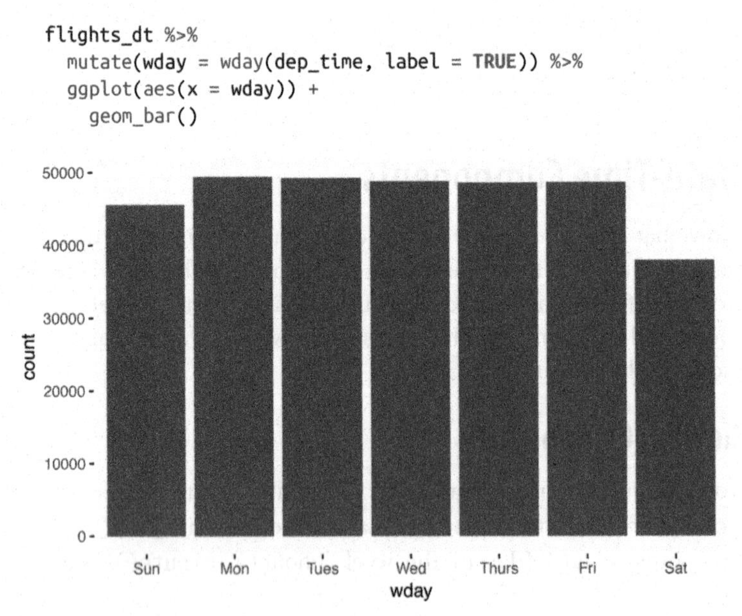

There's an interesting pattern if we look at the average departure delay by minute within the hour. It looks like flights leaving in minutes 20–30 and 50–60 have much lower delays than the rest of the hour!

```
flights_dt %>%
  mutate(minute = minute(dep_time)) %>%
  group_by(minute) %>%
  summarize(
    avg_delay = mean(arr_delay, na.rm = TRUE),
    n = n()) %>%
  ggplot(aes(minute, avg_delay)) +
    geom_line()
```

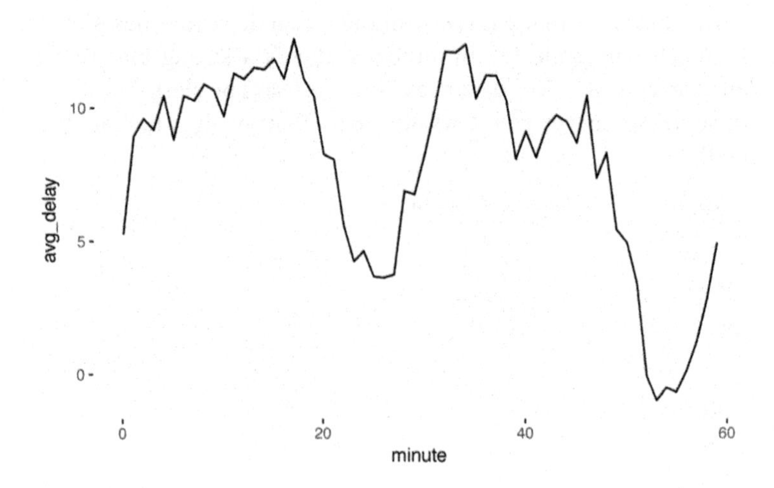

Interestingly, if we look at the *scheduled* departure time we don't see such a strong pattern:

```
sched_dep <- flights_dt %>%
  mutate(minute = minute(sched_dep_time)) %>%
  group_by(minute) %>%
  summarize(
    avg_delay = mean(arr_delay, na.rm = TRUE),
    n = n())

ggplot(sched_dep, aes(minute, avg_delay)) +
  geom_line()
```

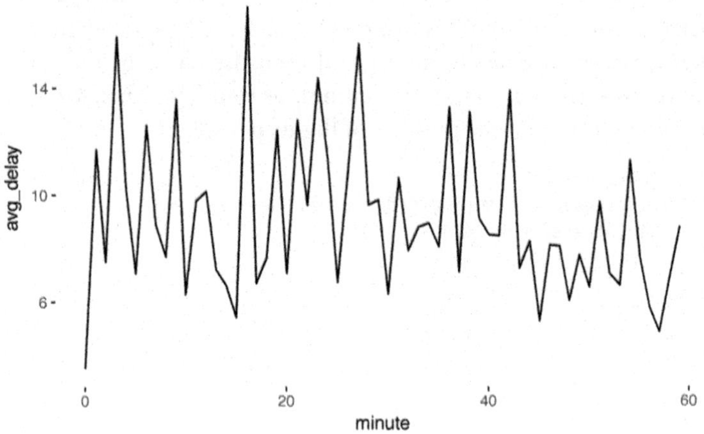

So why do we see that pattern with the actual departure times? Well, like much data collected by humans, there's a strong bias toward flights leaving at "nice" departure times. Always be alert for this sort of pattern whenever you work with data that involves human judgment!

```
ggplot(sched_dep, aes(minute, n)) +
  geom_line()
```

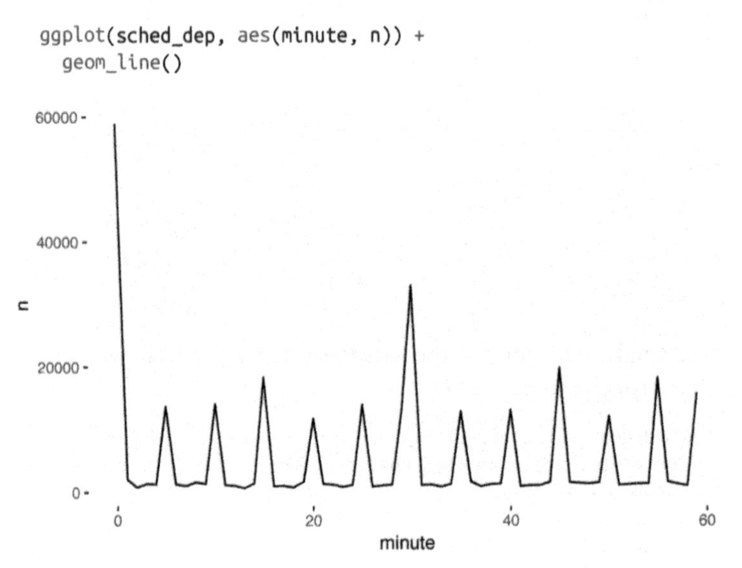

Rounding

An alternative approach to plotting individual components is to round the date to a nearby unit of time, with `floor_date()`, `round_date()`, and `ceiling_date()`. Each ceiling_date()function takes a vector of dates to adjust and then the name of the unit to round down (floor), round up (ceiling), or round to. This, for example, allows us to plot the number of flights per week:

```
flights_dt %>%
  count(week = floor_date(dep_time, "week")) %>%
  ggplot(aes(week, n)) +
    geom_line()
```

Computing the difference between a rounded and unrounded date can be particularly useful.

Setting Components

You can also use each accessor function to set the components of a date/time:

```
(datetime <- ymd_hms("2016-07-08 12:34:56"))
#> [1] "2016-07-08 12:34:56 UTC"

year(datetime) <- 2020
datetime
#> [1] "2020-07-08 12:34:56 UTC"
month(datetime) <- 01
datetime
#> [1] "2020-01-08 12:34:56 UTC"
hour(datetime) <- hour(datetime) + 1
```

Alternatively, rather than modifying in place, you can create a new date-time with `update()`. This also allows you to set multiple values at once:

```
update(datetime, year = 2020, month = 2, mday = 2, hour = 2)
#> [1] "2020-02-02 02:34:56 UTC"
```

If values are too big, they will roll over:

```
ymd("2015-02-01") %>%
  update(mday = 30)
#> [1] "2015-03-02"
ymd("2015-02-01") %>%
```

```
  update(hour = 400)
#> [1] "2015-02-17 16:00:00 UTC"
```

You can use `update()` to show the distribution of flights across the course of the day for every day of the year:

```
flights_dt %>%
  mutate(dep_hour = update(dep_time, yday = 1)) %>%
  ggplot(aes(dep_hour)) +
    geom_freqpoly(binwidth = 300)
```

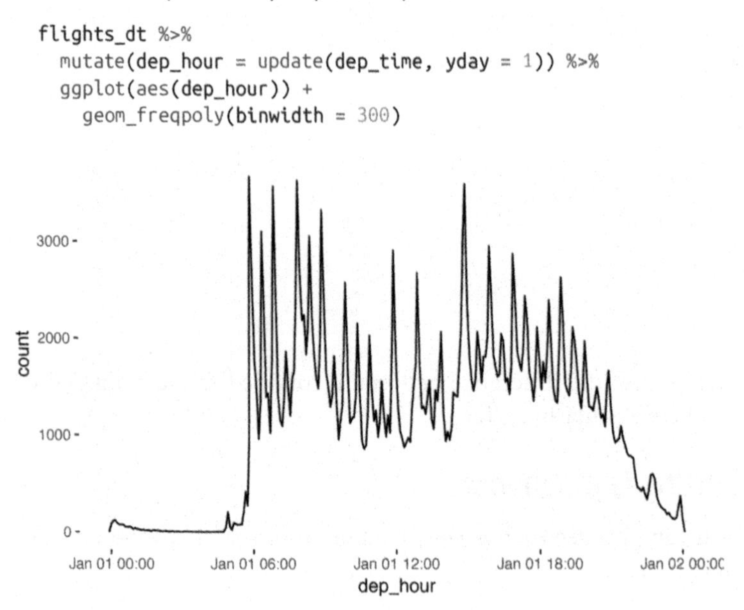

Setting larger components of a date to a constant is a powerful technique that allows you to explore patterns in the smaller components.

Exercises

1. How does the distribution of flight times within a day change over the course of the year?
2. Compare `dep_time`, `sched_dep_time`, and `dep_delay`. Are they consistent? Explain your findings.
3. Compare `air_time` with the duration between the departure and arrival. Explain your findings. (Hint: consider the location of the airport.)
4. How does the average delay time change over the course of a day? Should you use `dep_time` or `sched_dep_time`? Why?
5. On what day of the week should you leave if you want to minimize the chance of a delay?

6. What makes the distribution of `diamonds$carat` and `flights$sched_dep_time` similar?
7. Confirm my hypothesis that the early departures of flights in minutes 20–30 and 50–60 are caused by scheduled flights that leave early. Hint: create a binary variable that tells you whether or not a flight was delayed.

Time Spans

Next you'll learn about how arithmetic with dates works, including subtraction, addition, and division. Along the way, you'll learn about three important classes that represent time spans:

- *Durations*, which represent an exact number of seconds.
- *Periods*, which represent human units like weeks and months.
- *Intervals*, which represent a starting and ending point.

Durations

In R, when you subtract two dates, you get a difftime object:

```
# How old is Hadley?
h_age <- today() - ymd(19791014)
h_age
#> Time difference of 13511 days
```

A difftime class object records a time span of seconds, minutes, hours, days, or weeks. This ambiguity can make difftimes a little painful to work with, so **lubridate** provides an alternative that always uses seconds—the *duration*:

```
as.duration(h_age)
#> [1] "1167350400s (~36.99 years)"
```

Durations come with a bunch of convenient constructors:

```
dseconds(15)
#> [1] "15s"
dminutes(10)
#> [1] "600s (~10 minutes)"
dhours(c(12, 24))
#> [1] "43200s (~12 hours)" "86400s (~1 days)"
ddays(0:5)
#> [1] "0s"                 "86400s (~1 days)"
#> [3] "172800s (~2 days)" "259200s (~3 days)"
```

```
#> [5] "345600s (~4 days)" "432000s (~5 days)"
dweeks(3)
#> [1] "1814400s (~3 weeks)"
dyears(1)
#> [1] "31536000s (~52.14 weeks)"
```

Durations always record the time span in seconds. Larger units are created by converting minutes, hours, days, weeks, and years to seconds at the standard rate (60 seconds in a minute, 60 minutes in an hour, 24 hours in day, 7 days in a week, 365 days in a year).

You can add and multiply durations:

```
2 * dyears(1)
#> [1] "63072000s (~2 years)"
dyears(1) + dweeks(12) + dhours(15)
#> [1] "38847600s (~1.23 years)"
```

You can add and subtract durations to and from days:

```
tomorrow <- today() + ddays(1)
last_year <- today() - dyears(1)
```

However, because durations represent an exact number of seconds, sometimes you might get an unexpected result:

```
one_pm <- ymd_hms(
  "2016-03-12 13:00:00",
  tz = "America/New_York"
)

one_pm
#> [1] "2016-03-12 13:00:00 EST"
one_pm + ddays(1)
#> [1] "2016-03-13 14:00:00 EDT"
```

Why is one day after 1 p.m. on March 12, 2 p.m. on March 13?! If you look carefully at the date you might also notice that the time zones have changed. Because of DST, March 12 only has 23 hours, so if we add a full day's worth of seconds we end up with a different time.

Periods

To solve this problem, **lubridate** provides *periods*. Periods are time spans but don't have a fixed length in seconds; instead they work with "human" times, like days and months. That allows them to work in a more intuitive way:

```
one_pm
#> [1] "2016-03-12 13:00:00 EST"
one_pm + days(1)
#> [1] "2016-03-13 13:00:00 EDT"
```

Like durations, periods can be created with a number of friendly constructor functions:

```
seconds(15)
#> [1] "15S"
minutes(10)
#> [1] "10M 0S"
hours(c(12, 24))
#> [1] "12H 0M 0S" "24H 0M 0S"
days(7)
#> [1] "7d 0H 0M 0S"
months(1:6)
#> [1] "1m 0d 0H 0M 0S" "2m 0d 0H 0M 0S" "3m 0d 0H 0M 0S"
#> [4] "4m 0d 0H 0M 0S" "5m 0d 0H 0M 0S" "6m 0d 0H 0M 0S"
weeks(3)
#> [1] "21d 0H 0M 0S"
years(1)
#> [1] "1y 0m 0d 0H 0M 0S"
```

You can add and multiply periods:

```
10 * (months(6) + days(1))
#> [1] "60m 10d 0H 0M 0S"
days(50) + hours(25) + minutes(2)
#> [1] "50d 25H 2M 0S"
```

And of course, add them to dates. Compared to durations, periods are more likely to do what you expect:

```
# A leap year
ymd("2016-01-01") + dyears(1)
#> [1] "2016-12-31"
ymd("2016-01-01") + years(1)
#> [1] "2017-01-01"

# Daylight Savings Time
one_pm + ddays(1)
#> [1] "2016-03-13 14:00:00 EDT"
one_pm + days(1)
#> [1] "2016-03-13 13:00:00 EDT"
```

Let's use periods to fix an oddity related to our flight dates. Some planes appear to have arrived at their destination *before* they departed from New York City:

```
flights_dt %>%
  filter(arr_time < dep_time)
```

```
#> # A tibble: 10,633 × 9
#>   origin  dest dep_delay arr_delay            dep_time
#>    <chr> <chr>     <dbl>     <dbl>              <dttm>
#> 1    EWR   BQN         9        -4 2013-01-01 19:29:00
#> 2    JFK   DFW        59        NA 2013-01-01 19:39:00
#> 3    EWR   TPA        -2         9 2013-01-01 20:58:00
#> 4    EWR   SJU        -6       -12 2013-01-01 21:02:00
#> 5    EWR   SFO        11       -14 2013-01-01 21:08:00
#> 6    LGA   FLL       -10        -2 2013-01-01 21:20:00
#> # ... with 1.063e+04 more rows, and 4 more variables:
#> #   sched_dep_time <dttm>, arr_time <dttm>,
#> #   sched_arr_time <dttm>, air_time <dbl>
```

These are overnight flights. We used the same date information for both the departure and the arrival times, but these flights arrived on the following day. We can fix this by adding `days(1)` to the arrival time of each overnight flight:

```
flights_dt <- flights_dt %>%
  mutate(
    overnight = arr_time < dep_time,
    arr_time = arr_time + days(overnight * 1),
    sched_arr_time = sched_arr_time + days(overnight * 1)
  )
```

Now all of our flights obey the laws of physics:

```
flights_dt %>%
  filter(overnight, arr_time < dep_time)
#> # A tibble: 0 × 10
#> # ... with 10 variables: origin <chr>, dest <chr>,
#> #   dep_delay <dbl>, arr_delay <dbl>, dep_time <dttm>,
#> #   sched_dep_time <dttm>, arr_time <dttm>,
#> #   sched_arr_time <dttm>, air_time <dbl>, overnight <lgl>
```

Intervals

It's obvious what `dyears(1) / ddays(365)` should return: one, because durations are always represented by a number of seconds, and a duration of a year is defined as 365 days' worth of seconds.

What should `years(1) / days(1)` return? Well, if the year was 2015 it should return 365, but if it was 2016, it should return 366! There's not quite enough information for **lubridate** to give a single clear answer. What it does instead is give an estimate, with a warning:

```
years(1) / days(1)
#> estimate only: convert to intervals for accuracy
#> [1] 365
```

If you want a more accurate measurement, you'll have to use an *interval*. An interval is a duration with a starting point; that makes it precise so you can determine exactly how long it is:

```
next_year <- today() + years(1)
(today() %--% next_year) / ddays(1)
#> [1] 365
```

To find out how many periods fall into an interval, you need to use integer division:

```
(today() %--% next_year) %/% days(1)
#> [1] 365
```

Summary

How do you pick between duration, periods, and intervals? As always, pick the simplest data structure that solves your problem. If you only care about physical time, use a duration; if you need to add human times, use a period; if you need to figure out how long a span is in human units, use an interval.

Figure 13-1 summarizes permitted arithmetic operations between the different data types.

	date	date time	duration	period	interval	number
date	-		- +	- +		- +
date time		-	- +	- +		- +
duration	- +	- +	- + /			- + × /
period	- +	- +		- +		- + × /
interval			/	/		
number	- +	- +	- + ×	- + ×	- + ×	- + × /

Figure 13-1. The allowed arithmetic operations between pairs of date/time classes

Exercises

1. Why is there `months()` but no `dmonths()`?
2. Explain `days(overnight * 1)` to someone who has just started learning R. How does it work?

3. Create a vector of dates giving the first day of every month in 2015. Create a vector of dates giving the first day of every month in the *current* year.
4. Write a function that, given your birthday (as a date), returns how old you are in years.
5. Why can't `(today() %--% (today() + years(1)) / months(1)` work?

Time Zones

Time zones are an enormously complicated topic because of their interaction with geopolitical entities. Fortunately we don't need to dig into all the details as they're not all that important for data analysis, but there are a few challenges we'll need to tackle head on.

The first challenge is that everyday names of time zones tend to be ambiguous. For example, if you're American you're probably familiar with EST, or Eastern Standard Time. However, both Australia and Canada also have EST! To avoid confusion, R uses the international standard IANA time zones. These use a consistent naming scheme with "/", typically in the form "<continent>/<city>" (there are a few exceptions because not every country lies on a continent). Examples include "America/New_York," "Europe/Paris," and "Pacific/Auckland."

You might wonder why the time zone uses a city, when typically you think of time zones as associated with a country or region within a country. This is because the IANA database has to record decades' worth of time zone rules. In the course of decades, countries change names (or break apart) fairly frequently, but city names tend to stay the same. Another problem is that name needs to reflect not only to the current behavior, but also the complete history. For example, there are time zones for both "America/New_York" and "America/Detroit." These cities both currently use Eastern Standard Time but in 1969–1972, Michigan (the state in which Detroit is located) did not follow DST, so it needs a different name. It's worth reading the raw time zone database (available at *http://www.iana.org/time-zones*) just to read some of these stories!

You can find out what R thinks your current time zone is with `Sys.timezone()`:

```
Sys.timezone()
#> [1] "America/Los_Angeles"
```

(If R doesn't know, you'll get an NA.)

And see the complete list of all time zone names with `OlsonNames()`:

```
length(OlsonNames())
#> [1] 589
head(OlsonNames())
#> [1] "Africa/Abidjan"     "Africa/Accra"
#> [3] "Africa/Addis_Ababa" "Africa/Algiers"
#> [5] "Africa/Asmara"      "Africa/Asmera"
```

In R, the time zone is an attribute of the date-time that only controls printing. For example, these three objects represent the same instant in time:

```
(x1 <- ymd_hms("2015-06-01 12:00:00", tz = "America/New_York"))
#> [1] "2015-06-01 12:00:00 EDT"
(x2 <- ymd_hms("2015-06-01 18:00:00", tz = "Europe/Copenhagen"))
#> [1] "2015-06-01 18:00:00 CEST"
(x3 <- ymd_hms("2015-06-02 04:00:00", tz = "Pacific/Auckland"))
#> [1] "2015-06-02 04:00:00 NZST"
```

You can verify that they're the same time using subtraction:

```
x1 - x2
#> Time difference of 0 secs
x1 - x3
#> Time difference of 0 secs
```

Unless otherwise specified, **lubridate** always uses UTC. UTC (Coordinated Universal Time) is the standard time zone used by the scientific community and is roughly equivalent to its predecessor GMT (Greenwich Mean Time). It does not have DST, which makes it a convenient representation for computation. Operations that combine date-times, like `c()`, will often drop the time zone. In that case, the date-times will display in your local time zone:

```
x4 <- c(x1, x2, x3)
x4
#> [1] "2015-06-01 09:00:00 PDT" "2015-06-01 09:00:00 PDT"
#> [3] "2015-06-01 09:00:00 PDT"
```

You can change the time zone in two ways:

- Keep the instant in time the same, and change how it's displayed. Use this when the instant is correct, but you want a more natural display:

```
x4a <- with_tz(x4, tzone = "Australia/Lord_Howe")
x4a
#> [1] "2015-06-02 02:30:00 LHST"
#> [2] "2015-06-02 02:30:00 LHST"
#> [3] "2015-06-02 02:30:00 LHST"
x4a - x4
#> Time differences in secs
#> [1] 0 0 0
```

(This also illustrates another challenge of times zones: they're not all integer hour offsets!)

- Change the underlying instant in time. Use this when you have an instant that has been labeled with the incorrect time zone, and you need to fix it:

```
x4b <- force_tz(x4, tzone = "Australia/Lord_Howe")
x4b
#> [1] "2015-06-01 09:00:00 LHST"
#> [2] "2015-06-01 09:00:00 LHST"
#> [3] "2015-06-01 09:00:00 LHST"
x4b - x4
#> Time differences in hours
#> [1] -17.5 -17.5 -17.5
```

PART III
Program

In this part of the book, you'll improve your programming skills. Programming is a cross-cutting skill needed for all data science work: you must use a computer to do data science; you cannot do it in your head, or with pencil and paper.

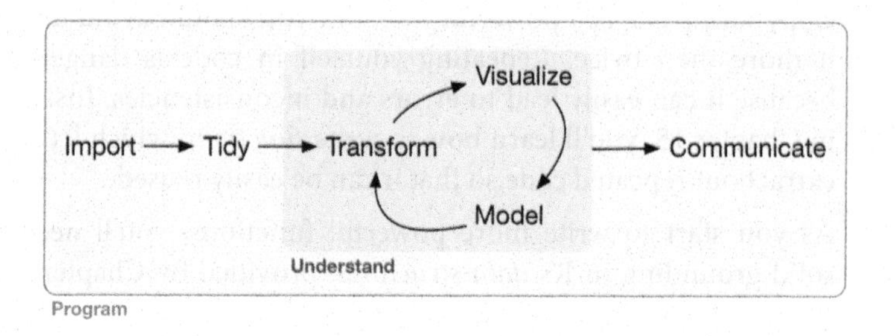

Programming produces code, and code is a tool of communication. Obviously code tells the computer what you want it to do. But it also communicates meaning to other humans. Thinking about code as a vehicle for communication is important because every project you do is fundamentally collaborative. Even if you're not working with other people, you'll definitely be working with future-you! Writing clear code is important so that others (like future-you) can understand why you tackled an analysis in the way you did. That means getting better at programming also involves getting better at communicating. Over time, you want your code to become not just easier to write, but easier for others to read.

Writing code is similar in many ways to writing prose. One parallel that I find particularly useful is that in both cases rewriting is the key to clarity. The first expression of your ideas is unlikely to be particularly clear, and you may need to rewrite multiple times. After solving a data analysis challenge, it's often worth looking at your code and thinking about whether or not it's obvious what you've done. If you spend a little time rewriting your code while the ideas are fresh, you can save a lot of time later trying to re-create what your code did. But this doesn't mean you should rewrite every function: you need to balance what you need to achieve now with saving time in the long run. (But the more you rewrite your functions the more likely your first attempt will be clear.)

In the following four chapters, you'll learn skills that will allow you to both tackle new programs and solve existing problems with greater clarity and ease:

- In Chapter 14, you will dive deep into the *pipe*, `%>%`, and learn more about how it works, what the alternatives are, and when not to use it.
- Copy-and-paste is a powerful tool, but you should avoid doing it more than twice. Repeating yourself in code is dangerous because it can easily lead to errors and inconsistencies. Instead, in Chapter 15, you'll learn how to write *functions*, which let you extract out repeated code so that it can be easily reused.
- As you start to write more powerful functions, you'll need a solid grounding in R's *data structures*, provided by Chapter 16. You must master the four common atomic vectors and the three important S3 classes built on top of them, and understand the mysteries of the list and data frame.
- Functions extract out repeated code, but you often need to repeat the same actions on different inputs. You need tools for *iteration* that let you do similar things again and again. These tools include for loops and functional programming, which you'll learn about in Chapter 17.

Learning More

The goal of these chapters is to teach you the minimum about programming that you need to practice data science, which turns out to be a reasonable amount. Once you have mastered the material in this book, I strongly believe you should invest further in your programming skills. Learning more about programming is a long-term investment: it won't pay off immediately, but in the long term it will allow you to solve new problems more quickly, and let you reuse your insights from previous problems in new scenarios.

To learn more you need to study R as a programming language, not just an interactive environment for data science. We have written two books that will help you do so:

- *Hands-On Programming with R*, by Garrett Grolemund. This is an introduction to R as a programming language and is a great place to start if R is your first programming language. It covers similar material to these chapters, but with a different style and different motivation examples (based in the casino). It's a useful complement if you find that these four chapters go by too quickly.
- *Advanced R* by Hadley Wickham. This dives into the details of R the programming language. This is a great place to start if you have existing programming experience. It's also a great next step once you've internalized the ideas in these chapters. You can read it online at *http://adv-r.had.co.nz*.

CHAPTER 14

Pipes with magrittr

Introduction

Pipes are a powerful tool for clearly expressing a sequence of multiple operations. So far, you've been using them without knowing how they work, or what the alternatives are. Now, in this chapter, it's time to explore the pipe in more detail. You'll learn the alternatives to the pipe, when you shouldn't use the pipe, and some useful related tools.

Prerequisites

The pipe, `%>%`, comes from the **magrittr** package by Stefan Milton Bache. Packages in the tidyverse load `%>%` for you automatically, so you don't usually load **magrittr** explicitly. Here, however, we're focusing on piping, and we aren't loading any other packages, so we will load it explicitly.

```
library(magrittr)
```

Piping Alternatives

The point of the pipe is to help you write code in a way that is easier to read and understand. To see why the pipe is so useful, we're going to explore a number of ways of writing the same code. Let's use code to tell a story about a little bunny named Foo Foo:

> Little bunny Foo Foo
> Went hopping through the forest
> Scooping up the field mice
> And bopping them on the head

This is a popular children's poem that is accompanied by hand actions.

We'll start by defining an object to represent little bunny Foo Foo:

```
foo_foo <- little_bunny()
```

And we'll use a function for each key verb: `hop()`, `scoop()`, and `bop()`. Using this object and these verbs, there are (at least) four ways we could retell the story in code:

- Save each intermediate step as a new object.
- Overwrite the original object many times.
- Compose functions.
- Use the pipe.

We'll work through each approach, showing you the code and talking about the advantages and disadvantages.

Intermediate Steps

The simplest approach is to save each step as a new object:

```
foo_foo_1 <- hop(foo_foo, through = forest)
foo_foo_2 <- scoop(foo_foo_1, up = field_mice)
foo_foo_3 <- bop(foo_foo_2, on = head)
```

The main downside of this form is that it forces you to name each intermediate element. If there are natural names, this is a good idea, and you should do it. But many times, like in this example, there aren't natural names, and you add numeric suffixes to make the names unique. That leads to two problems:

- The code is cluttered with unimportant names.
- You have to carefully increment the suffix on each line.

Whenever I write code like this, I invariably use the wrong number on one line and then spend 10 minutes scratching my head and trying to figure out what went wrong with my code.

You may also worry that this form creates many copies of your data and takes up a lot of memory. Surprisingly, that's not the case. First, note that proactively worrying about memory is not a useful way to spend your time: worry about it when it becomes a problem (i.e., you run out of memory), not before. Second, R isn't stupid, and it will share columns across data frames, where possible. Let's take a look at an actual data manipulation pipeline where we add a new column to `ggplot2::diamonds`:

```
diamonds <- ggplot2::diamonds
diamonds2 <- diamonds %>%
  dplyr::mutate(price_per_carat = price / carat)

pryr::object_size(diamonds)
#> 3.46 MB
pryr::object_size(diamonds2)
#> 3.89 MB
pryr::object_size(diamonds, diamonds2)
#> 3.89 MB
```

`pryr::object_size()` gives the memory occupied by all of its arguments. The results seem counterintuitive at first:

- `diamonds` takes up 3.46 MB.
- `diamonds2` takes up 3.89 MB.
- `diamonds` and `diamonds2` together take up 3.89 MB!

How can that work? Well, `diamonds2` has 10 columns in common with `diamonds`: there's no need to duplicate all that data, so the two data frames have variables in common. These variables will only get copied if you modify one of them. In the following example, we modify a single value in `diamonds$carat`. That means the `carat` variable can no longer be shared between the two data frames, and a copy must be made. The size of each data frame is unchanged, but the collective size increases:

```
diamonds$carat[1] <- NA
pryr::object_size(diamonds)
#> 3.46 MB
pryr::object_size(diamonds2)
#> 3.89 MB
pryr::object_size(diamonds, diamonds2)
#> 4.32 MB
```

(Note that we use `pryr::object_size()` here, not the built-in `object.size()`. `object.size()` only takes a single object so it can't compute how data is shared across multiple objects.)

Overwrite the Original

Instead of creating intermediate objects at each step, we could overwrite the original object:

```
foo_foo <- hop(foo_foo, through = forest)
foo_foo <- scoop(foo_foo, up = field_mice)
foo_foo <- bop(foo_foo, on = head)
```

This is less typing (and less thinking), so you're less likely to make mistakes. However, there are two problems:

- Debugging is painful. If you make a mistake you'll need to rerun the complete pipeline from the beginning.
- The repetition of the object being transformed (we've written `foo_foo` six times!) obscures what's changing on each line.

Function Composition

Another approach is to abandon assignment and just string the function calls together:

```
bop(
  scoop(
    hop(foo_foo, through = forest),
    up = field_mice
  ),
  on = head
)
```

Here the disadvantage is that you have to read from inside-out, from right-to-left, and that the arguments end up spread far apart (evocatively called the Dagwood sandwich (*https://en.wikipedia.org/wiki/Dagwood_sandwich*) problem). In short, this code is hard for a human to consume.

Use the Pipe

Finally, we can use the pipe:

```
foo_foo %>%
  hop(through = forest) %>%
```

```
scoop(up = field_mouse) %>%
bop(on = head)
```

This is my favorite form, because it focuses on verbs, not nouns. You can read this series of function compositions like it's a set of imperative actions. Foo Foo hops, then scoops, then bops. The downside, of course, is that you need to be familiar with the pipe. If you've never seen %>% before, you'll have no idea what this code does. Fortunately, most people pick up the idea very quickly, so when you share your code with others who aren't familiar with the pipe, you can easily teach them.

The pipe works by performing a "lexical transformation": behind the scenes, **magrittr** reassembles the code in the pipe to a form that works by overwriting an intermediate object. When you run a pipe like the preceding one, **magrittr** does something like this:

```
my_pipe <- function(.) {
  . <- hop(., through = forest)
  . <- scoop(., up = field_mice)
  bop(., on = head)
}
my_pipe(foo_foo)
```

This means that the pipe won't work for two classes of functions:

- Functions that use the current environment. For example, `assign()` will create a new variable with the given name in the current environment:

  ```
  assign("x", 10)
  x
  #> [1] 10

  "x" %>% assign(100)
  x
  #> [1] 10
  ```

  The use of `assign` with the pipe does not work because it assigns it to a temporary environment used by %>%. If you do want to use `assign` with the pipe, you must be explicit about the environment:

  ```
  env <- environment()
  "x" %>% assign(100, envir = env)
  x
  #> [1] 100
  ```

  Other functions with this problem include `get()` and `load()`.

- Functions that use lazy evaluation. In R, function arguments are only computed when the function uses them, not prior to calling the function. The pipe computes each element in turn, so you can't rely on this behavior.

  One place that this is a problem is `tryCatch()`, which lets you capture and handle errors:

  ```
  tryCatch(stop("!"), error = function(e) "An error")
  #> [1] "An error"

  stop("!") %>%
    tryCatch(error = function(e) "An error")
  #> Error in eval(expr, envir, enclos): !
  ```

  There are a relatively wide class of functions with this behavior, including `try()`, `suppressMessages()`, and `suppressWarnings()` in base R.

When Not to Use the Pipe

The pipe is a powerful tool, but it's not the only tool at your disposal, and it doesn't solve every problem! Pipes are most useful for rewriting a fairly short linear sequence of operations. I think you should reach for another tool when:

- Your pipes are longer than (say) 10 steps. In that case, create intermediate objects with meaningful names. That will make debugging easier, because you can more easily check the intermediate results, and it makes it easier to understand your code, because the variable names can help communicate intent.
- You have multiple inputs or outputs. If there isn't one primary object being transformed, but two or more objects being combined together, don't use the pipe.
- You are starting to think about a directed graph with a complex dependency structure. Pipes are fundamentally linear and expressing complex relationships with them will typically yield confusing code.

Other Tools from magrittr

All packages in the tidyverse automatically make `%>%` available for you, so you don't normally load **magrittr** explicitly. However, there are some other useful tools inside **magrittr** that you might want to try out:

- When working with more complex pipes, it's sometimes useful to call a function for its side effects. Maybe you want to print out the current object, or plot it, or save it to disk. Many times, such functions don't return anything, effectively terminating the pipe.

  To work around this problem, you can use the "tee" pipe. `%T>%` works like `%>%` except that it returns the lefthand side instead of the righthand side. It's called "tee" because it's like a literal T-shaped pipe:

  ```
  rnorm(100) %>%
    matrix(ncol = 2) %>%
    plot() %>%
    str()
  #>  NULL
  ```

  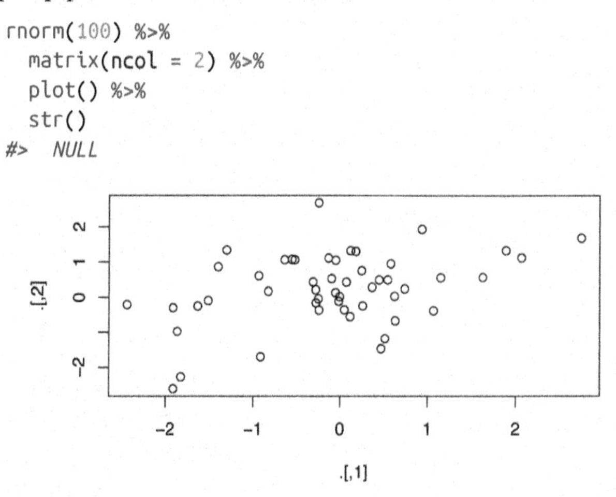


  ```
  rnorm(100) %>%
    matrix(ncol = 2) %T>%
    plot() %>%
    str()
  #>  num [1:50, 1:2] -0.387 -0.785 -1.057 -0.796 -1.756 ...
  ```

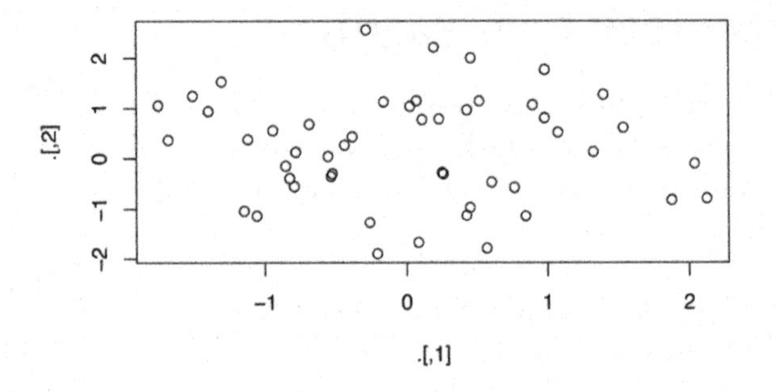

- If you're working with functions that don't have a data frame–based API (i.e., you pass them individual vectors, not a data frame and expressions to be evaluated in the context of that data frame), you might find `%$%` useful. It "explodes" out the variables in a data frame so that you can refer to them explicitly. This is useful when working with many functions in base R:

  ```
  mtcars %$%
    cor(disp, mpg)
  #> [1] -0.848
  ```

- For assignment **magrittr** provides the `%<>%` operator, which allows you to replace code like:

  ```
  mtcars <- mtcars %>%
    transform(cyl = cyl * 2)
  ```

  with:

  ```
  mtcars %<>% transform(cyl = cyl * 2)
  ```

  I'm not a fan of this operator because I think assignment is such a special operation that it should always be clear when it's occurring. In my opinion, a little bit of duplication (i.e., repeating the name of the object twice) is fine in return for making assignment more explicit.

CHAPTER 15

Functions

Introduction

One of the best ways to improve your reach as a data scientist is to write functions. Functions allow you to automate common tasks in a more powerful and general way than copying and pasting. Writing a function has three big advantages over using copy-and-paste:

- You can give a function an evocative name that makes your code easier to understand.
- As requirements change, you only need to update code in one place, instead of many.
- You eliminate the chance of making incidental mistakes when you copy and paste (i.e., updating a variable name in one place, but not in another).

Writing good functions is a lifetime journey. Even after using R for many years I still learn new techniques and better ways of approaching old problems. The goal of this chapter is not to teach you every esoteric detail of functions but to get you started with some pragmatic advice that you can apply immediately.

As well as practical advice for writing functions, this chapter also gives you some suggestions for how to style your code. Good code style is like correct punctuation. Youcanmanagewithoutit, but it sure makes things easier to read! As with styles of punctuation, there are many possible variations. Here we present the style we use in our code, but the most important thing is to be consistent.

Prerequisites

The focus of this chapter is on writing functions in base R, so you won't need any extra packages.

When Should You Write a Function?

You should consider writing a function whenever you've copied and pasted a block of code more than twice (i.e., you now have three copies of the same code). For example, take a look at this code. What does it do?

```
df <- tibble::tibble(
  a = rnorm(10),
  b = rnorm(10),
  c = rnorm(10),
  d = rnorm(10)
)

df$a <- (df$a - min(df$a, na.rm = TRUE)) /
  (max(df$a, na.rm = TRUE) - min(df$a, na.rm = TRUE))
df$b <- (df$b - min(df$b, na.rm = TRUE)) /
  (max(df$b, na.rm = TRUE) - min(df$a, na.rm = TRUE))
df$c <- (df$c - min(df$c, na.rm = TRUE)) /
  (max(df$c, na.rm = TRUE) - min(df$c, na.rm = TRUE))
df$d <- (df$d - min(df$d, na.rm = TRUE)) /
  (max(df$d, na.rm = TRUE) - min(df$d, na.rm = TRUE))
```

You might be able to puzzle out that this rescales each column to have a range from 0 to 1. But did you spot the mistake? I made an error when copying and pasting the code for `df$b`: I forgot to change an `a` to a `b`. Extracting repeated code out into a function is a good idea because it prevents you from making this type of mistake.

To write a function you need to first analyze the code. How many inputs does it have?

```
(df$a - min(df$a, na.rm = TRUE)) /
  (max(df$a, na.rm = TRUE) - min(df$a, na.rm = TRUE))
```

This code only has one input: `df$a`. (If you're surprised that `TRUE` is not an input, you can explore why in the following exercise.) To make the inputs more clear, it's a good idea to rewrite the code using temporary variables with general names. Here this code only requires a single numeric vector, so I'll call it `x`:

```
x <- df$a
(x - min(x, na.rm = TRUE)) /
```

```
(max(x, na.rm = TRUE) - min(x, na.rm = TRUE))
#>  [1] 0.289 0.751 0.000 0.678 0.853 1.000 0.172 0.611 0.612
#>  [10] 0.601
```

There is some duplication in this code. We're computing the range of the data three times, but it makes sense to do it in one step:

```
rng <- range(x, na.rm = TRUE)
(x - rng[1]) / (rng[2] - rng[1])
#>  [1] 0.289 0.751 0.000 0.678 0.853 1.000 0.172 0.611 0.612
#>  [10] 0.601
```

Pulling out intermediate calculations into named variables is a good practice because it makes it more clear what the code is doing. Now that I've simplified the code, and checked that it still works, I can turn it into a function:

```
rescale01 <- function(x) {
  rng <- range(x, na.rm = TRUE)
  (x - rng[1]) / (rng[2] - rng[1])
}
rescale01(c(0, 5, 10))
#> [1] 0.0 0.5 1.0
```

There are three key steps to creating a new function:

1. You need to pick a *name* for the function. Here I've used `rescale01` because this function rescales a vector to lie between 0 and 1.

2. You list the inputs, or *arguments*, to the function inside `function`. Here we have just one argument. If we had more the call would look like `function(x, y, z)`.

3. You place the code you have developed in the *body* of the function, a { block that immediately follows `function(...)`.

Note the overall process: I only made the function after I'd figured out how to make it work with a simple input. It's easier to start with working code and turn it into a function; it's harder to create a function and then try to make it work.

At this point it's a good idea to check your function with a few different inputs:

```
rescale01(c(-10, 0, 10))
#> [1] 0.0 0.5 1.0
rescale01(c(1, 2, 3, NA, 5))
#> [1] 0.00 0.25 0.50   NA 1.00
```

As you write more and more functions you'll eventually want to convert these informal, interactive tests into formal, automated tests. That process is called unit testing. Unfortunately, it's beyond the scope of this book, but you can learn about it at *http://r-pkgs.had.co.nz/tests.html*.

We can simplify the original example now that we have a function:

```
df$a <- rescale01(df$a)
df$b <- rescale01(df$b)
df$c <- rescale01(df$c)
df$d <- rescale01(df$d)
```

Compared to the original, this code is easier to understand and we've eliminated one class of copy-and-paste errors. There is still quite a bit of duplication since we're doing the same thing to multiple columns. We'll learn how to eliminate that duplication in Chapter 17, once you've learned more about R's data structures in Chapter 16.

Another advantage of functions is that if our requirements change, we only need to make the change in one place. For example, we might discover that some of our variables include infinite values, and `rescale01()` fails:

```
x <- c(1:10, Inf)
rescale01(x)
#>  [1]   0   0   0   0   0   0   0   0   0   0 NaN
```

Because we've extracted the code into a function, we only need to make the fix in one place:

```
rescale01 <- function(x) {
  rng <- range(x, na.rm = TRUE, finite = TRUE)
  (x - rng[1]) / (rng[2] - rng[1])
}
rescale01(x)
#>  [1] 0.000 0.111 0.222 0.333 0.444 0.556 0.667 0.778 0.889
#> [10] 1.000   Inf
```

This is an important part of the "do not repeat yourself" (or DRY) principle. The more repetition you have in your code, the more places you need to remember to update when things change (and they always do!), and the more likely you are to create bugs over time.

Exercises

1. Why is `TRUE` not a parameter to `rescale01()`? What would happen if `x` contained a single missing value, and `na.rm` was `FALSE`?

2. In the second variant of `rescale01()`, infinite values are left unchanged. Rewrite `rescale01()` so that `-Inf` is mapped to 0, and `Inf` is mapped to 1.

3. Practice turning the following code snippets into functions. Think about what each function does. What would you call it? How many arguments does it need? Can you rewrite it to be more expressive or less duplicative?

   ```
   mean(is.na(x))

   x / sum(x, na.rm = TRUE)

   sd(x, na.rm = TRUE) / mean(x, na.rm = TRUE)
   ```

4. Follow *http://nicercode.github.io/intro/writing-functions.html* to write your own functions to compute the variance and skew of a numeric vector.

5. Write `both_na()`, a function that takes two vectors of the same length and returns the number of positions that have an `NA` in both vectors.

6. What do the following functions do? Why are they useful even though they are so short?

   ```
   is_directory <- function(x) file.info(x)$isdir
   is_readable <- function(x) file.access(x, 4) == 0
   ```

7. Read the complete lyrics (*http://bit.ly/littlebunnyfoofoo*) to "Little Bunny Foo Foo." There's a lot of duplication in this song. Extend the initial piping example to re-create the complete song, and use functions to reduce the duplication.

Functions Are for Humans and Computers

It's important to remember that functions are not just for the computer, but are also for humans. R doesn't care what your function is called, or what comments it contains, but these are important for human readers. This section discusses some things that you should bear in mind when writing functions that humans can understand.

The name of a function is important. Ideally, the name of your function will be short, but clearly evoke what the function does. That's hard! But it's better to be clear than short, as RStudio's autocomplete makes it easy to type long names.

Generally, function names should be verbs, and arguments should be nouns. There are some exceptions: nouns are OK if the function computes a very well known noun (i.e., `mean()` is better than `compute_mean()`), or is accessing some property of an object (i.e., `coef()` is better than `get_coefficients()`). A good sign that a noun might be a better choice is if you're using a very broad verb like "get," "compute," "calculate," or "determine." Use your best judgment and don't be afraid to rename a function if you figure out a better name later:

```
# Too short
f()

# Not a verb, or descriptive
my_awesome_function()

# Long, but clear
impute_missing()
collapse_years()
```

If your function name is composed of multiple words, I recommend using "snake_case," where each lowercase word is separated by an underscore. camelCase is a popular alternative. It doesn't really matter which one you pick; the important thing is to be consistent: pick one or the other and stick with it. R itself is not very consistent, but there's nothing you can do about that. Make sure you don't fall into the same trap by making your code as consistent as possible:

```
# Never do this!
col_mins <- function(x, y) {}
rowMaxes <- function(y, x) {}
```

If you have a family of functions that do similar things, make sure they have consistent names and arguments. Use a common prefix to indicate that they are connected. That's better than a common suffix because autocomplete allows you to type the prefix and see all the members of the family:

```
# Good
input_select()
input_checkbox()
input_text()
```

```
# Not so good
select_input()
checkbox_input()
text_input()
```

A good example of this design is the **stringr** package: if you don't remember exactly which function you need, you can type `str_` and jog your memory.

Where possible, avoid overriding existing functions and variables. It's impossible to do in general because so many good names are already taken by other packages, but avoiding the most common names from base R will avoid confusion:

```
# Don't do this!
T <- FALSE
c <- 10
mean <- function(x) sum(x)
```

Use comments, lines starting with `#`, to explain the "why" of your code. You generally should avoid comments that explain the "what" or the "how." If you can't understand what the code does from reading it, you should think about how to rewrite it to be more clear. Do you need to add some intermediate variables with useful names? Do you need to break out a subcomponent of a large function so you can name it? However, your code can never capture the reasoning behind your decisions: why did you choose this approach instead of an alternative? What else did you try that didn't work? It's a great idea to capture that sort of thinking in a comment.

Another important use of comments is to break up your file into easily readable chunks. Use long lines of `-` or `=` to make it easy to spot the breaks:

```
# Load data --------------------------------------

# Plot data --------------------------------------
```

RStudio provides a keyboard shortcut to create these headers (Cmd/Ctrl-Shift-R), and will display them in the code navigation drop-down at the bottom-left of the editor:

Load data
Plot data
Load data

Exercises

1. Read the source code for each of the following three functions, puzzle out what they do, and then brainstorm better names:

```
f1 <- function(string, prefix) {
  substr(string, 1, nchar(prefix)) == prefix
}
f2 <- function(x) {
  if (length(x) <= 1) return(NULL)
  x[-length(x)]
}
f3 <- function(x, y) {
  rep(y, length.out = length(x))
}
```

2. Take a function that you've written recently and spend five minutes brainstorming a better name for it and its arguments.

3. Compare and contrast `rnorm()` and `MASS::mvrnorm()`. How could you make them more consistent?

4. Make a case for why `norm_r()`, `norm_d()`, etc., would be better than `rnorm()`, `dnorm()`. Make a case for the opposite.

Conditional Execution

An `if` statement allows you to conditionally execute code. It looks like this:

```
if (condition) {
  # code executed when condition is TRUE
} else {
  # code executed when condition is FALSE
}
```

To get help on `if` you need to surround it in backticks: ``?`if` ``. The help isn't particularly helpful if you're not already an experienced programmer, but at least you know how to get to it!

Here's a simple function that uses an `if` statement. The goal of this function is to return a logical vector describing whether or not each element of a vector is named:

```
has_name <- function(x) {
  nms <- names(x)
  if (is.null(nms)) {
    rep(FALSE, length(x))
```

```
  } else {
    !is.na(nms) & nms != ""
  }
}
```

This function takes advantage of the standard return rule: a function returns the last value that it computed. Here that is either one of the two branches of the `if` statement.

Conditions

The `condition` must evaluate to either `TRUE` or `FALSE`. If it's a vector, you'll get a warning message; if it's an `NA`, you'll get an error. Watch out for these messages in your own code:

```
if (c(TRUE, FALSE)) {}
#> Warning in if (c(TRUE, FALSE)) {:
#> the condition has length > 1 and only the
#> first element will be used
#> NULL

if (NA) {}
#> Error in if (NA) {: missing value where TRUE/FALSE needed
```

You can use `||` (or) and `&&` (and) to combine multiple logical expressions. These operators are "short-circuiting": as soon as `||` sees the first `TRUE` it returns `TRUE` without computing anything else. As soon as `&&` sees the first `FALSE` it returns `FALSE`. You should never use `|` or `&` in an `if` statement: these are vectorized operations that apply to multiple values (that's why you use them in `filter()`). If you do have a logical vector, you can use `any()` or `all()` to collapse it to a single value.

Be careful when testing for equality. `==` is vectorized, which means that it's easy to get more than one output. Either check the length is already 1, collapse with `all()` or `any()`, or use the nonvectorized `identical()`. `identical()` is very strict: it always returns either a single `TRUE` or a single `FALSE`, and doesn't coerce types. This means that you need to be careful when comparing integers and doubles:

```
identical(0L, 0)
#> [1] FALSE
```

You also need to be wary of floating-point numbers:

```
x <- sqrt(2) ^ 2
x
#> [1] 2
```

```
x == 2
#> [1] FALSE
x - 2
#> [1] 4.44e-16
```

Instead use `dplyr::near()` for comparisons, as described in "Comparisons" on page 46.

And remember, `x == NA` doesn't do anything useful!

Multiple Conditions

You can chain multiple `if` statements together:

```
if (this) {
  # do that
} else if (that) {
  # do something else
} else {
  #
}
```

But if you end up with a very long series of chained `if` statements, you should consider rewriting. One useful technique is the `switch()` function. It allows you to evaluate selected code based on position or name:

```
#> function(x, y, op) {
#>   switch(op,
#>     plus = x + y,
#>     minus = x - y,
#>     times = x * y,
#>     divide = x / y,
#>     stop("Unknown op!")
#>   )
#> }
```

Another useful function that can often eliminate long chains of `if` statements is `cut()`. It's used to discretize continuous variables.

Code Style

Both `if` and `function` should (almost) always be followed by squiggly brackets (`{}`), and the contents should be indented by two spaces. This makes it easier to see the hierarchy in your code by skimming the lefthand margin.

An opening curly brace should never go on its own line and should always be followed by a new line. A closing curly brace should always go on its own line, unless it's followed by `else`. Always indent the code inside curly braces:

```
# Good
if (y < 0 && debug) {
  message("Y is negative")
}

if (y == 0) {
  log(x)
} else {
  y ^ x
}

# Bad
if (y < 0 && debug)
message("Y is negative")

if (y == 0) {
  log(x)
}
else {
  y ^ x
}
```

It's OK to drop the curly braces if you have a very short `if` statement that can fit on one line:

```
y <- 10
x <- if (y < 20) "Too low" else "Too high"
```

I recommend this only for very brief `if` statements. Otherwise, the full form is easier to read:

```
if (y < 20) {
  x <- "Too low"
} else {
  x <- "Too high"
}
```

Exercises

1. What's the difference between `if` and `ifelse()`? Carefully read the help and construct three examples that illustrate the key differences.

2. Write a greeting function that says "good morning," "good afternoon," or "good evening," depending on the time of day. (Hint: use a time argument that defaults to `lubridate::now()`. That will make it easier to test your function.)

3. Implement a `fizzbuzz` function. It takes a single number as input. If the number is divisible by three, it returns "fizz". If it's divisible by five it returns "buzz". If it's divisible by three and five, it returns "fizzbuzz". Otherwise, it returns the number. Make sure you first write working code before you create the function.

4. How could you use `cut()` to simplify this set of nested if-else statements?

```
if (temp <= 0) {
  "freezing"
} else if (temp <= 10) {
  "cold"
} else if (temp <= 20) {
  "cool"
} else if (temp <= 30) {
  "warm"
} else {
  "hot"
}
```

   How would you change the call to `cut()` if I'd used < instead of <=? What is the other chief advantage of `cut()` for this problem? (Hint: what happens if you have many values in `temp`?)

5. What happens if you use `switch()` with numeric values?

6. What does this `switch()` call do? What happens if `x` is "e"?

```
switch(x,
  a = ,
  b = "ab",
  c = ,
  d = "cd"
)
```

   Experiment, then carefully read the documentation.

Function Arguments

The arguments to a function typically fall into two broad sets: one set supplies the *data* to compute on, and the other supplies arguments that control the *details* of the computation. For example:

- In `log()`, the data is `x`, and the detail is the `base` of the logarithm.

- In `mean()`, the data is `x`, and the details are how much data to trim from the ends (`trim`) and how to handle missing values (`na.rm`).
- In `t.test()`, the data are `x` and `y`, and the details of the test are `alternative`, `mu`, `paired`, `var.equal`, and `conf.level`.
- In `str_c()` you can supply any number of strings to `...`, and the details of the concatenation are controlled by `sep` and `collapse`.

Generally, data arguments should come first. Detail arguments should go on the end, and usually should have default values. You specify a default value in the same way you call a function with a named argument:

```
# Compute confidence interval around
# mean using normal approximation
mean_ci <- function(x, conf = 0.95) {
  se <- sd(x) / sqrt(length(x))
  alpha <- 1 - conf
  mean(x) + se * qnorm(c(alpha / 2, 1 - alpha / 2))
}

x <- runif(100)
mean_ci(x)
#> [1] 0.498 0.610
mean_ci(x, conf = 0.99)
#> [1] 0.480 0.628
```

The default value should almost always be the most common value. The few exceptions to this rule have to do with safety. For example, it makes sense for `na.rm` to default to `FALSE` because missing values are important. Even though `na.rm = TRUE` is what you usually put in your code, it's a bad idea to silently ignore missing values by default.

When you call a function, you typically omit the names of the data arguments, because they are used so commonly. If you override the default value of a detail argument, you should use the full name:

```
# Good
mean(1:10, na.rm = TRUE)

# Bad
mean(x = 1:10, , FALSE)
mean(, TRUE, x = c(1:10, NA))
```

You can refer to an argument by its unique prefix (e.g., `mean(x, n = TRUE)`), but this is generally best avoided given the possibilities for confusion.

Notice that when you call a function, you should place a space around `=` in function calls, and always put a space after a comma, not before (just like in regular English). Using whitespace makes it easier to skim the function for the important components:

```
# Good
average <- mean(feet / 12 + inches, na.rm = TRUE)

# Bad
average<-mean(feet/12+inches,na.rm=TRUE)
```

Choosing Names

The names of the arguments are also important. R doesn't care, but the readers of your code (including future-you!) will. Generally you should prefer longer, more descriptive names, but there are a handful of very common, very short names. It's worth memorizing these:

- `x`, `y`, `z`: vectors.
- `w`: a vector of weights.
- `df`: a data frame.
- `i`, `j`: numeric indices (typically rows and columns).
- `n`: length, or number of rows.
- `p`: number of columns.

Otherwise, consider matching names of arguments in existing R functions. For example, use `na.rm` to determine if missing values should be removed.

Checking Values

As you start to write more functions, you'll eventually get to the point where you don't remember exactly how your function works. At this point it's easy to call your function with invalid inputs. To avoid this problem, it's often useful to make constraints explicit. For example, imagine you've written some functions for computing weighted summary statistics:

```
wt_mean <- function(x, w) {
  sum(x * w) / sum(x)
}
wt_var <- function(x, w) {
  mu <- wt_mean(x, w)
  sum(w * (x - mu) ^ 2) / sum(w)
}
wt_sd <- function(x, w) {
  sqrt(wt_var(x, w))
}
```

What happens if x and w are not the same length?

```
wt_mean(1:6, 1:3)
#> [1] 2.19
```

In this case, because of R's vector recycling rules, we don't get an error.

It's good practice to check important preconditions, and throw an error (with `stop()`) if they are not true:

```
wt_mean <- function(x, w) {
  if (length(x) != length(w)) {
    stop("`x` and `w` must be the same length", call. = FALSE)
  }
  sum(w * x) / sum(x)
}
```

Be careful not to take this too far. There's a trade-off between how much time you spend making your function robust, versus how long you spend writing it. For example, if you also added a `na.rm` argument, I probably wouldn't check it carefully:

```
wt_mean <- function(x, w, na.rm = FALSE) {
  if (!is.logical(na.rm)) {
    stop("`na.rm` must be logical")
  }
  if (length(na.rm) != 1) {
    stop("`na.rm` must be length 1")
  }
  if (length(x) != length(w)) {
    stop("`x` and `w` must be the same length", call. = FALSE)
  }

  if (na.rm) {
    miss <- is.na(x) | is.na(w)
    x <- x[!miss]
    w <- w[!miss]
  }
  sum(w * x) / sum(x)
}
```

This is a lot of extra work for little additional gain. A useful compromise is the built-in `stopifnot()`; it checks that each argument is `TRUE`, and produces a generic error message if not:

```
wt_mean <- function(x, w, na.rm = FALSE) {
  stopifnot(is.logical(na.rm), length(na.rm) == 1)
  stopifnot(length(x) == length(w))

  if (na.rm) {
    miss <- is.na(x) | is.na(w)
    x <- x[!miss]
    w <- w[!miss]
  }
  sum(w * x) / sum(x)
}
wt_mean(1:6, 6:1, na.rm = "foo")
#> Error: is.logical(na.rm) is not TRUE
```

Note that when using `stopifnot()` you assert what should be true rather than checking for what might be wrong.

Dot-Dot-Dot (...)

Many functions in R take an arbitrary number of inputs:

```
sum(1, 2, 3, 4, 5, 6, 7, 8, 9, 10)
#> [1] 55
stringr::str_c("a", "b", "c", "d", "e", "f")
#> [1] "abcdef"
```

How do these functions work? They rely on a special argument: `...` (pronounced dot-dot-dot). This special argument captures any number of arguments that aren't otherwise matched.

It's useful because you can then send those `...` on to another function. This is a useful catch-all if your function primarily wraps another function. For example, I commonly create these helper functions that wrap around `str_c()`:

```
commas <- function(...) stringr::str_c(..., collapse = ", ")
commas(letters[1:10])
#> [1] "a, b, c, d, e, f, g, h, i, j"

rule <- function(..., pad = "-") {
  title <- paste0(...)
  width <- getOption("width") - nchar(title) - 5
  cat(title, " ", stringr::str_dup(pad, width), "\n", sep = "")
}
rule("Important output")
#> Important output ------------------------------------
```

Here ... lets me forward on any arguments that I don't want to deal with to `str_c()`. It's a very convenient technique. But it does come at a price: any misspelled arguments will not raise an error. This makes it easy for typos to go unnoticed:

```
x <- c(1, 2)
sum(x, na.mr = TRUE)
#> [1] 4
```

If you just want to capture the values of the ..., use `list(...)`.

Lazy Evaluation

Arguments in R are lazily evaluated: they're not computed until they're needed. That means if they're never used, they're never called. This is an important property of R as a programming language, but is generally not important when you're writing your own functions for data analysis. You can read more about lazy evaluation at *http://adv-r.had.co.nz/Functions.html#lazy-evaluation*.

Exercises

1. What does `commas(letters, collapse = "-")` do? Why?
2. It'd be nice if you could supply multiple characters to the `pad` argument, e.g., `rule("Title", pad = "-+")`. Why doesn't this currently work? How could you fix it?
3. What does the `trim` argument to `mean()` do? When might you use it?
4. The default value for the `method` argument to `cor()` is `c("pearson", "kendall", "spearman")`. What does that mean? What value is used by default?

Return Values

Figuring out what your function should return is usually straightforward: it's why you created the function in the first place! There are two things you should consider when returning a value:

- Does returning early make your function easier to read?
- Can you make your function pipeable?

Explicit Return Statements

The value returned by the function is usually the last statement it evaluates, but you can choose to return early by using `return()`. I think it's best to save the use of `return()` to signal that you can return early with a simpler solution. A common reason to do this is because the inputs are empty:

```
complicated_function <- function(x, y, z) {
  if (length(x) == 0 || length(y) == 0) {
    return(0)
  }

  # Complicated code here
}
```

Another reason is because you have a `if` statement with one complex block and one simple block. For example, you might write an `if` statement like this:

```
f <- function() {
  if (x) {
    # Do
    # something
    # that
    # takes
    # many
    # lines
    # to
    # express
  } else {
    # return something short
  }
}
```

But if the first block is very long, by the time you get to the `else`, you've forgotten the `condition`. One way to rewrite it is to use an early return for the simple case:

```
f <- function() {
  if (!x) {
    return(something_short)
  }

  # Do
  # something
  # that
  # takes
  # many
  # lines
```

```
    # to
    # express
}
```

This tends to make the code easier to understand, because you don't need quite so much context to understand it.

Writing Pipeable Functions

If you want to write your own pipeable functions, thinking about the return value is important. There are two main types of pipeable functions: transformation and side-effect.

In *transformation* functions, there's a clear "primary" object that is passed in as the first argument, and a modified version is returned by the function. For example, the key objects for **dplyr** and **tidyr** are data frames. If you can identify what the object type is for your domain, you'll find that your functions just work with the pipe.

Side-effect functions are primarily called to perform an action, like drawing a plot or saving a file, not transforming an object. These functions should "invisibly" return the first argument, so they're not printed by default, but can still be used in a pipeline. For example, this simple function prints out the number of missing values in a data frame:

```
show_missings <- function(df) {
  n <- sum(is.na(df))
  cat("Missing values: ", n, "\n", sep = "")

  invisible(df)
}
```

If we call it interactively, the `invisible()` means that the input `df` doesn't get printed out:

```
show_missings(mtcars)
#> Missing values: 0
```

But it's still there, it's just not printed by default:

```
x <- show_missings(mtcars)
#> Missing values: 0
class(x)
#> [1] "data.frame"
dim(x)
#> [1] 32 11
```

And we can still use it in a pipe:

```
mtcars %>%
  show_missings() %>%
  mutate(mpg = ifelse(mpg < 20, NA, mpg)) %>%
  show_missings()
#> Missing values: 0
#> Missing values: 18
```

Environment

The last component of a function is its environment. This is not something you need to understand deeply when you first start writing functions. However, it's important to know a little bit about environments because they are crucial to how functions work. The environment of a function controls how R finds the value associated with a name. For example, take this function:

```
f <- function(x) {
  x + y
}
```

In many programming languages, this would be an error, because y is not defined inside the function. In R, this is valid code because R uses rules called *lexical scoping* to find the value associated with a name. Since y is not defined inside the function, R will look in the *environment* where the function was defined:

```
y <- 100
f(10)
#> [1] 110

y <- 1000
f(10)
#> [1] 1010
```

This behavior seems like a recipe for bugs, and indeed you should avoid creating functions like this deliberately, but by and large it doesn't cause too many problems (especially if you regularly restart R to get to a clean slate).

The advantage of this behavior is that from a language standpoint it allows R to be very consistent. Every name is looked up using the same set of rules. For f() that includes the behavior of two things that you might not expect: { and +. This allows you to do devious things like:

```
`+` <- function(x, y) {
  if (runif(1) < 0.1) {
    sum(x, y)
  } else {
    sum(x, y) * 1.1
  }
}
table(replicate(1000, 1 + 2))
#>
#>   3 3.3
#> 100 900
rm(`+`)
```

This is a common phenomenon in R. R places few limits on your power. You can do many things that you can't do in other programming languages. You can do many things that 99% of the time are extremely ill-advised (like overriding how addition works!). But this power and flexibility is what makes tools like **ggplot2** and **dplyr** possible. Learning how to make best use of this flexibility is beyond the scope of this book, but you can read about in *Advanced R* (*http://adv-r.had.co.nz*).

CHAPTER 16
Vectors

Introduction

So far this book has focused on tibbles and packages that work with them. But as you start to write your own functions, and dig deeper into R, you need to learn about vectors, the objects that underlie tibbles. If you've learned R in a more traditional way, you're probably already familiar with vectors, as most R resources start with vectors and work their way up to tibbles. I think it's better to start with tibbles because they're immediately useful, and then work your way down to the underlying components.

Vectors are particularly important as most of the functions you will write will work with vectors. It is possible to write functions that work with tibbles (like in **ggplot2**, **dplyr**, and **tidyr**), but the tools you need to write such functions are currently idiosyncratic and immature. I am working on a better approach, *https://github.com/hadley/lazyeval*, but it will not be ready in time for the publication of the book. Even when complete, you'll still need to understand vectors; it'll just make it easier to write a user-friendly layer on top.

Prerequisites

The focus of this chapter is on base R data structures, so it isn't essential to load any packages. We will, however, use a handful of functions from the **purrr** package to avoid some inconsistences in base R.

```
library(tidyverse)
#> Loading tidyverse: ggplot2
#> Loading tidyverse: tibble
#> Loading tidyverse: tidyr
#> Loading tidyverse: readr
#> Loading tidyverse: purrr
#> Loading tidyverse: dplyr
#> Conflicts with tidy packages ----------------------------------
#> filter(): dplyr, stats
#> lag():    dplyr, stats
```

Vector Basics

There are two types of vectors:

- *Atomic* vectors, of which there are six types: *logical*, *integer*, *double*, *character*, *complex*, and *raw*. Integer and double vectors are collectively known as *numeric* vectors.
- *Lists*, which are sometimes called recursive vectors because lists can contain other lists.

The chief difference between atomic vectors and lists is that atomic vectors are *homogeneous*, while lists can be *heterogeneous*. There's one other related object: `NULL`. `NULL` is often used to represent the absence of a vector (as opposed to `NA`, which is used to represent the absence of a value in a vector). `NULL` typically behaves like a vector of length 0. Figure 16-1 summarizes the interrelationships.

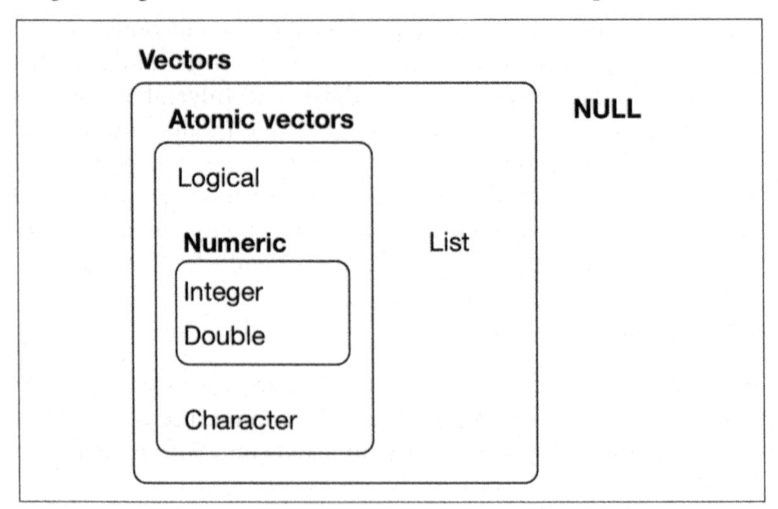

Figure 16-1. The hierarchy of R's vector types

Every vector has two key properties:

- Its *type*, which you can determine with `typeof()`:

```
typeof(letters)
#> [1] "character"
typeof(1:10)
#> [1] "integer"
```

- Its *length*, which you can determine with `length()`:

```
x <- list("a", "b", 1:10)
length(x)
#> [1] 3
```

Vectors can also contain arbitrary additional metadata in the form of attributes. These attributes are used to create *augmented vectors*, which build on additional behavior. There are four important types of augmented vector:

- Factors are built on top of integer vectors.
- Dates and date-times are built on top of numeric vectors.
- Data frames and tibbles are built on top of lists.

This chapter will introduce you to these important vectors from simplest to most complicated. You'll start with atomic vectors, then build up to lists, and finish off with augmented vectors.

Important Types of Atomic Vector

The four most important types of atomic vector are logical, integer, double, and character. Raw and complex are rarely used during a data analysis, so I won't discuss them here.

Logical

Logical vectors are the simplest type of atomic vector because they can take only three possible values: `FALSE`, `TRUE`, and `NA`. Logical vectors are usually constructed with comparison operators, as described in "Comparisons" on page 46. You can also create them by hand with `c()`:

```
1:10 %% 3 == 0
#>  [1] FALSE FALSE  TRUE FALSE FALSE
#>  [2] TRUE FALSE FALSE  TRUE FALSE
```

```
c(TRUE, TRUE, FALSE, NA)
#> [1]  TRUE  TRUE FALSE    NA
```

Numeric

Integer and double vectors are known collectively as numeric vectors. In R, numbers are doubles by default. To make an integer, place a `L` after the number:

```
typeof(1)
#> [1] "double"
typeof(1L)
#> [1] "integer"
1.5L
#> [1] 1.5
```

The distinction between integers and doubles is not usually important, but there are two important differences that you should be aware of:

- Doubles are approximations. Doubles represent floating-point numbers that cannot always be precisely represented with a fixed amount of memory. This means that you should consider all doubles to be approximations. For example, what is square of the square root of two?

  ```
  x <- sqrt(2) ^ 2
  x
  #> [1] 2
  x - 2
  #> [1] 4.44e-16
  ```

  This behavior is common when working with floating-point numbers: most calculations include some approximation error. Instead of comparing floating-point numbers using `==`, you should use `dplyr::near()`, which allows for some numerical tolerance.

- Integers have one special value, `NA`, while doubles have four, `NA`, `NaN`, `Inf`, and `-Inf`. All three special values can arise during division:

  ```
  c(-1, 0, 1) / 0
  #> [1] -Inf  NaN  Inf
  ```

  Avoid using `==` to check for these other special values. Instead use the helper functions `is.finite()`, `is.infinite()`, and `is.nan()`:

	0	Inf	NA	NaN
is.finite()	X			
is.infinite()		X		
is.na()			X	X
is.nan()				X

Character

Character vectors are the most complex type of atomic vector, because each element of a character vector is a string, and a string can contain an arbitrary amount of data.

You've already learned a lot about working with strings in Chapter 11. Here I want to mention one important feature of the underlying string implementation: R uses a global string pool. This means that each unique string is only stored in memory once, and every use of the string points to that representation. This reduces the amount of memory needed by duplicated strings. You can see this behavior in practice with `pryr::object_size()`:

```
x <- "This is a reasonably long string."
pryr::object_size(x)
#> 136 B

y <- rep(x, 1000)
pryr::object_size(y)
#> 8.13 kB
```

`y` doesn't take up 1000x as much memory as `x`, because each element of `y` is just a pointer to that same string. A pointer is 8 bytes, so 1000 pointers to a 136 B string is 8 * 1000 + 136 = 8.13 kB.

Missing Values

Note that each type of atomic vector has its own missing value:

```
NA            # logical
#> [1] NA
NA_integer_   # integer
#> [1] NA
NA_real_      # double
#> [1] NA
NA_character_ # character
#> [1] NA
```

Normally you don't need to know about these different types because you can always use `NA` and it will be converted to the correct type using the implicit coercion rules described next. However, there are some functions that are strict about their inputs, so it's useful to have this knowledge sitting in your back pocket so you can be specific when needed.

Exercises

1. Describe the difference between `is.finite(x)` and `!is.infinite(x)`.
2. Read the source code for `dplyr::near()` (Hint: to see the source code, drop the `()`). How does it work?
3. A logical vector can take three possible values. How many possible values can an integer vector take? How many possible values can a double take? Use Google to do some research.
4. Brainstorm at least four functions that allow you to convert a double to an integer. How do they differ? Be precise.
5. What functions from the **readr** package allow you to turn a string into a logical, integer, and double vector?

Using Atomic Vectors

Now that you understand the different types of atomic vector, it's useful to review some of the important tools for working with them. These include:

- How to convert from one type to another, and when that happens automatically.
- How to tell if an object is a specific type of vector.
- What happens when you work with vectors of different lengths.
- How to name the elements of a vector.
- How to pull out elements of interest.

Coercion

There are two ways to convert, or coerce, one type of vector to another:

- Explicit coercion happens when you call a function like `as.logical()`, `as.integer()`, `as.double()`, or `as.character()`. Whenever you find yourself using explicit coercion, you should always check whether you can make the fix upstream, so that the vector never had the wrong type in the first place. For example, you may need to tweak your **readr** `col_types` specification.
- Implicit coercion happens when you use a vector in a specific context that expects a certain type of vector. For example, when you use a logical vector with a numeric summary function, or when you use a double vector where an integer vector is expected.

Because explicit coercion is used relatively rarely, and is largely easy to understand, I'll focus on implicit coercion here.

You've already seen the most important type of implicit coercion: using a logical vector in a numeric context. In this case `TRUE` is converted to `1` and `FALSE` is converted to 0. That means the sum of a logical vector is the number of trues, and the mean of a logical vector is the proportion of trues:

```
x <- sample(20, 100, replace = TRUE)
y <- x > 10
sum(y)  # how many are greater than 10?
#> [1] 44
mean(y) # what proportion are greater than 10?
#> [1] 0.44
```

You may see some code (typically older) that relies on implicit coercion in the opposite direction, from integer to logical:

```
if (length(x)) {
  # do something
}
```

In this case, 0 is converted to `FALSE` and everything else is converted to `TRUE`. I think this makes it harder to understand your code, and I don't recommend it. Instead be explicit: `length(x) > 0`.

It's also important to understand what happens when you try and create a vector containing multiple types with `c()`—the most complex type always wins:

```
typeof(c(TRUE, 1L))
#> [1] "integer"
typeof(c(1L, 1.5))
```

```
#> [1] "double"
typeof(c(1.5, "a"))
#> [1] "character"
```

An atomic vector cannot have a mix of different types because the type is a property of the complete vector, not the individual elements. If you need to mix multiple types in the same vector, you should use a list, which you'll learn about shortly.

Test Functions

Sometimes you want to do different things based on the type of vector. One option is to use `typeof()`. Another is to use a test function that returns a `TRUE` or `FALSE`. Base R provides many functions like `is.vector()` and `is.atomic()`, but they often return surprising results. Instead, it's safer to use the `is_*` functions provided by **purrr**, which are summarized in the following table.

	lgl	int	dbl	chr	list
`is_logical()`	x				
`is_integer()`		x			
`is_double()`			x		
`is_numeric()`		x	x		
`is_character()`				x	
`is_atomic()`	x	x	x	x	
`is_list()`					x
`is_vector()`	x	x	x	x	x

Each predicate also comes with a "scalar" version, like `is_scalar_atomic()`, which checks that the length is 1. This is useful, for example, if you want to check that an argument to your function is a single logical value.

Scalars and Recycling Rules

As well as implicitly coercing the types of vectors to be compatible, R will also implicitly coerce the length of vectors. This is called vector *recycling*, because the shorter vector is repeated, or recycled, to the same length as the longer vector.

This is generally most useful when you are mixing vectors and "scalars." I put scalars in quotes because R doesn't actually have scalars: instead, a single number is a vector of length 1. Because there are no scalars, most built-in functions are *vectorized*, meaning that they will operate on a vector of numbers. That's why, for example, this code works:

```
sample(10) + 100
#>  [1] 109 108 104 102 103 110 106 107 105 101
runif(10) > 0.5
#>  [1]  TRUE  TRUE FALSE  TRUE  TRUE  TRUE FALSE  TRUE  TRUE
#> [10]  TRUE
```

In R, basic mathematical operations work with vectors. That means that you should never need to perform explicit iteration when performing simple mathematical computations.

It's intuitive what should happen if you add two vectors of the same length, or a vector and a "scalar," but what happens if you add two vectors of different lengths?

```
1:10 + 1:2
#>  [1]  2  4  4  6  6  8  8 10 10 12
```

Here, R will expand the shortest vector to the same length as the longest, so-called recycling. This is silent except when the length of the longer is not an integer multiple of the length of the shorter:

```
1:10 + 1:3
#> Warning in 1:10 + 1:3:
#> longer object length is not a multiple of shorter
#> object length
#>  [1]  2  4  6  5  7  9  8 10 12 11
```

While vector recycling can be used to create very succinct, clever code, it can also silently conceal problems. For this reason, the vectorized functions in tidyverse will throw errors when you recycle anything other than a scalar. If you do want to recycle, you'll need to do it yourself with `rep()`:

```
tibble(x = 1:4, y = 1:2)
#> Error: Variables must be length 1 or 4.
#> Problem variables: 'y'

tibble(x = 1:4, y = rep(1:2, 2))
#> # A tibble: 4 × 2
#>       x     y
#>   <int> <int>
#> 1     1     1
#> 2     2     2
#> 3     3     1
#> 4     4     2
```

```
tibble(x = 1:4, y = rep(1:2, each = 2))
#> # A tibble: 4 × 2
#>       x     y
#>   <int> <int>
#> 1     1     1
#> 2     2     1
#> 3     3     2
#> 4     4     2
```

Naming Vectors

All types of vectors can be named. You can name them during creation with `c()`:

```
c(x = 1, y = 2, z = 4)
#> x y z
#> 1 2 4
```

Or after the fact with `purrr::set_names()`:

```
set_names(1:3, c("a", "b", "c"))
#> a b c
#> 1 2 3
```

Named vectors are most useful for subsetting, described next.

Subsetting

So far we've used `dplyr::filter()` to filter the rows in a tibble. `filter()` only works with tibble, so we'll need a new tool for vectors: `[`. `[` is the subsetting function, and is called like `x[a]`. There are four types of things that you can subset a vector with:

- A numeric vector containing only integers. The integers must either be all positive, all negative, or zero.

  Subsetting with positive integers keeps the elements at those positions:

  ```
  x <- c("one", "two", "three", "four", "five")
  x[c(3, 2, 5)]
  #> [1] "three" "two"   "five"
  ```

  By repeating a position, you can actually make a longer output than input:

  ```
  x[c(1, 1, 5, 5, 5, 2)]
  #> [1] "one"  "one"  "five" "five" "five" "two"
  ```

Negative values drop the elements at the specified positions:

```
x[c(-1, -3, -5)]
#> [1] "two"  "four"
```

It's an error to mix positive and negative values:

```
x[c(1, -1)]
#> Error in x[c(1, -1)]:
#> only 0's may be mixed with negative subscripts
```

The error message mentions subsetting with zero, which returns no values:

```
x[0]
#> character(0)
```

This is not useful very often, but it can be helpful if you want to create unusual data structures to test your functions with.

- Subsetting with a logical vector keeps all values corresponding to a `TRUE` value. This is most often useful in conjunction with the comparison functions:

```
x <- c(10, 3, NA, 5, 8, 1, NA)

# All non-missing values of x
x[!is.na(x)]
#> [1] 10  3  5  8  1

# All even (or missing!) values of x
x[x %% 2 == 0]
#> [1] 10 NA  8 NA
```

- If you have a named vector, you can subset it with a character vector:

```
x <- c(abc = 1, def = 2, xyz = 5)
x[c("xyz", "def")]
#> xyz def
#>   5   2
```

Like with positive integers, you can also use a character vector to duplicate individual entries.

- The simplest type of subsetting is nothing, `x[]`, which returns the complete `x`. This is not useful for subsetting vectors, but it is useful when subsetting matrices (and other high-dimensional structures) because it lets you select all the rows or all the columns, by leaving that index blank. For example, if `x` is 2D, x[1,] selects the first row and all the columns, and x[, -1] selects all rows and all columns except the first.

To learn more about the applications of subsetting, read the "Subsetting" chapter of *Advanced R* (*http://bit.ly/subsetadvR*).

There is an important variation of [called [[. [[only ever extracts a single element, and always drops names. It's a good idea to use it whenever you want to make it clear that you're extracting a single item, as in a for loop. The distinction between [and [[is most important for lists, as we'll see shortly.

Exercises

1. What does mean(is.na(x)) tell you about a vector x? What about sum(!is.finite(x))?
2. Carefully read the documentation of is.vector(). What does it actually test for? Why does is.atomic() not agree with the definition of atomic vectors above?
3. Compare and contrast setNames() with purrr::set_names().
4. Create functions that take a vector as input and return:
   a. The last value. Should you use [or [[?
   b. The elements at even numbered positions.
   c. Every element except the last value.
   d. Only even numbers (and no missing values).
5. Why is x[-which(x > 0)] not the same as x[x <= 0]?
6. What happens when you subset with a positive integer that's bigger than the length of the vector? What happens when you subset with a name that doesn't exist?

Recursive Vectors (Lists)

Lists are a step up in complexity from atomic vectors, because lists can contain other lists. This makes them suitable for representing hierarchical or tree-like structures. You create a list with list():

```
x <- list(1, 2, 3)
x
#> [[1]]
```

```
#> [1] 1
#>
#> [[2]]
#> [1] 2
#>
#> [[3]]
#> [1] 3
```

A very useful tool for working with lists is `str()` because it focuses on the *str*ucture, not the contents:

```
str(x)
#> List of 3
#>  $ : num 1
#>  $ : num 2
#>  $ : num 3

x_named <- list(a = 1, b = 2, c = 3)
str(x_named)
#> List of 3
#>  $ a: num 1
#>  $ b: num 2
#>  $ c: num 3
```

Unlike atomic vectors, `lists()` can contain a mix of objects:

```
y <- list("a", 1L, 1.5, TRUE)
str(y)
#> List of 4
#>  $ : chr "a"
#>  $ : int 1
#>  $ : num 1.5
#>  $ : logi TRUE
```

Lists can even contain other lists!

```
z <- list(list(1, 2), list(3, 4))
str(z)
#> List of 2
#>  $ :List of 2
#>   ..$ : num 1
#>   ..$ : num 2
#>  $ :List of 2
#>   ..$ : num 3
#>   ..$ : num 4
```

Visualizing Lists

To explain more complicated list manipulation functions, it's helpful to have a visual representation of lists. For example, take these three lists:

```
x1 <- list(c(1, 2), c(3, 4))
x2 <- list(list(1, 2), list(3, 4))
x3 <- list(1, list(2, list(3)))
```

I'll draw them as follows:

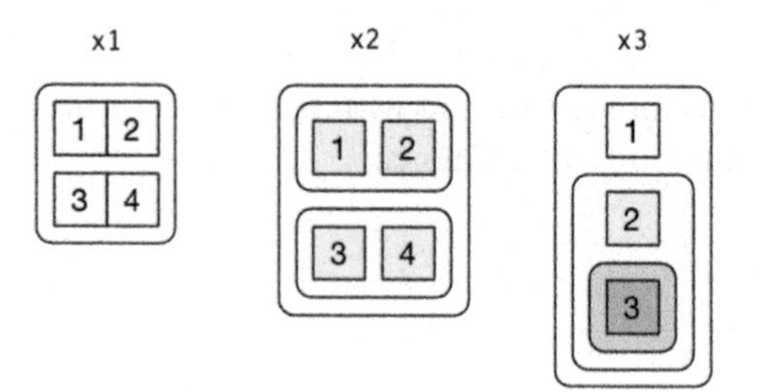

There are three principles:

- Lists have rounded corners. Atomic vectors have square corners.
- Children are drawn inside their parent, and have a slightly darker background to make it easier to see the hierarchy.
- The orientation of the children (i.e., rows or columns) isn't important, so I'll pick a row or column orientation to either save space or illustrate an important property in the example.

Subsetting

There are three ways to subset a list, which I'll illustrate with `a`:

```
a <- list(a = 1:3, b = "a string", c = pi, d = list(-1, -5))
```

- `[` extracts a sublist. The result will always be a list:

  ```
  str(a[1:2])
  #> List of 2
  #>  $ a: int [1:3] 1 2 3
  #>  $ b: chr "a string"
  str(a[4])
  #> List of 1
  #>  $ d:List of 2
  #>   ..$ : num -1
  #>   ..$ : num -5
  ```

  Like with vectors, you can subset with a logical, integer, or character vector.

- `[[` extracts a single component from a list. It removes a level of hierarchy from the list:

```
str(y[[1]])
#>  chr "a"
str(y[[4]])
#>  logi TRUE
```

- `$` is a shorthand for extracting named elements of a list. It works similarly to `[[` except that you don't need to use quotes:

```
a$a
#> [1] 1 2 3
a[["a"]]
#> [1] 1 2 3
```

The distinction between `[` and `[[` is really important for lists, because `[[` drills down into the list while `[` returns a new, smaller list. Compare the preceding code and output with the visual representation in Figure 16-2.

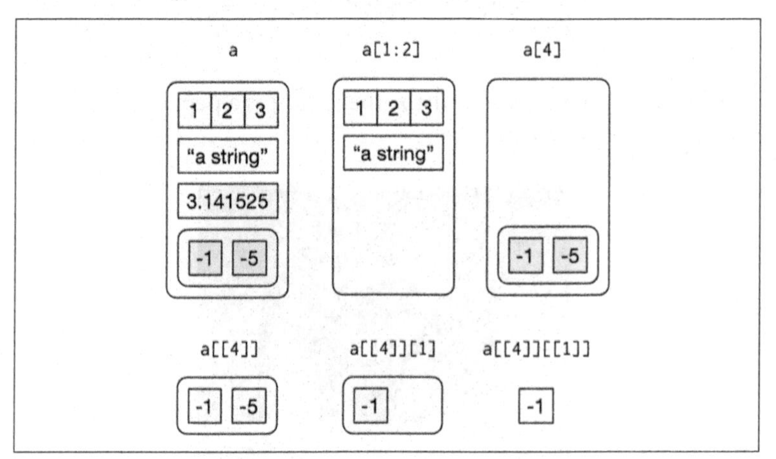

Figure 16-2. Subsetting a list, visually

Lists of Condiments

The difference between `[` and `[[` is very important, but it's easy to get confused. To help you remember, let me show you an unusual pepper shaker:

If this pepper shaker is your list x, then, x[1] is a pepper shaker containing a single pepper packet:

x[2] would look the same, but would contain the second packet. x[1:2] would be a pepper shaker containing two pepper packets.

x[[1]] is:

If you wanted to get the content of the pepper package, you'd need x[[1]][[1]]:

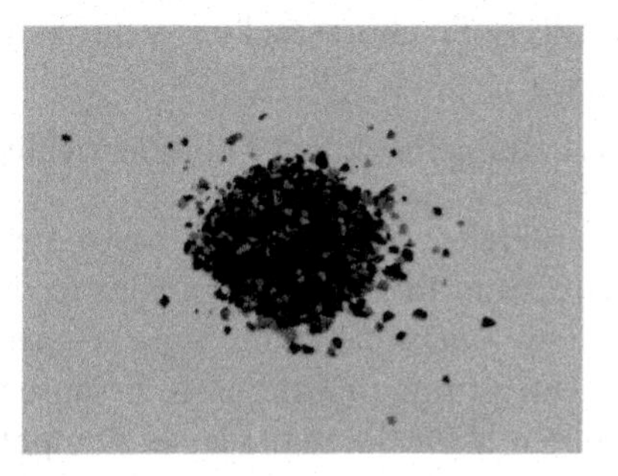

Exercises

1. Draw the following lists as nested sets:

   a. `list(a, b, list(c, d), list(e, f))`

   b. `list(list(list(list(list(list(a))))))`

2. What happens if you subset a tibble as if you're subsetting a list? What are the key differences between a list and a tibble?

Attributes

Any vector can contain arbitrary additional metadata through its *attributes*. You can think of attributes as a named list of vectors that can be attached to any object. You can get and set individual attribute values with `attr()` or see them all at once with `attributes()`:

```
x <- 1:10
attr(x, "greeting")
#> NULL
attr(x, "greeting") <- "Hi!"
attr(x, "farewell") <- "Bye!"
attributes(x)
#> $greeting
#> [1] "Hi!"
#>
#> $farewell
#> [1] "Bye!"
```

There are three very important attributes that are used to implement fundamental parts of R:

- *Names* are used to name the elements of a vector.
- *Dimensions* (dims, for short) make a vector behave like a matrix or array.
- *Class* is used to implement the S3 object-oriented system.

You've seen names earlier, and we won't cover dimensions because we don't use matrices in this book. It remains to describe the class, which controls how *generic functions* work. Generic functions are key to object-oriented programming in R, because they make functions behave differently for different classes of input. A detailed discussion of object-oriented programming is beyond the scope of this book, but you can read more about it in *Advanced R* (*http://bit.ly/OOproadvR*).

Here's what a typical generic function looks like:

```
as.Date
#> function (x, ...)
#> UseMethod("as.Date")
#> <bytecode: 0x7fa61e0590d8>
#> <environment: namespace:base>
```

The call to "UseMethod" means that this is a generic function, and it will call a specific *method*, a function, based on the class of the first argument. (All methods are functions; not all functions are methods.) You can list all the methods for a generic with `methods()`:

```
methods("as.Date")
#> [1] as.Date.character as.Date.date      as.Date.dates
#> [4] as.Date.default as.Date.factor    as.Date.numeric
#> [7] as.Date.POSIXct   as.Date.POSIXlt
#> see '?methods' for accessing help and source code
```

For example, if `x` is a character vector, `as.Date()` will call `as.Date.character()`; if it's a factor, it'll call `as.Date.factor()`.

You can see the specific implementation of a method with `getS3method()`:

```
getS3method("as.Date", "default")
#> function (x, ...)
#> {
#>     if (inherits(x, "Date"))
#>         return(x)
#>     if (is.logical(x) && all(is.na(x)))
#>         return(structure(as.numeric(x), class = "Date"))
#>     stop(
#>       gettextf("do not know how to convert '%s' to class %s",
#>       deparse(substitute(x)), dQuote("Date")), domain = NA)
#> }
#> <bytecode: 0x7fa61dd47e78>
#> <environment: namespace:base>
getS3method("as.Date", "numeric")
#> function (x, origin, ...)
#> {
#>     if (missing(origin))
#>         stop("'origin' must be supplied")
#>     as.Date(origin, ...) + x
#> }
#> <bytecode: 0x7fa61dd463b8>
#> <environment: namespace:base>
```

The most important S3 generic is `print()`: it controls how the object is printed when you type its name at the console. Other important generics are the subsetting functions `[`, `[[`, and `$`.

Augmented Vectors

Atomic vectors and lists are the building blocks for other important vector types like factors and dates. I call these *augmented vectors*, because they are vectors with additional *attributes*, including class. Because augmented vectors have a class, they behave differently to the atomic vector on which they are built. In this book, we make use of four important augmented vectors:

- Factors
- Date-times and times
- Tibbles

These are described next.

Factors

Factors are designed to represent categorical data that can take a fixed set of possible values. Factors are built on top of integers, and have a levels attribute:

```
x <- factor(c("ab", "cd", "ab"), levels = c("ab", "cd", "ef"))
typeof(x)
#> [1] "integer"
attributes(x)
#> $levels
#> [1] "ab" "cd" "ef"
#>
#> $class
#> [1] "factor"
```

Dates and Date-Times

Dates in R are numeric vectors that represent the number of days since 1 January 1970:

```
x <- as.Date("1971-01-01")
unclass(x)
#> [1] 365

typeof(x)
#> [1] "double"
attributes(x)
#> $class
#> [1] "Date"
```

Date-times are numeric vectors with class `POSIXct` that represent the number of seconds since 1 January 1970. (In case you were wondering, "POSIXct" stands for "Portable Operating System Interface," calendar time.)

```
x <- lubridate::ymd_hm("1970-01-01 01:00")
unclass(x)
#> [1] 3600
#> attr(,"tzone")
#> [1] "UTC"
```

```
typeof(x)
#> [1] "double"
attributes(x)
#> $tzone
#> [1] "UTC"
#>
#> $class
#> [1] "POSIXct" "POSIXt"
```

The `tzone` attribute is optional. It controls how the time is printed, not what absolute time it refers to:

```
attr(x, "tzone") <- "US/Pacific"
x
#> [1] "1969-12-31 17:00:00 PST"

attr(x, "tzone") <- "US/Eastern"
x
#> [1] "1969-12-31 20:00:00 EST"
```

There is another type of date-times called `POSIXlt`. These are built on top of named lists:

```
y <- as.POSIXlt(x)
typeof(y)
#> [1] "list"
attributes(y)
#> $names
#>  [1] "sec"    "min"    "hour"   "mday"   "mon"    "year"
#>  [7] "wday"   "yday"   "isdst"  "zone"   "gmtoff"
#>
#> $class
#> [1] "POSIXlt" "POSIXt"
#>
#> $tzone
#> [1] "US/Eastern" "EST"        "EDT"
```

`POSIXlt`s are rare inside the tidyverse. They do crop up in base R, because they are needed to extract specific components of a date, like the year or month. Since **lubridate** provides helpers for you to do this instead, you don't need them. `POSIXct`'s are always easier to work with, so if you find you have a `POSIXlt`, you should always convert it to a regular date-time with `lubridate::as_date_time()`.

Tibbles

Tibbles are augmented lists. They have three classes: `tbl_df`, `tbl`, and `data.frame`. They have two attributes: (column) `names` and `row.names`.

```
tb <- tibble::tibble(x = 1:5, y = 5:1)
typeof(tb)
#> [1] "list"
attributes(tb)
#> $names
#> [1] "x" "y"
#>
#> $class
#> [1] "tbl_df"     "tbl"        "data.frame"
#>
#> $row.names
#> [1] 1 2 3 4 5
```

Traditional `data.frames` have a very similar structure:

```
df <- data.frame(x = 1:5, y = 5:1)
typeof(df)
#> [1] "list"
attributes(df)
#> $names
#> [1] "x" "y"
#>
#> $row.names
#> [1] 1 2 3 4 5
#>
#> $class
#> [1] "data.frame"
```

The main difference is the class. The class of tibble includes "data.frame," which means tibbles inherit the regular data frame behavior by default.

The difference between a tibble or a data frame and a list is that all of the elements of a tibble or data frame must be vectors with the same length. All functions that work with tibbles enforce this constraint.

Exercises

1. What does `hms::hms(3600)` return? How does it print? What primitive type is the augmented vector built on top of? What attributes does it use?
2. Try and make a tibble that has columns with different lengths. What happens?
3. Based of the previous definition, is it OK to have a list as a column of a tibble?

CHAPTER 17

Iteration with purrr

Introduction

In Chapter 15, we talked about how important it is to reduce duplication in your code by creating functions instead of copying and pasting. Reducing code duplication has three main benefits:

- It's easier to see the intent of your code, because your eyes are drawn to what's different, not what stays the same.
- It's easier to respond to changes in requirements. As your needs change, you only need to make changes in one place, rather than remembering to change every place that you copied and pasted the code.
- You're likely to have fewer bugs because each line of code is used in more places.

One tool for reducing duplication is functions, which reduce duplication by identifying repeated patterns of code and extracting them out into independent pieces that can be easily reused and updated. Another tool for reducing duplication is *iteration*, which helps you when you need to do the same thing to multiple inputs: repeating the same operation on different columns, or on different datasets. In this chapter you'll learn about two important iteration paradigms: imperative programming and functional programming. On the imperative side you have tools like for loops and while loops, which are a great place to start because they make iteration very explicit, so it's obvious what's happening. However, for loops are quite verbose, and require quite a bit of bookkeeping code that is duplicated for every for loop. Functional programming (FP) offers tools to extract out this duplicated code, so each common for loop pattern gets its own function. Once you master the vocabulary of FP, you can solve many common iteration problems with less code, more ease, and fewer errors.

Prerequisites

Once you've mastered the for loops provided by base R, you'll learn some of the powerful programming tools provided by **purrr**, one of the tidyverse core packages.

```
library(tidyverse)
```

For Loops

Imagine we have this simple tibble:

```
df <- tibble(
  a = rnorm(10),
  b = rnorm(10),
  c = rnorm(10),
  d = rnorm(10)
)
```

We want to compute the median of each column. You *could* do it with copy-and-paste:

```
median(df$a)
#> [1] -0.246
median(df$b)
#> [1] -0.287
median(df$c)
#> [1] -0.0567
median(df$d)
#> [1] 0.144
```

But that breaks our rule of thumb: never copy and paste more than twice. Instead, we could use a for loop:

```
output <- vector("double", ncol(df))  # 1. output
for (i in seq_along(df)) {            # 2. sequence
  output[[i]] <- median(df[[i]])      # 3. body
}
output
#> [1] -0.2458 -0.2873 -0.0567  0.1443
```

Every for loop has three components:

output `output <- vector("double", length(x))`
: Before you start the loop, you must always allocate sufficient space for the output. This is very important for efficiency: if you grow the for loop at each iteration using `c()` (for example), your for loop will be very slow.

  A general way of creating an empty vector of given length is the `vector()` function. It has two arguments: the type of the vector ("logical," "integer," "double," "character," etc.) and the length of the vector.

sequence `i in seq_along(df)`
: This determines what to loop over: each run of the for loop will assign `i` to a different value from `seq_along(df)`. It's useful to think of `i` as a pronoun, like "it."

  You might not have seen `seq_along()` before. It's a safe version of the familiar `1:length(l)`, with an important difference; if you have a zero-length vector, `seq_along()` does the right thing:

```
y <- vector("double", 0)
seq_along(y)
#> integer(0)
1:length(y)
#> [1] 1 0
```

  You probably won't create a zero-length vector deliberately, but it's easy to create them accidentally. If you use `1:length(x)` instead of `seq_along(x)`, you're likely to get a confusing error message.

body `output[[i]] <- median(df[[i]])`
: This is the code that does the work. It's run repeatedly, each time with a different value for `i`. The first iteration will run `output[[1]] <- median(df[[1]])`, the second will run `output[[2]] <- median(df[[2]])`, and so on.

That's all there is to the for loop! Now is a good time to practice creating some basic (and not so basic) for loops using the following exercises. Then we'll move on to some variations of the for loop that help you solve other problems that will crop up in practice.

Exercises

1. Write for loops to:

    a. Compute the mean of every column in `mtcars`.

    b. Determine the type of each column in `nycflights13::flights`.

    c. Compute the number of unique values in each column of `iris`.

    d. Generate 10 random normals for each of $\mu = -10, 0, 10$, and 100.

    Think about the output, sequence, and body *before* you start writing the loop.

2. Eliminate the for loop in each of the following examples by taking advantage of an existing function that works with vectors:

```
out <- ""
for (x in letters) {
  out <- stringr::str_c(out, x)
}

x <- sample(100)
sd <- 0
for (i in seq_along(x)) {
  sd <- sd + (x[i] - mean(x)) ^ 2
}
sd <- sqrt(sd / (length(x) - 1))

x <- runif(100)
out <- vector("numeric", length(x))
out[1] <- x[1]
for (i in 2:length(x)) {
  out[i] <- out[i - 1] + x[i]
}
```

3. Combine your function writing and for loop skills:

    a. Write a for loop that `prints()` the lyrics to the children's song "Alice the Camel."

    b. Convert the nursery rhyme "Ten in the Bed" to a function. Generalize it to any number of people in any sleeping structure.

c. Convert the song "99 Bottles of Beer on the Wall" to a function. Generalize to any number of any vessel containing any liquid on any surface.

4. It's common to see for loops that don't preallocate the output and instead increase the length of a vector at each step:

```
output <- vector("integer", 0)
for (i in seq_along(x)) {
  output <- c(output, lengths(x[[i]]))
}
output
```

How does this affect performance? Design and execute an experiment.

For Loop Variations

Once you have the basic for loop under your belt, there are some variations that you should be aware of. These variations are important regardless of how you do iteration, so don't forget about them once you've mastered the FP techniques you'll learn about in the next section.

There are four variations on the basic theme of the for loop:

- Modifying an existing object, instead of creating a new object.
- Looping over names or values, instead of indices.
- Handling outputs of unknown length.
- Handling sequences of unknown length.

Modifying an Existing Object

Sometimes you want to use a for loop to modify an existing object. For example, remember our challenge from Chapter 15. We wanted to rescale every column in a data frame:

```
df <- tibble(
  a = rnorm(10),
  b = rnorm(10),
  c = rnorm(10),
  d = rnorm(10)
)
rescale01 <- function(x) {
```

```
  rng <- range(x, na.rm = TRUE)
  (x - rng[1]) / (rng[2] - rng[1])
}

df$a <- rescale01(df$a)
df$b <- rescale01(df$b)
df$c <- rescale01(df$c)
df$d <- rescale01(df$d)
```

To solve this with a for loop we again think about the three components:

Output
: We already have the output—it's the same as the input!

Sequence
: We can think about a data frame as a list of columns, so we can iterate over each column with `seq_along(df)`.

Body
: Apply `rescale01()`.

This gives us:

```
for (i in seq_along(df)) {
  df[[i]] <- rescale01(df[[i]])
}
```

Typically you'll be modifying a list or data frame with this sort of loop, so remember to use `[[`, not `[`. You might have spotted that I used `[[` in all my for loops: I think it's better to use `[[` even for atomic vectors because it makes it clear that I want to work with a single element.

Looping Patterns

There are three basic ways to loop over a vector. So far I've shown you the most general: looping over the numeric indices with `for (i in seq_along(xs))`, and extracting the value with `x[[i]]`. There are two other forms:

- Loop over the elements: `for (x in xs)`. This is most useful if you only care about side effects, like plotting or saving a file, because it's difficult to save the output efficiently.
- Loop over the names: `for (nm in names(xs))`. This gives you a name, which you can use to access the value with `x[[nm]]`. This is useful if you want to use the name in a plot title or a filename.

If you're creating named output, make sure to name the results vector like so:

```
results <- vector("list", length(x))
names(results) <- names(x)
```

Iteration over the numeric indices is the most general form, because given the position you can extract both the name and the value:

```
for (i in seq_along(x)) {
  name <- names(x)[[i]]
  value <- x[[i]]
}
```

Unknown Output Length

Sometimes you might not know how long the output will be. For example, imagine you want to simulate some random vectors of random lengths. You might be tempted to solve this problem by progressively growing the vector:

```
means <- c(0, 1, 2)

output <- double()
for (i in seq_along(means)) {
  n <- sample(100, 1)
  output <- c(output, rnorm(n, means[[i]]))
}
str(output)
#>  num [1:202] 0.912 0.205 2.584 -0.789 0.588 ...
```

But this is not very efficient because in each iteration, R has to copy all the data from the previous iterations. In technical terms you get "quadratic" ($O(n^2)$) behavior, which means that a loop with three times as many elements would take nine (3^2) times as long to run.

A better solution is to save the results in a list, and then combine into a single vector after the loop is done:

```
out <- vector("list", length(means))
for (i in seq_along(means)) {
  n <- sample(100, 1)
  out[[i]] <- rnorm(n, means[[i]])
}
str(out)
#> List of 3
#>  $ : num [1:83] 0.367 1.13 -0.941 0.218 1.415 ...
#>  $ : num [1:21] -0.485 -0.425 2.937 1.688 1.324 ...
#>  $ : num [1:40] 2.34 1.59 2.93 3.84 1.3 ...
```

```
str(unlist(out))
#>  num [1:144] 0.367 1.13 -0.941 0.218 1.415 ...
```

Here I've used `unlist()` to flatten a list of vectors into a single vector. A stricter option is to use `purrr::flatten_dbl()`—it will throw an error if the input isn't a list of doubles.

This pattern occurs in other places too:

- You might be generating a long string. Instead of `paste()`ing together each iteration with the previous, save the output in a character vector and then combine that vector into a single string with `paste(output, collapse = "")`.
- You might be generating a big data frame. Instead of sequentially `rbind()`ing in each iteration, save the output in a list, then use `dplyr::bind_rows(output)` to combine the output into a single data frame.

Watch out for this pattern. Whenever you see it, switch to a more complex result object, and then combine in one step at the end.

Unknown Sequence Length

Sometimes you don't even know how long the input sequence should be. This is common when doing simulations. For example, you might want to loop until you get three heads in a row. You can't do that sort of iteration with the for loop. Instead, you can use a while loop. A while loop is simpler than a for loop because it only has two components, a condition and a body:

```
while (condition) {
  # body
}
```

A while loop is also more general than a for loop, because you can rewrite any for loop as a while loop, but you can't rewrite every while loop as a for loop:

```
for (i in seq_along(x)) {
  # body
}

# Equivalent to
i <- 1
while (i <= length(x)) {
  # body
```

```
  i <- i + 1
}
```

Here's how we could use a while loop to find how many tries it takes to get three heads in a row:

```
flip <- function() sample(c("T", "H"), 1)

flips <- 0
nheads <- 0

while (nheads < 3) {
  if (flip() == "H") {
    nheads <- nheads + 1
  } else {
    nheads <- 0
  }
  flips <- flips + 1
}
flips
#> [1] 3
```

I mention while loops only briefly, because I hardly ever use them. They're most often used for simulation, which is outside the scope of this book. However, it is good to know they exist so that you're prepared for problems where the number of iterations is not known in advance.

Exercises

1. Imagine you have a directory full of CSV files that you want to read in. You have their paths in a vector, `files <- dir("data/", pattern = "\\.csv$", full.names = TRUE)`, and now want to read each one with `read_csv()`. Write the for loop that will load them into a single data frame.

2. What happens if you use `for (nm in names(x))` and `x` has no names? What if only some of the elements are named? What if the names are not unique?

3. Write a function that prints the mean of each numeric column in a data frame, along with its name. For example, `show_mean(iris)` would print:

   ```
   show_mean(iris)
   #> Sepal.Length: 5.84
   #> Sepal.Width:  3.06
   #> Petal.Length: 3.76
   #> Petal.Width:  1.20
   ```

(Extra challenge: what function did I use to make sure that the numbers lined up nicely, even though the variable names had different lengths?)

4. What does this code do? How does it work?

```
trans <- list(
  disp = function(x) x * 0.0163871,
  am = function(x) {
    factor(x, labels = c("auto", "manual"))
  }
)
for (var in names(trans)) {
  mtcars[[var]] <- trans[[var]](mtcars[[var]])
}
```

For Loops Versus Functionals

For loops are not as important in R as they are in other languages because R is a functional programming language. This means that it's possible to wrap up for loops in a function, and call that function instead of using the for loop directly.

To see why this is important, consider (again) this simple data frame:

```
df <- tibble(
  a = rnorm(10),
  b = rnorm(10),
  c = rnorm(10),
  d = rnorm(10)
)
```

Imagine you want to compute the mean of every column. You could do that with a for loop:

```
output <- vector("double", length(df))
for (i in seq_along(df)) {
  output[[i]] <- mean(df[[i]])
}
output
#> [1]  0.2026 -0.2068  0.1275 -0.0917
```

You realize that you're going to want to compute the means of every column pretty frequently, so you extract it out into a function:

```
col_mean <- function(df) {
  output <- vector("double", length(df))
  for (i in seq_along(df)) {
```

```
    output[i] <- mean(df[[i]])
  }
  output
}
```

But then you think it'd also be helpful to be able to compute the median, and the standard deviation, so you copy and paste your `col_mean()` function and replace the `mean()` with `median()` and `sd()`:

```
col_median <- function(df) {
  output <- vector("double", length(df))
  for (i in seq_along(df)) {
    output[i] <- median(df[[i]])
  }
  output
}
col_sd <- function(df) {
  output <- vector("double", length(df))
  for (i in seq_along(df)) {
    output[i] <- sd(df[[i]])
  }
  output
}
```

Uh oh! You've copied and pasted this code twice, so it's time to think about how to generalize it. Notice that most of this code is for-loop boilerplate and it's hard to see the one thing (`mean()`, `median()`, `sd()`) that is different between the functions.

What would you do if you saw a set of functions like this?

```
f1 <- function(x) abs(x - mean(x)) ^ 1
f2 <- function(x) abs(x - mean(x)) ^ 2
f3 <- function(x) abs(x - mean(x)) ^ 3
```

Hopefully, you'd notice that there's a lot of duplication, and extract it out into an additional argument:

```
f <- function(x, i) abs(x - mean(x)) ^ i
```

You've reduced the chance of bugs (because you now have 1/3 less code), and made it easy to generalize to new situations.

We can do exactly the same thing with `col_mean()`, `col_median()`, and `col_sd()` by adding an argument that supplies the function to apply to each column:

```
col_summary <- function(df, fun) {
  out <- vector("double", length(df))
  for (i in seq_along(df)) {
```

```
    out[i] <- fun(df[[i]])
  }
  out
}
col_summary(df, median)
#> [1]  0.237 -0.218  0.254 -0.133
col_summary(df, mean)
#> [1]  0.2026 -0.2068  0.1275 -0.0917
```

The idea of passing a function to another function is an extremely powerful idea, and it's one of the behaviors that makes R a functional programming language. It might take you a while to wrap your head around the idea, but it's worth the investment. In the rest of the chapter, you'll learn about and use the **purrr** package, which provides functions that eliminate the need for many common for loops. The apply family of functions in base R (`apply()`, `lapply()`, `tapply()`, etc.) solve a similar problem, but **purrr** is more consistent and thus is easier to learn.

The goal of using **purrr** functions instead of for loops is to allow you to break common list manipulation challenges into independent pieces:

- How can you solve the problem for a single element of the list? Once you've solved that problem, **purrr** takes care of generalizing your solution to every element in the list.
- If you're solving a complex problem, how can you break it down into bite-sized pieces that allow you to advance one small step toward a solution? With **purrr**, you get lots of small pieces that you can compose together with the pipe.

This structure makes it easier to solve new problems. It also makes it easier to understand your solutions to old problems when you reread your old code.

Exercises

1. Read the documentation for `apply()`. In the second case, what two for loops does it generalize?
2. Adapt `col_summary()` so that it only applies to numeric columns. You might want to start with an `is_numeric()` function that returns a logical vector that has a `TRUE` corresponding to each numeric column.

The Map Functions

The pattern of looping over a vector, doing something to each element, and saving the results is so common that the **purrr** package provides a family of functions to do it for you. There is one function for each type of output:

- `map()` makes a list.
- `map_lgl()` makes a logical vector.
- `map_int()` makes an integer vector.
- `map_dbl()` makes a double vector.
- `map_chr()` makes a character vector.

Each function takes a vector as input, applies a function to each piece, and then returns a new vector that's the same length (and has the same names) as the input. The type of the vector is determined by the suffix to the map function.

Once you master these functions, you'll find it takes much less time to solve iteration problems. But you should never feel bad about using a for loop instead of a map function. The map functions are a step up a tower of abstraction, and it can take a long time to get your head around how they work. The important thing is that you solve the problem that you're working on, not write the most concise and elegant code (although that's definitely something you want to strive toward!).

Some people will tell you to avoid for loops because they are slow. They're wrong! (Well at least they're rather out of date, as for loops haven't been slow for many years). The chief benefit of using functions like `map()` is not speed, but clarity: they make your code easier to write and to read.

We can use these functions to perform the same computations as the last for loop. Those summary functions returned doubles, so we need to use `map_dbl()`:

```
map_dbl(df, mean)
#>       a       b       c       d
#>  0.2026 -0.2068  0.1275 -0.0917
map_dbl(df, median)
#>      a      b      c      d
#>  0.237 -0.218  0.254 -0.133
```

```
map_dbl(df, sd)
#>     a     b     c     d
#> 0.796 0.759 1.164 1.062
```

Compared to using a for loop, focus is on the operation being performed (i.e., `mean()`, `median()`, `sd()`), not the bookkeeping required to loop over every element and store the output. This is even more apparent if we use the pipe:

```
df %>% map_dbl(mean)
#>       a       b       c       d
#>  0.2026 -0.2068  0.1275 -0.0917
df %>% map_dbl(median)
#>      a      b      c      d
#>  0.237 -0.218  0.254 -0.133
df %>% map_dbl(sd)
#>     a     b     c     d
#> 0.796 0.759 1.164 1.062
```

There are a few differences between `map_*()` and `col_summary()`:

- All **purrr** functions are implemented in C. This makes them a little faster at the expense of readability.
- The second argument, `.f`, the function to apply, can be a formula, a character vector, or an integer vector. You'll learn about those handy shortcuts in the next section.
- `map_*()` uses ... ("Dot-Dot-Dot (...)" on page 284) to pass along additional arguments to `.f` each time it's called:

  ```
  map_dbl(df, mean, trim = 0.5)
  #>      a      b      c      d
  #>  0.237 -0.218  0.254 -0.133
  ```

- The map functions also preserve names:

  ```
  z <- list(x = 1:3, y = 4:5)
  map_int(z, length)
  #> x y
  #> 3 2
  ```

Shortcuts

There are a few shortcuts that you can use with `.f` in order to save a little typing. Imagine you want to fit a linear model to each group in a dataset. The following toy example splits up the `mtcars` dataset into three pieces (one for each value of cylinder) and fits the same linear model to each piece:

```
models <- mtcars %>%
  split(.$cyl) %>%
  map(function(df) lm(mpg ~ wt, data = df))
```

The syntax for creating an anonymous function in R is quite verbose so **purrr** provides a convenient shortcut—a one-sided formula:

```
models <- mtcars %>%
  split(.$cyl) %>%
  map(~lm(mpg ~ wt, data = .))
```

Here I've used `.` as a pronoun: it refers to the current list element (in the same way that `i` referred to the current index in the for loop).

When you're looking at many models, you might want to extract a summary statistic like the R^2. To do that we need to first run `summary()` and then extract the component called `r.squared`. We could do that using the shorthand for anonymous functions:

```
models %>%
  map(summary) %>%
  map_dbl(~.$r.squared)
#>     4     6     8
#> 0.509 0.465 0.423
```

But extracting named components is a common operation, so **purrr** provides an even shorter shortcut: you can use a string.

```
models %>%
  map(summary) %>%
  map_dbl("r.squared")
#>     4     6     8
#> 0.509 0.465 0.423
```

You can also use an integer to select elements by position:

```
x <- list(list(1, 2, 3), list(4, 5, 6), list(7, 8, 9))
x %>% map_dbl(2)
#> [1] 2 5 8
```

Base R

If you're familiar with the apply family of functions in base R, you might have noticed some similarities with the **purrr** functions:

- `lapply()` is basically identical to `map()`, except that `map()` is consistent with all the other functions in **purrr**, and you can use the shortcuts for `.f`.

- Base `sapply()` is a wrapper around `lapply()` that automatically simplifies the output. This is useful for interactive work but is problematic in a function because you never know what sort of output you'll get:

```
x1 <- list(
  c(0.27, 0.37, 0.57, 0.91, 0.20),
  c(0.90, 0.94, 0.66, 0.63, 0.06),
  c(0.21, 0.18, 0.69, 0.38, 0.77)
)
x2 <- list(
  c(0.50, 0.72, 0.99, 0.38, 0.78),
  c(0.93, 0.21, 0.65, 0.13, 0.27),
  c(0.39, 0.01, 0.38, 0.87, 0.34)
)

threshold <- function(x, cutoff = 0.8) x[x > cutoff]
x1 %>% sapply(threshold) %>% str()
#> List of 3
#>  $ : num 0.91
#>  $ : num [1:2] 0.9 0.94
#>  $ : num(0)
x2 %>% sapply(threshold) %>% str()
#>  num [1:3] 0.99 0.93 0.87
```

- `vapply()` is a safe alternative to `sapply()` because you supply an additional argument that defines the type. The only problem with `vapply()` is that it's a lot of typing: `vapply(df, is.numeric, logical(1))` is equivalent to `map_lgl(df, is.numeric)`. One advantage of `vapply()` over **purrr**'s map functions is that it can also produce matrices—the map functions only ever produce vectors.

I focus on **purrr** functions here because they have more consistent names and arguments, helpful shortcuts, and in the future will provide easy parallelism and progress bars.

Exercises

1. Write code that uses one of the map functions to:

   a. Compute the mean of every column in `mtcars`.

   b. Determine the type of each column in `nycflights13::flights`.

c. Compute the number of unique values in each column of `iris`.

d. Generate 10 random normals for each of $\mu = -10$, 0, 10, and 100.

2. How can you create a single vector that for each column in a data frame indicates whether or not it's a factor?

3. What happens when you use the map functions on vectors that aren't lists? What does `map(1:5, runif)` do? Why?

4. What does `map(-2:2, rnorm, n = 5)` do? Why? What does `map_dbl(-2:2, rnorm, n = 5)` do? Why?

5. Rewrite `map(x, function(df) lm(mpg ~ wt, data = df))` to eliminate the anonymous function.

Dealing with Failure

When you use the map functions to repeat many operations, the chances are much higher that one of those operations will fail. When this happens, you'll get an error message, and no output. This is annoying: why does one failure prevent you from accessing all the other successes? How do you ensure that one bad apple doesn't ruin the whole barrel?

In this section you'll learn how to deal with this situation with a new function: `safely()`. `safely()` is an adverb: it takes a function (a verb) and returns a modified version. In this case, the modified function will never throw an error. Instead, it always returns a list with two elements:

`result`
: The original result. If there was an error, this will be `NULL`.

`error`
: An error object. If the operation was successful, this will be `NULL`.

(You might be familiar with the `try()` function in base R. It's similar, but because it sometimes returns the original result and it sometimes returns an error object it's more difficult to work with.)

Let's illustrate this with a simple example, `log()`:

```
safe_log <- safely(log)
str(safe_log(10))
#> List of 2
#>  $ result: num 2.3
#>  $ error : NULL
str(safe_log("a"))
#> List of 2
#>  $ result: NULL
#>  $ error :List of 2
#>   ..$ message: chr "non-numeric argument to mathematical ..."
#>   ..$ call   : language .f(...)
#>   ..- attr(*, "class")= chr [1:3] "simpleError" "error" ...
```

When the function succeeds the `result` element contains the result and the `error` element is `NULL`. When the function fails, the `result` element is `NULL` and the `error` element contains an error object.

`safely()` is designed to work with `map`:

```
x <- list(1, 10, "a")
y <- x %>% map(safely(log))
str(y)
#> List of 3
#>  $ :List of 2
#>   ..$ result: num 0
#>   ..$ error : NULL
#>  $ :List of 2
#>   ..$ result: num 2.3
#>   ..$ error : NULL
#>  $ :List of 2
#>   ..$ result: NULL
#>   ..$ error :List of 2
#>   .. ..$ message: chr "non-numeric argument to ..."
#>   .. ..$ call   : language .f(...)
#>   .. ..- attr(*, "class")=chr [1:3] "simpleError" "error" ...
```

This would be easier to work with if we had two lists: one of all the errors and one of all the output. That's easy to get with `purrr::transpose()`:

```
y <- y %>% transpose()
str(y)
#> List of 2
#>  $ result:List of 3
#>   ..$ : num 0
#>   ..$ : num 2.3
#>   ..$ : NULL
#>  $ error :List of 3
#>   ..$ : NULL
```

```
#>    ..$ : NULL
#>    ..$ :List of 2
#>    .. ..$ message: chr "non-numeric argument to ..."
#>    .. ..$ call   : language .f(...)
#>    .. ..- attr(*, "class")=chr [1:3] "simpleError" "error" ...
```

It's up to you how to deal with the errors, but typically you'll either look at the values of x where y is an error, or work with the values of y that are OK:

```
is_ok <- y$error %>% map_lgl(is_null)
x[!is_ok]
#> [[1]]
#> [1] "a"
y$result[is_ok] %>% flatten_dbl()
#> [1] 0.0 2.3
```

purrr provides two other useful adverbs:

- Like `safely()`, `possibly()` always succeeds. It's simpler than `safely()`, because you give it a default value to return when there is an error:

  ```
  x <- list(1, 10, "a")
  x %>% map_dbl(possibly(log, NA_real_))
  #> [1] 0.0 2.3  NA
  ```

- `quietly()` performs a similar role to `safely()`, but instead of capturing errors, it captures printed output, messages, and warnings:

  ```
  x <- list(1, -1)
  x %>% map(quietly(log)) %>% str()
  #> List of 2
  #>  $ :List of 4
  #>   ..$ result  : num 0
  #>   ..$ output  : chr ""
  #>   ..$ warnings: chr(0)
  #>   ..$ messages: chr(0)
  #>  $ :List of 4
  #>   ..$ result  : num NaN
  #>   ..$ output  : chr ""
  #>   ..$ warnings: chr "NaNs produced"
  #>   ..$ messages: chr(0)
  ```

Mapping over Multiple Arguments

So far we've mapped along a single input. But often you have multiple related inputs that you need to iterate along in parallel. That's the job of the `map2()` and `pmap()` functions. For example, imagine you want to simulate some random normals with different means. You know how to do that with `map()`:

```
mu <- list(5, 10, -3)
mu %>%
  map(rnorm, n = 5) %>%
  str()
#> List of 3
#>  $ : num [1:5] 5.45 5.5 5.78 6.51 3.18
#>  $ : num [1:5] 10.79 9.03 10.89 10.76 10.65
#>  $ : num [1:5] -3.54 -3.08 -5.01 -3.51 -2.9
```

What if you also want to vary the standard deviation? One way to do that would be to iterate over the indices and index into vectors of means and sds:

```
sigma <- list(1, 5, 10)
seq_along(mu) %>%
  map(~rnorm(5, mu[[.]], sigma[[.]])) %>%
  str()
#> List of 3
#>  $ : num [1:5] 4.94 2.57 4.37 4.12 5.29
#>  $ : num [1:5] 11.72 5.32 11.46 10.24 12.22
#>  $ : num [1:5] 3.68 -6.12 22.24 -7.2 10.37
```

But that obfuscates the intent of the code. Instead we could use `map2()`, which iterates over two vectors in parallel:

```
map2(mu, sigma, rnorm, n = 5) %>% str()
#> List of 3
#>  $ : num [1:5] 4.78 5.59 4.93 4.3 4.47
#>  $ : num [1:5] 10.85 10.57 6.02 8.82 15.93
#>  $ : num [1:5] -1.12 7.39 -7.5 -10.09 -2.7
```

`map2()` generates this series of function calls:

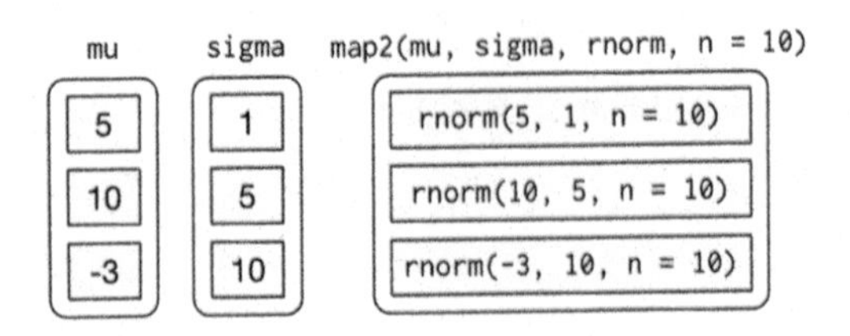

Note that the arguments that vary for each call come *before* the function; arguments that are the same for every call come *after*.

Like map(), map2() is just a wrapper around a for loop:

```
map2 <- function(x, y, f, ...) {
  out <- vector("list", length(x))
  for (i in seq_along(x)) {
    out[[i]] <- f(x[[i]], y[[i]], ...)
  }
  out
}
```

You could also imagine map3(), map4(), map5(), map6(), etc., but that would get tedious quickly. Instead, **purrr** provides pmap(), which takes a list of arguments. You might use that if you wanted to vary the mean, standard deviation, and number of samples:

```
n <- list(1, 3, 5)
args1 <- list(n, mu, sigma)
args1 %>%
  pmap(rnorm) %>%
  str()
#> List of 3
#>  $ : num 4.55
#>  $ : num [1:3] 13.4 18.8 13.2
#>  $ : num [1:5] 0.685 10.801 -11.671 21.363 -2.562
```

That looks like:

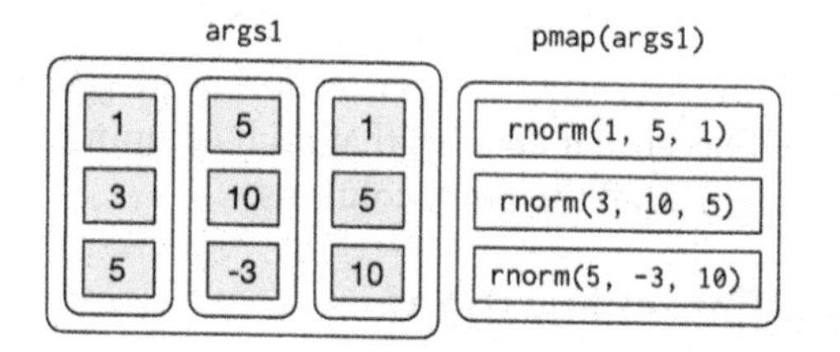

If you don't name the elements of list, pmap() will use positional matching when calling the function. That's a little fragile, and makes the code harder to read, so it's better to name the arguments:

```
args2 <- list(mean = mu, sd = sigma, n = n)
args2 %>%
  pmap(rnorm) %>%
  str()
```

That generates longer, but safer, calls:

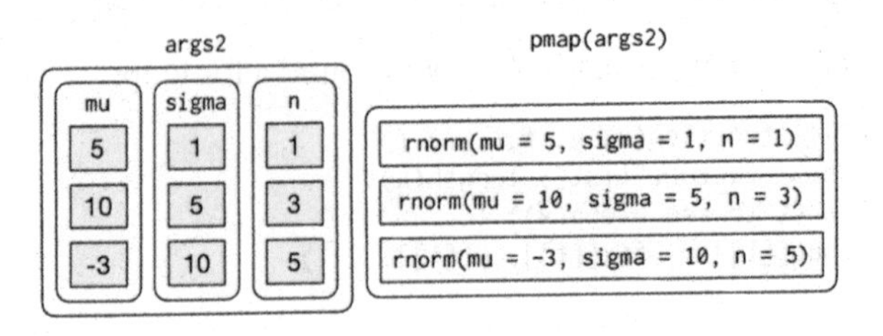

Since the arguments are all the same length, it makes sense to store them in a data frame:

```
params <- tribble(
  ~mean, ~sd, ~n,
    5,     1,  1,
   10,     5,  3,
   -3,    10,  5
)
params %>%
  pmap(rnorm)
#> [[1]]
#> [1] 4.68
#>
#> [[2]]
#> [1] 23.44 12.85  7.28
#>
#> [[3]]
#> [1]  -5.34 -17.66   0.92   6.06   9.02
```

As soon as your code gets complicated, I think a data frame is a good approach because it ensures that each column has a name and is the same length as all the other columns.

Invoking Different Functions

There's one more step up in complexity—as well as varying the arguments to the function you might also vary the function itself:

```
f <- c("runif", "rnorm", "rpois")
param <- list(
  list(min = -1, max = 1),
  list(sd = 5),
  list(lambda = 10)
)
```

To handle this case, you can use `invoke_map()`:

```
invoke_map(f, param, n = 5) %>% str()
#> List of 3
#>  $ : num [1:5] 0.762 0.36 -0.714 0.531 0.254
#>  $ : num [1:5] 3.07 -3.09 1.1 5.64 9.07
#>  $ : int [1:5] 9 14 8 9 7
```

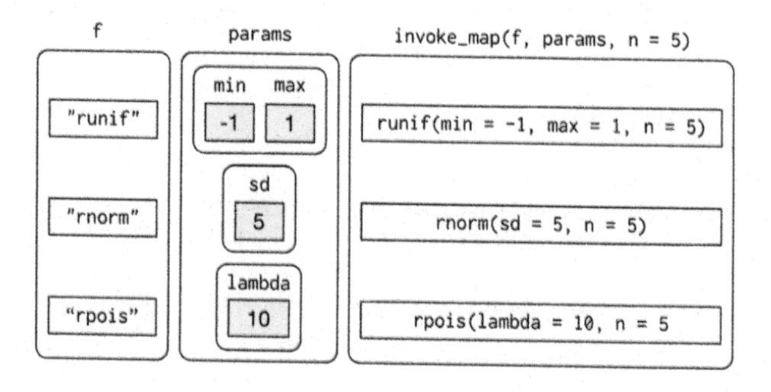

The first argument is a list of functions or a character vector of function names. The second argument is a list of lists giving the arguments that vary for each function. The subsequent arguments are passed on to every function.

And again, you can use `tribble()` to make creating these matching pairs a little easier:

```
sim <- tribble(
  ~f,      ~params,
  "runif", list(min = -1, max = 1),
  "rnorm", list(sd = 5),
  "rpois", list(lambda = 10)
)
sim %>%
  mutate(sim = invoke_map(f, params, n = 10))
```

Walk

Walk is an alternative to map that you use when you want to call a function for its side effects, rather than for its return value. You typically do this because you want to render output to the screen or save files to disk—the important thing is the action, not the return value. Here's a very simple example:

```
x <- list(1, "a", 3)

x %>%
```

```
  walk(print)
#> [1] 1
#> [1] "a"
#> [1] 3
```

`walk()` is generally not that useful compared to `walk2()` or `pwalk()`. For example, if you had a list of plots and a vector of filenames, you could use `pwalk()` to save each file to the corresponding location on disk:

```
library(ggplot2)
plots <- mtcars %>%
  split(.$cyl) %>%
  map(~ggplot(., aes(mpg, wt)) + geom_point())
paths <- stringr::str_c(names(plots), ".pdf")

pwalk(list(paths, plots), ggsave, path = tempdir())
```

`walk()`, `walk2()`, and `pwalk()` all invisibly return `.x`, the first argument. This makes them suitable for use in the middle of pipelines.

Other Patterns of For Loops

purrr provides a number of other functions that abstract over other types of for loops. You'll use them less frequently than the map functions, but they're useful to know about. The goal here is to briefly illustrate each function, so hopefully it will come to mind if you see a similar problem in the future. Then you can go look up the documentation for more details.

Predicate Functions

A number of functions work with *predicate* functions that return either a single `TRUE` or `FALSE`.

`keep()` and `discard()` keep elements of the input where the predicate is `TRUE` or `FALSE`, respectively:

```
iris %>%
  keep(is.factor) %>%
  str()
#> 'data.frame':    150 obs. of  1 variable:
#>  $ Species: Factor w/ 3 levels "setosa","versicolor",..: ...

iris %>%
  discard(is.factor) %>%
  str()
#> 'data.frame':    150 obs. of  4 variables:
```

```
#>  $ Sepal.Length: num  5.1 4.9 4.7 4.6 5 5.4 4.6 5 4.4 4.9 ...
#>  $ Sepal.Width : num  3.5 3 3.2 3.1 3.6 3.9 3.4 3.4 2.9 3 ...
#>  $ Petal.Length: num  1.4 1.4 1.3 1.5 1.4 1.7 1.4 1.5 1.4 ...
#>  $ Petal.Width : num  0.2 0.2 0.2 0.2 0.2 0.4 0.3 0.2 0.2 ...
```

`some()` and `every()` determine if the predicate is true for any or for all of the elements:

```
x <- list(1:5, letters, list(10))

x %>%
  some(is_character)
#> [1] TRUE

x %>%
  every(is_vector)
#> [1] TRUE
```

`detect()` finds the first element where the predicate is true; `detect_index()` returns its position:

```
x <- sample(10)
x
#>  [1]  8  7  5  6  9  2 10  1  3  4

x %>%
  detect(~ . > 5)
#> [1] 8

x %>%
  detect_index(~ . > 5)
#> [1] 1
```

`head_while()` and `tail_while()` take elements from the start or end of a vector while a predicate is true:

```
x %>%
  head_while(~ . > 5)
#> [1] 8 7

x %>%
  tail_while(~ . > 5)
#> integer(0)
```

Reduce and Accumulate

Sometimes you have a complex list that you want to reduce to a simple list by repeatedly applying a function that reduces a pair to a singleton. This is useful if you want to apply a two-table **dplyr** verb to multiple tables. For example, you might have a list of data frames, and you want to reduce to a single data frame by joining the elements together:

```
dfs <- list(
  age = tibble(name = "John", age = 30),
  sex = tibble(name = c("John", "Mary"), sex = c("M", "F")),
  trt = tibble(name = "Mary", treatment = "A")
)

dfs %>% reduce(full_join)
#> Joining, by = "name"
#> Joining, by = "name"
#> # A tibble: 2 × 4
#>    name   age   sex treatment
#>   <chr> <dbl> <chr>     <chr>
#> 1  John    30     M      <NA>
#> 2  Mary    NA     F         A
```

Or maybe you have a list of vectors, and want to find the intersection:

```
vs <- list(
  c(1, 3, 5, 6, 10),
  c(1, 2, 3, 7, 8, 10),
  c(1, 2, 3, 4, 8, 9, 10)
)

vs %>% reduce(intersect)
#> [1]  1  3 10
```

The `reduce` function takes a "binary" function (i.e., a function with two primary inputs), and applies it repeatedly to a list until there is only a single element left.

Accumulate is similar but it keeps all the interim results. You could use it to implement a cumulative sum:

```
x <- sample(10)
x
#>  [1]  6  9  8  5  2  4  7  1 10  3
x %>% accumulate(`+`)
#>  [1]  6 15 23 28 30 34 41 42 52 55
```

Exercises

1. Implement your own version of `every()` using a for loop. Compare it with `purrr::every()`. What does **purrr**'s version do that your version doesn't?

2. Create an enhanced `col_sum()` that applies a summary function to every numeric column in a data frame.

3. A possible base R equivalent of `col_sum()` is:

```
col_sum3 <- function(df, f) {
  is_num <- sapply(df, is.numeric)
  df_num <- df[, is_num]

  sapply(df_num, f)
}
```

But it has a number of bugs as illustrated with the following inputs:

```
df <- tibble(
  x = 1:3,
  y = 3:1,
  z = c("a", "b", "c")
)
# OK
col_sum3(df, mean)
# Has problems: don't always return numeric vector
col_sum3(df[1:2], mean)
col_sum3(df[1], mean)
col_sum3(df[0], mean)
```

What causes the bugs?

PART IV
Model

Now that you are equipped with powerful programming tools we can finally return to modeling. You'll use your new tools of data wrangling and programming to fit many models and understand how they work. The focus of this book is on exploration, not confirmation or formal inference. But you'll learn a few basic tools that help you understand the variation within your models.

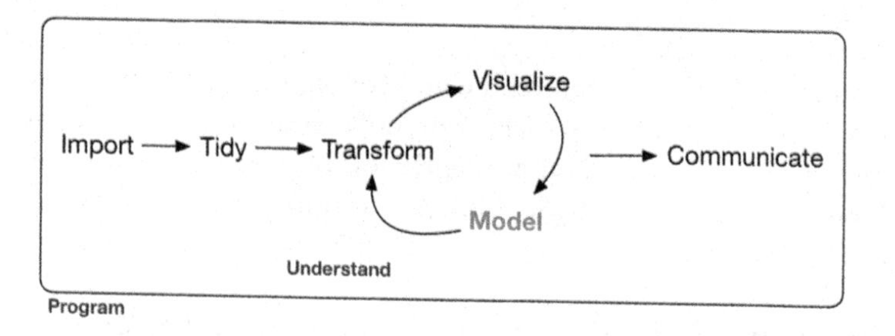

The goal of a model is to provide a simple low-dimensional summary of a dataset. Ideally, the model will capture true "signals" (i.e., patterns generated by the phenomenon of interest), and ignore "noise" (i.e., random variation that you're not interested in). Here we only cover "predictive" models, which, as the name suggests, generate predictions. There is another type of model that we're not going to discuss: "data discovery" models. These models don't make predictions, but instead help you discover interesting relationships within your data. (These two categories of models are sometimes called supervised and unsupervised, but I don't think that terminology is particularly illuminating.)

This book is not going to give you a deep understanding of the mathematical theory that underlies models. It will, however, build your intuition about how statistical models work, and give you a family of useful tools that allow you to use models to better understand your data:

- In Chapter 18, you'll learn how models work mechanistically, focusing on the important family of linear models. You'll learn general tools for gaining insight into what a predictive model tells you about your data, focusing on simple simulated datasets.
- In Chapter 19, you'll learn how to use models to pull out known patterns in real data. Once you have recognized an important pattern it's useful to make it explicit in a model, because then you can more easily see the subtler signals that remain.
- In Chapter 20, you'll learn how to use many simple models to help understand complex datasets. This is a powerful technique, but to access it you'll need to combine modeling and programming tools.

These topics are notable because of what they don't include: any tools for quantitatively assessing models. That is deliberate: precisely quantifying a model requires a couple of big ideas that we just don't have the space to cover here. For now, you'll rely on qualitative assessment and your natural skepticism. In "Learning More About Models" on page 396, we'll point you to other resources where you can learn more.

Hypothesis Generation Versus Hypothesis Confirmation

In this book, we are going to use models as a tool for exploration, completing the trifecta of the tools for EDA that were introduced in Part I. This is not how models are usually taught, but as you will see, models are an important tool for exploration. Traditionally, the focus of modeling is on inference, or for confirming that a hypothesis is true. Doing this correctly is not complicated, but it is hard.

There is a pair of ideas that you must understand in order to do inference correctly:

- Each observation can either be used for exploration or confirmation, not both.
- You can use an observation as many times as you like for exploration, but you can only use it once for confirmation. As soon as you use an observation twice, you've switched from confirmation to exploration.

This is necessary because to confirm a hypothesis you must use data independent of the data that you used to generate the hypothesis. Otherwise you will be overoptimistic. There is absolutely nothing wrong with exploration, but you should never sell an exploratory analysis as a confirmatory analysis because it is fundamentally misleading.

If you are serious about doing a confirmatory analysis, one approach is to split your data into three pieces before you begin the analysis:

- 60% of your data goes into a *training* (or exploration) set. You're allowed to do anything you like with this data: visualize it and fit tons of models to it.
- 20% goes into a *query* set. You can use this data to compare models or visualizations by hand, but you're not allowed to use it as part of an automated process.
- 20% is held back for a *test* set. You can only use this data ONCE, to test your final model.

This partitioning allows you to explore the training data, occasionally generating candidate hypotheses that you check with the query set. When you are confident you have the right model, you can check it once with the test data.

(Note that even when doing confirmatory modeling, you will still need to do EDA. If you don't do any EDA you will remain blind to the quality problems with your data.)

CHAPTER 18

Model Basics with modelr

Introduction

The goal of a model is to provide a simple low-dimensional summary of a dataset. In the context of this book we're going to use models to partition data into patterns and residuals. Strong patterns will hide subtler trends, so we'll use models to help peel back layers of structure as we explore a dataset.

However, before we can start using models on interesting, real datasets, you need to understand the basics of how models work. For that reason, this chapter of the book is unique because it uses only simulated datasets. These datasets are very simple, and not at all interesting, but they will help you understand the essence of modeling before you apply the same techniques to real data in the next chapter.

There are two parts to a model:

1. First, you define a *family of models* that express a precise, but generic, pattern that you want to capture. For example, the pattern might be a straight line, or a quadatric curve. You will express the model family as an equation like `y = a_1 * x + a_2` or `y = a_1 * x ^ a_2`. Here, `x` and `y` are known variables from your data, and `a_1` and `a_2` are parameters that can vary to capture different patterns.
2. Next, you generate a *fitted model* by finding the model from the family that is the closest to your data. This takes the generic

model family and makes it specific, like `y = 3 * x + 7` or `y = 9 * x ^ 2`.

It's important to understand that a fitted model is just the closest model from a family of models. That implies that you have the "best" model (according to some criteria); it doesn't imply that you have a good model and it certainly doesn't imply that the model is "true." George Box puts this well in his famous aphorism:

> All models are wrong, but some are useful.

It's worth reading the fuller context of the quote:

> Now it would be very remarkable if any system existing in the real world could be exactly represented by any simple model. However, cunningly chosen parsimonious models often do provide remarkably useful approximations. For example, the law PV = RT relating pressure P, volume V and temperature T of an "ideal" gas via a constant R is not exactly true for any real gas, but it frequently provides a useful approximation and furthermore its structure is informative since it springs from a physical view of the behavior of gas molecules.
>
> For such a model there is no need to ask the question "Is the model true?" If "truth" is to be the "whole truth" the answer must be "No." The only question of interest is "Is the model illuminating and useful?"

The goal of a model is not to uncover truth, but to discover a simple approximation that is still useful.

Prerequisites

In this chapter we'll use the **modelr** package, which wraps around base R's modeling functions to make them work naturally in a pipe.

```
library(tidyverse)

library(modelr)
options(na.action = na.warn)
```

A Simple Model

Let's take a look at the simulated dataset `sim1`. It contains two continuous variables, `x` and `y`. Let's plot them to see how they're related:

```
ggplot(sim1, aes(x, y)) +
  geom_point()
```

You can see a strong pattern in the data. Let's use a model to capture that pattern and make it explicit. It's our job to supply the basic form of the model. In this case, the relationship looks linear, i.e., `y = a_0 + a_1 * x`. Let's start by getting a feel for what models from that family look like by randomly generating a few and overlaying them on the data. For this simple case, we can use `geom_abline()`, which takes a slope and intercept as parameters. Later on we'll learn more general techniques that work with any model:

```
models <- tibble(
  a1 = runif(250, -20, 40),
  a2 = runif(250, -5, 5)
)

ggplot(sim1, aes(x, y)) +
  geom_abline(
    aes(intercept = a1, slope = a2),
    data = models, alpha = 1/4
  ) +
  geom_point()
```

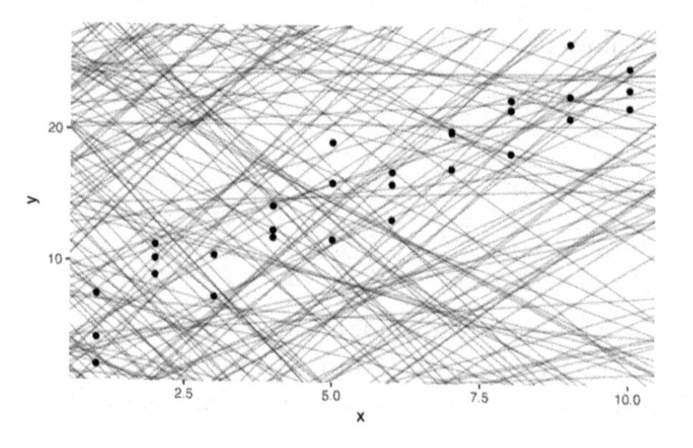

There are 250 models on this plot, but a lot are really bad! We need to find the good models by making precise our intuition that a good model is "close" to the data. We need a way to quantify the distance between the data and a model. Then we can fit the model by finding the values of `a_0` and `a_1` that generate the model with the smallest distance from this data.

One easy place to start is to find the vertical distance between each point and the model, as in the following diagram. (Note that I've shifted the x values slightly so you can see the individual distances.)

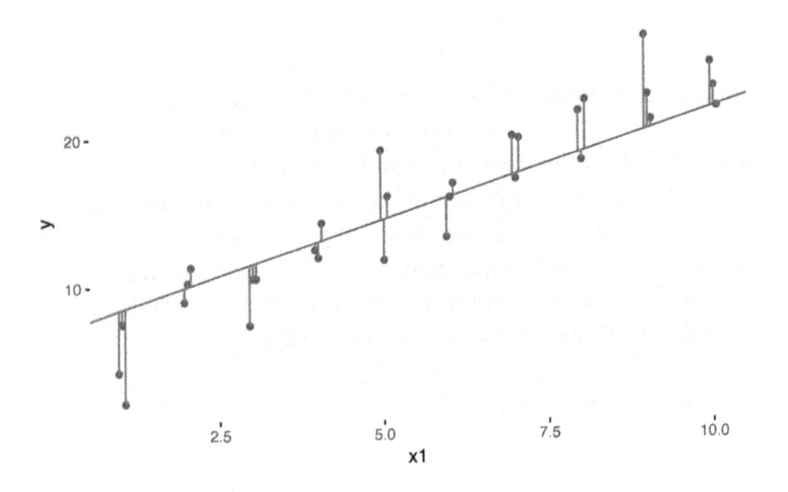

This distance is just the difference between the `y` value given by the model (the *prediction*), and the actual `y` value in the data (the *response*).

To compute this distance, we first turn our model family into an R function. This takes the model parameters and the data as inputs, and gives values predicted by the model as output:

```
model1 <- function(a, data) {
  a[1] + data$x * a[2]
}
model1(c(7, 1.5), sim1)
#>  [1]  8.5  8.5  8.5 10.0 10.0 10.0 11.5 11.5 11.5 13.0 13.0
#> [12] 13.0 14.5 14.5 14.5 16.0 16.0 16.0 17.5 17.5 17.5 19.0
#> [23] 19.0 19.0 20.5 20.5 20.5 22.0 22.0 22.0
```

Next, we need some way to compute an overall distance between the predicted and actual values. In other words, the plot shows 30 distances: how do we collapse that into a single number?

One common way to do this in statistics is to use the "root-mean-squared deviation." We compute the difference between actual and predicted, square them, average them, and then take the square root. This distance has lots of appealing mathematical properties, which we're not going to talk about here. You'll just have to take my word for it!

```
measure_distance <- function(mod, data) {
  diff <- data$y - model1(mod, data)
  sqrt(mean(diff ^ 2))
}
measure_distance(c(7, 1.5), sim1)
#> [1] 2.67
```

Now we can use **purrr** to compute the distance for all the models defined previously. We need a helper function because our distance function expects the model as a numeric vector of length 2:

```
sim1_dist <- function(a1, a2) {
  measure_distance(c(a1, a2), sim1)
}

models <- models %>%
  mutate(dist = purrr::map2_dbl(a1, a2, sim1_dist))
models
#> # A tibble: 250 × 3
#>        a1      a2  dist
#>     <dbl>   <dbl> <dbl>
#> 1 -15.15  0.0889  30.8
#> 2  30.06 -0.8274  13.2
#> 3  16.05  2.2695  13.2
#> 4 -10.57  1.3769  18.7
#> 5 -19.56 -1.0359  41.8
#> 6   7.98  4.5948  19.3
#> # ... with 244 more rows
```

Next, let's overlay the 10 best models on to the data. I've colored the models by `-dist`: this is an easy way to make sure that the best models (i.e., the ones with the smallest distance) get the brighest colors:

```
ggplot(sim1, aes(x, y)) +
  geom_point(size = 2, color = "grey30") +
  geom_abline(
    aes(intercept = a1, slope = a2, color = -dist),
    data = filter(models, rank(dist) <= 10)
  )
```

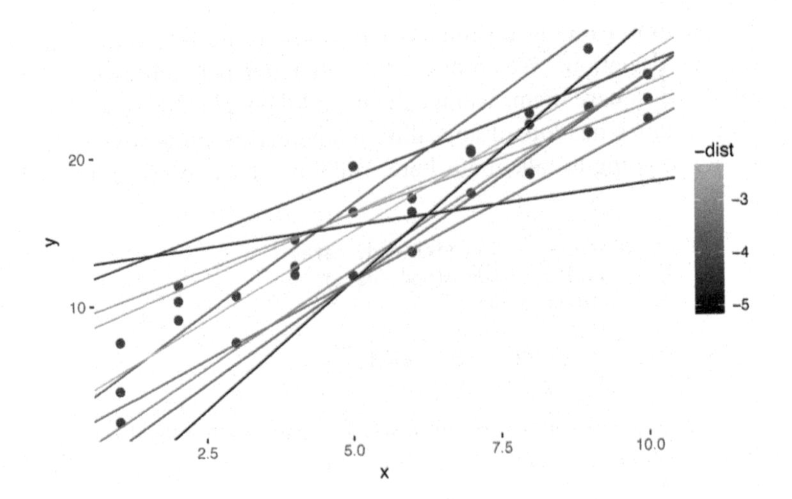

We can also think about these models as observations, and visualize them with a scatterplot of `a1` versus `a2`, again colored by `-dist`. We can no longer directly see how the model compares to the data, but we can see many models at once. Again, I've highlighted the 10 best models, this time by drawing red circles underneath them:

```
ggplot(models, aes(a1, a2)) +
  geom_point(
    data = filter(models, rank(dist) <= 10),
    size = 4, color = "red"
  ) +
  geom_point(aes(colour = -dist))
```

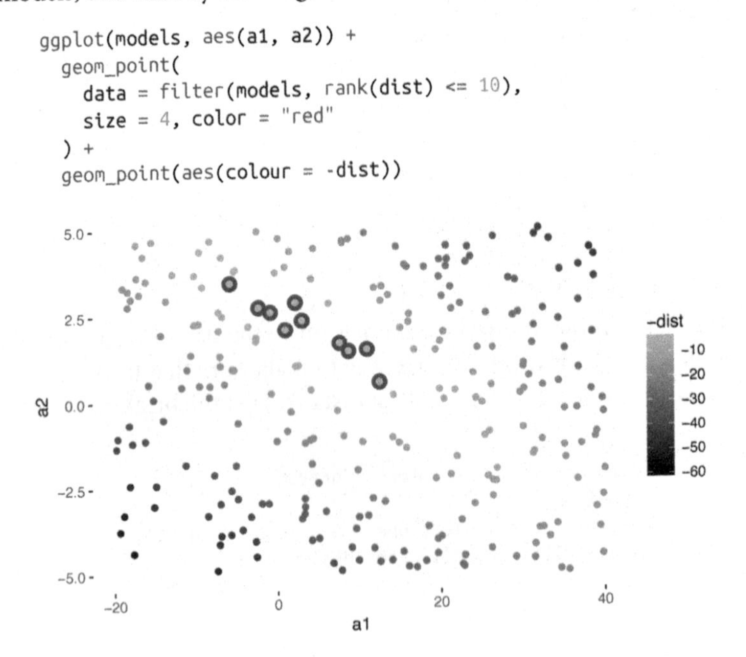

Instead of trying lots of random models, we could be more systematic and generate an evenly spaced grid of points (this is called a grid search). I picked the parameters of the grid roughly by looking at where the best models were in the preceding plot:

```
grid <- expand.grid(
  a1 = seq(-5, 20, length = 25),
  a2 = seq(1, 3, length = 25)
  ) %>%
  mutate(dist = purrr::map2_dbl(a1, a2, sim1_dist))

grid %>%
  ggplot(aes(a1, a2)) +
  geom_point(
    data = filter(grid, rank(dist) <= 10),
    size = 4, colour = "red"
  ) +
  geom_point(aes(color = -dist))
```

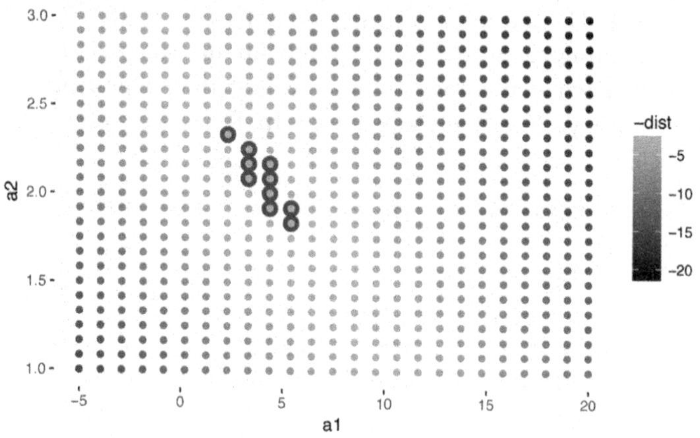

When you overlay the best 10 models back on the original data, they all look pretty good:

```
ggplot(sim1, aes(x, y)) +
  geom_point(size = 2, color = "grey30") +
  geom_abline(
    aes(intercept = a1, slope = a2, color = -dist),
    data = filter(grid, rank(dist) <= 10)
  )
```

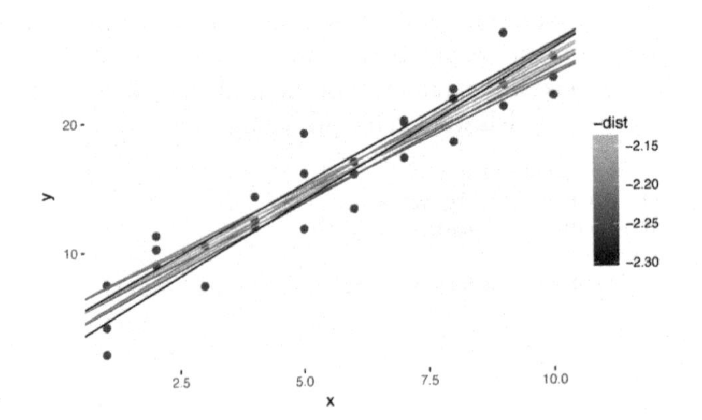

You could imagine iteratively making the grid finer and finer until you narrowed in on the best model. But there's a better way to tackle that problem: a numerical minimization tool called Newton–Raphson search. The intuition of Newton–Raphson is pretty simple: you pick a starting point and look around for the steepest slope. You then ski down that slope a little way, and then repeat again and again, until you can't go any lower. In R, we can do that with `optim()`:

```
best <- optim(c(0, 0), measure_distance, data = sim1)
best$par
#> [1] 4.22 2.05

ggplot(sim1, aes(x, y)) +
  geom_point(size = 2, color = "grey30") +
  geom_abline(intercept = best$par[1], slope = best$par[2])
```

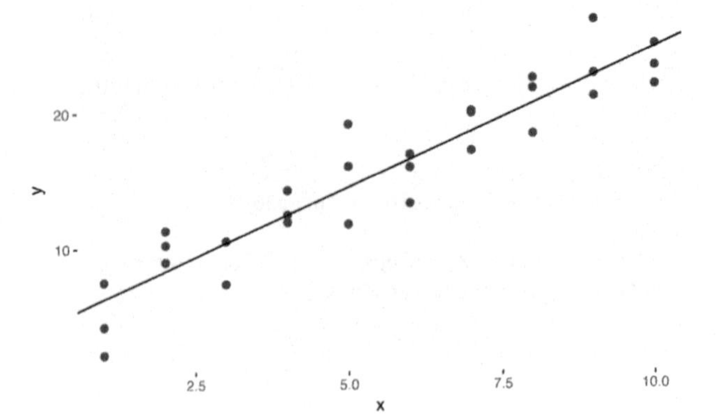

Don't worry too much about the details of how `optim()` works. It's the intuition that's important here. If you have a function that defines the distance between a model and a dataset, and an algorithm that can minimize that distance by modifying the parameters of the model, you can find the best model. The neat thing about this approach is that it will work for any family of models that you can write an equation for.

There's one more approach that we can use for this model, because it is a special case of a broader family: linear models. A linear model has the general form `y = a_1 + a_2 * x_1 + a_3 * x_2 + ... + a_n * x_(n - 1)`. So this simple model is equivalent to a general linear model where `n` is `2` and `x_1` is `x`. R has a tool specifically designed for fitting linear models called `lm()`. `lm()` has a special way to specify the model family: formulas. Formulas look like `y ~ x`, which `lm()` will translate to a function like `y = a_1 + a_2 * x`. We can fit the model and look at the output:

```
sim1_mod <- lm(y ~ x, data = sim1)
coef(sim1_mod)
#> (Intercept)           x
#>        4.22        2.05
```

These are exactly the same values we got with `optim()`! Behind the scenes `lm()` doesn't use `optim()` but instead takes advantage of the mathematical structure of linear models. Using some connections between geometry, calculus, and linear algebra, `lm()` actually finds the closest model in a single step, using a sophisticated algorithm. This approach is faster and guarantees that there is a global minimum.

Exercises

1. One downside of the linear model is that it is sensitive to unusual values because the distance incorporates a squared term. Fit a linear model to the following simulated data, and visualize the results. Rerun a few times to generate different simulated datasets. What do you notice about the model?

```
sim1a <- tibble(
  x = rep(1:10, each = 3),
  y = x * 1.5 + 6 + rt(length(x), df = 2)
)
```

2. One way to make linear models more robust is to use a different distance measure. For example, instead of root-mean-squared distance, you could use mean-absolute distance:

   ```
   measure_distance <- function(mod, data) {
     diff <- data$y - make_prediction(mod, data)
     mean(abs(diff))
   }
   ```

   Use optim() to fit this model to the previous simulated data and compare it to the linear model.

3. One challenge with performing numerical optimization is that it's only guaranteed to find one local optima. What's the problem with optimizing a three-parameter model like this?

   ```
   model1 <- function(a, data) {
     a[1] + data$x * a[2] + a[3]
   }
   ```

Visualizing Models

For simple models, like the one in the previous section, you can figure out what pattern the model captures by carefully studying the model family and the fitted coefficients. And if you ever take a statistics course on modeling, you're likely to spend a lot of time doing just that. Here, however, we're going to take a different tack. We're going to focus on understanding a model by looking at its predictions. This has a big advantage: every type of predictive model makes predictions (otherwise what use would it be?) so we can use the same set of techniques to understand any type of predictive model.

It's also useful to see what the model doesn't capture, the so-called residuals that are left after subtracting the predictions from the data. Residuals are powerful because they allow us to use models to remove striking patterns so we can study the subtler trends that remain.

Predictions

To visualize the predictions from a model, we start by generating an evenly spaced grid of values that covers the region where our data lies. The easiest way to do that is to use `modelr::data_grid()`. Its first argument is a data frame, and for each subsequent argument it finds the unique variables and then generates all combinations:

```
grid <- sim1 %>%
  data_grid(x)
grid
#> # A tibble: 10 × 1
#>       x
#>   <int>
#> 1     1
#> 2     2
#> 3     3
#> 4     4
#> 5     5
#> 6     6
#> # ... with 4 more rows
```

(This will get more interesting when we start to add more variables to our model.)

Next we add predictions. We'll use `modelr::add_predictions()`, which takes a data frame and a model. It adds the predictions from the model to a new column in the data frame:

```
grid <- grid %>%
  add_predictions(sim1_mod)
grid
#> # A tibble: 10 × 2
#>       x  pred
#>   <int> <dbl>
#> 1     1  6.27
#> 2     2  8.32
#> 3     3 10.38
#> 4     4 12.43
#> 5     5 14.48
#> 6     6 16.53
#> # ... with 4 more rows
```

(You can also use this function to add predictions to your original dataset.)

Next, we plot the predictions. You might wonder about all this extra work compared to just using `geom_abline()`. But the advantage of this approach is that it will work with *any* model in R, from the simplest to the most complex. You're only limited by your visualization skills. For more ideas about how to visualize more complex model types, you might try *http://vita.had.co.nz/papers/model-vis.html.*

```
ggplot(sim1, aes(x)) +
  geom_point(aes(y = y)) +
```

```
  geom_line(
    aes(y = pred),
    data = grid,
    colour = "red",
    size = 1
  )
```

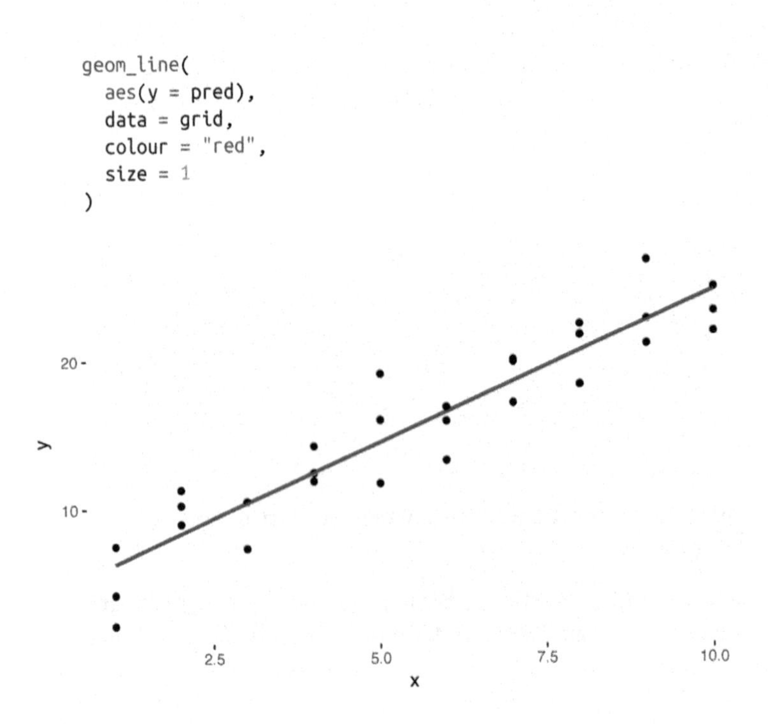

Residuals

The flip side of predictions are *residuals*. The predictions tell you the pattern that the model has captured, and the residuals tell you what the model has missed. The residuals are just the distances between the observed and predicted values that we computed earlier.

We add residuals to the data with `add_residuals()`, which works much like `add_predictions()`. Note, however, that we use the original dataset, not a manufactured grid. This is because to compute residuals we need actual y values:

```
sim1 <- sim1 %>%
  add_residuals(sim1_mod)
sim1
#> # A tibble: 30 × 3
#>         x     y  resid
#>     <int> <dbl>  <dbl>
#> 1       1  4.20 -2.072
#> 2       1  7.51  1.238
#> 3       1  2.13 -4.147
#> 4       2  8.99  0.665
#> 5       2 10.24  1.919
```

```
#> 6       2 11.30  2.973
#> # ... with 24 more rows
```

There are a few different ways to understand what the residuals tell us about the model. One way is to simply draw a frequency polygon to help us understand the spread of the residuals:

```
ggplot(sim1, aes(resid)) +
  geom_freqpoly(binwidth = 0.5)
```

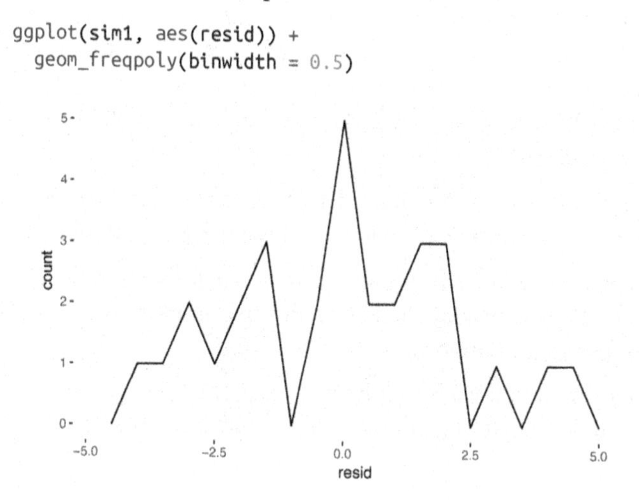

This helps you calibrate the quality of the model: how far away are the predictions from the observed values? Note that the average of the residual will always be 0.

You'll often want to re-create plots using the residuals instead of the original predictor. You'll see a lot of that in the next chapter:

```
ggplot(sim1, aes(x, resid)) +
  geom_ref_line(h = 0) +
  geom_point()
```

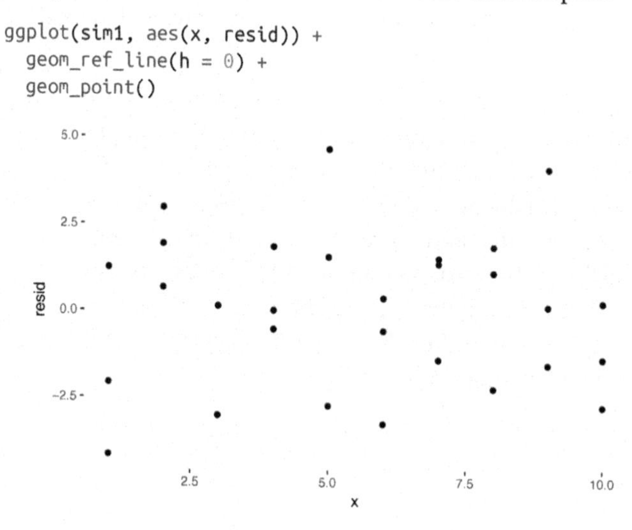

This looks like random noise, suggesting that our model has done a good job of capturing the patterns in the dataset.

Exercises

1. Instead of using `lm()` to fit a straight line, you can use `loess()` to fit a smooth curve. Repeat the process of model fitting, grid generation, predictions, and visualization on `sim1` using `loess()` instead of `lm()`. How does the result compare to `geom_smooth()`?
2. `add_predictions()` is paired with `gather_predictions()` and `spread_predictions()`. How do these three functions differ?
3. What does `geom_ref_line()` do? What package does it come from? Why is displaying a reference line in plots showing residuals useful and important?
4. Why might you want to look at a frequency polygon of absolute residuals? What are the pros and cons compared to looking at the raw residuals?

Formulas and Model Families

You've seen formulas before when using `facet_wrap()` and `facet_grid()`. In R, formulas provide a general way of getting "special behavior." Rather than evaluating the values of the variables right away, they capture them so they can be interpreted by the function.

The majority of modeling functions in R use a standard conversion from formulas to functions. You've seen one simple conversion already: `y ~ x` is translated to `y = a_1 + a_2 * x`. If you want to see what R actually does, you can use the `model_matrix()` function. It takes a data frame and a formula and returns a tibble that defines the model equation: each column in the output is associated with one coefficient in the model, and the function is always `y = a_1 * out1 + a_2 * out_2`. For the simplest case of `y ~ x1` this shows us something interesting:

```
df <- tribble(
  ~y, ~x1, ~x2,
  4, 2, 5,
  5, 1, 6
```

```
)
model_matrix(df, y ~ x1)
#> # A tibble: 2 × 2
#>   `(Intercept)`    x1
#>           <dbl> <dbl>
#> 1             1     2
#> 2             1     1
```

The way that R adds the intercept to the model is just by having a column that is full of ones. By default, R will always add this column. If you don't want that, you need to explicitly drop it with `-1`:

```
model_matrix(df, y ~ x1 - 1)
#> # A tibble: 2 × 1
#>      x1
#>   <dbl>
#> 1     2
#> 2     1
```

The model matrix grows in an unsurprising way when you add more variables to the model:

```
model_matrix(df, y ~ x1 + x2)
#> # A tibble: 2 × 3
#>   `(Intercept)`    x1    x2
#>           <dbl> <dbl> <dbl>
#> 1             1     2     5
#> 2             1     1     6
```

This formula notation is sometimes called "Wilkinson-Rogers notation," and was initially described in *Symbolic Description of Factorial Models for Analysis of Variance* (*http://bit.ly/wilkrog*), by G. N. Wilkinson and C. E. Rogers. It's worth digging up and reading the original paper if you'd like to understand the full details of the modeling algebra.

The following sections expand on how this formula notation works for categorcal variables, interactions, and transformation.

Categorical Variables

Generating a function from a formula is straightforward when the predictor is continuous, but things get a bit more complicated when the predictor is categorical. Imagine you have a formula like `y ~ sex`, where `sex` could either be male or female. It doesn't make sense to convert that to a formula like `y = x_0 + x_1 * sex` because `sex` isn't a number—you can't multiply it! Instead what R does is convert it to y = x_0 + x_1 * sex_male where sex_male is one if sex is male and zero otherwise:

```
df <- tribble(
  ~ sex, ~ response,
  "male", 1,
  "female", 2,
  "male", 1
)
model_matrix(df, response ~ sex)
#> # A tibble: 3 × 2
#>   `(Intercept)` sexmale
#>           <dbl>   <dbl>
#> 1             1       1
#> 2             1       0
#> 3             1       1
```

You might wonder why R also doesn't create a sexfemale column. The problem is that would create a column that is perfectly predictable based on the other columns (i.e., sexfemale = 1 - sexmale). Unfortunately the exact details of why this is a problem is beyond the scope of this book, but basically it creates a model family that is too flexible, and will have infinitely many models that are equally close to the data.

Fortunately, however, if you focus on visualizing predictions you don't need to worry about the exact parameterization. Let's look at some data and models to make that concrete. Here's the sim2 dataset from **modelr**:

```
ggplot(sim2) +
  geom_point(aes(x, y))
```

We can fit a model to it, and generate predictions:

```
mod2 <- lm(y ~ x, data = sim2)

grid <- sim2 %>%
  data_grid(x) %>%
  add_predictions(mod2)
grid
#> # A tibble: 4 × 2
#>       x  pred
#>   <chr> <dbl>
#> 1     a  1.15
#> 2     b  8.12
#> 3     c  6.13
#> 4     d  1.91
```

Effectively, a model with a categorical x will predict the mean value for each category. (Why? Because the mean minimizes the root-mean-squared distance.) That's easy to see if we overlay the predictions on top of the original data:

```
ggplot(sim2, aes(x)) +
  geom_point(aes(y = y)) +
  geom_point(
    data = grid,
    aes(y = pred),
    color = "red",
    size = 4
  )
```

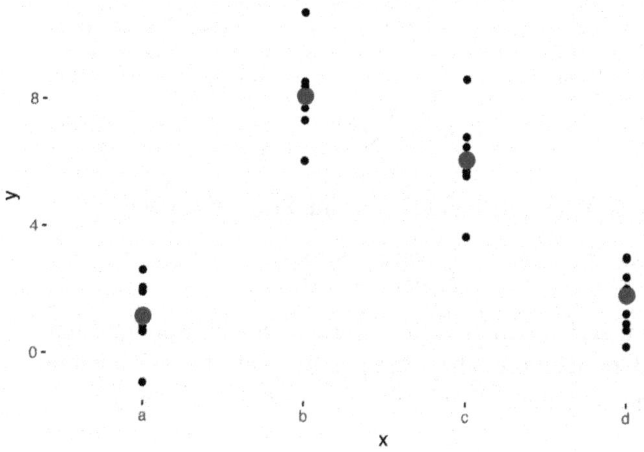

You can't make predictions about levels that you didn't observe. Sometimes you'll do this by accident so it's good to recognize this error message:

```
tibble(x = "e") %>%
  add_predictions(mod2)
#> Error in model.frame.default(Terms, newdata, na.action =
#> na.action, xlev = object$xlevels): factor x has new level e
```

Interactions (Continuous and Categorical)

What happens when you combine a continuous and a categorical variable? `sim3` contains a categorical predictor and a continuous predictor. We can visualize it with a simple plot:

```
ggplot(sim3, aes(x1, y)) +
  geom_point(aes(color = x2))
```

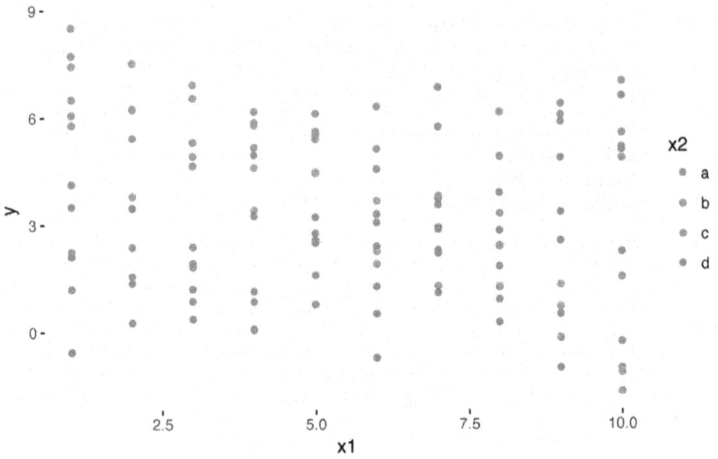

There are two possible models you could fit to this data:

```
mod1 <- lm(y ~ x1 + x2, data = sim3)
mod2 <- lm(y ~ x1 * x2, data = sim3)
```

When you add variables with `+`, the model will estimate each effect independent of all the others. It's possible to fit the so-called interaction by using `*`. For example, `y ~ x1 * x2` is translated to `y = a_0 + a_1 * a1 + a_2 * a2 + a_12 * a1 * a2`. Note that whenever you use `*`, both the interaction and the individual components are included in the model.

To visualize these models we need two new tricks:

- We have two predictors, so we need to give `data_grid()` both variables. It finds all the unique values of `x1` and `x2` and then generates all combinations.
- To generate predictions from both models simultaneously, we can use `gather_predictions()`, which adds each prediction as a row. The complement of `gather_predictions()` is `spread_predictions()`, which adds each prediction to a new column.

Together this gives us:

```
grid <- sim3 %>%
  data_grid(x1, x2) %>%
  gather_predictions(mod1, mod2)
grid
#> # A tibble: 80 × 4
#>    model    x1     x2  pred
#>    <chr> <int> <fctr> <dbl>
#> 1  mod1      1      a  1.67
#> 2  mod1      1      b  4.56
#> 3  mod1      1      c  6.48
#> 4  mod1      1      d  4.03
#> 5  mod1      2      a  1.48
#> 6  mod1      2      b  4.37
#> # ... with 74 more rows
```

We can visualize the results for both models on one plot using faceting:

```
ggplot(sim3, aes(x1, y, color = x2)) +
  geom_point() +
  geom_line(data = grid, aes(y = pred)) +
  facet_wrap(~ model)
```

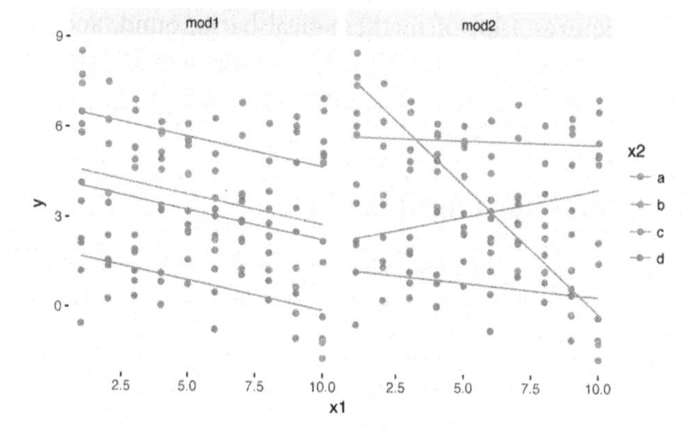

Note that the model that uses + has the same slope for each line, but different intercepts. The model that uses * has a different slope and intercept for each line.

Which model is better for this data? We can take look at the residuals. Here I've faceted by both model and x2 because it makes it easier to see the pattern within each group:

```
sim3 <- sim3 %>%
  gather_residuals(mod1, mod2)

ggplot(sim3, aes(x1, resid, color = x2)) +
  geom_point() +
  facet_grid(model ~ x2)
```

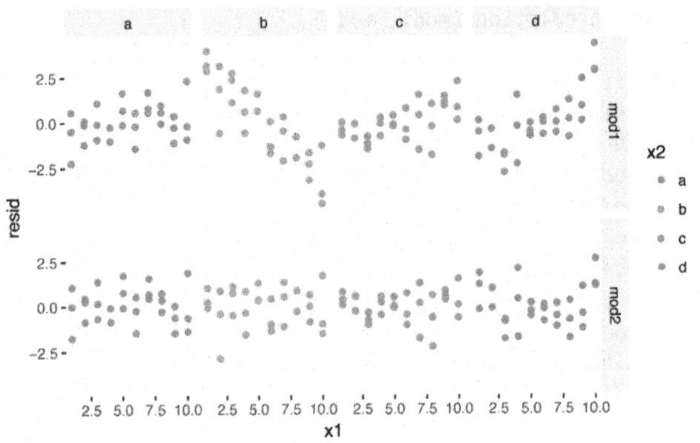

There is little obvious pattern in the residuals for mod2. The residuals for mod1 show that the model has clearly missed some pattern in b, and less so, but still present, is pattern in c, and d. You might wonder if there's a precise way to tell which of mod1 or mod2 is better. There is, but it requires a lot of mathematical background, and we don't really care. Here, we're interested in a qualitative assessment of whether or not the model has captured the pattern that we're interested in.

Interactions (Two Continuous)

Let's take a look at the equivalent model for two continuous variables. Initially things proceed almost identically to the previous example:

```
mod1 <- lm(y ~ x1 + x2, data = sim4)
mod2 <- lm(y ~ x1 * x2, data = sim4)
```

```
grid <- sim4 %>%
  data_grid(
    x1 = seq_range(x1, 5),
    x2 = seq_range(x2, 5)
  ) %>%
  gather_predictions(mod1, mod2)
grid
#> # A tibble: 50 × 4
#>    model    x1    x2   pred
#>    <chr> <dbl> <dbl>  <dbl>
#> 1  mod1   -1.0  -1.0  0.996
#> 2  mod1   -1.0  -0.5 -0.395
#> 3  mod1   -1.0   0.0 -1.786
#> 4  mod1   -1.0   0.5 -3.177
#> 5  mod1   -1.0   1.0 -4.569
#> 6  mod1   -0.5  -1.0  1.907
#> # ... with 44 more rows
```

Note my use of `seq_range()` inside `data_grid()`. Instead of using every unique value of `x`, I'm going to use a regularly spaced grid of five values between the minimum and maximum numbers. It's probably not super important here, but it's a useful technique in general. There are three other useful arguments to `seq_range()`:

- `pretty = TRUE` will generate a "pretty" sequence, i.e., something that looks nice to the human eye. This is useful if you want to produce tables of output:

  ```
  seq_range(c(0.0123, 0.923423), n = 5)
  #> [1] 0.0123 0.2401 0.4679 0.6956 0.9234
  seq_range(c(0.0123, 0.923423), n = 5, pretty = TRUE)
  #> [1] 0.0 0.2 0.4 0.6 0.8 1.0
  ```

- `trim = 0.1` will trim off 10% of the tail values. This is useful if the variable has a long-tailed distribution and you want to focus on generating values near the center:

  ```
  x1 <- rcauchy(100)
  seq_range(x1, n = 5)
  #> [1] -115.9  -83.5  -51.2  -18.8   13.5
  seq_range(x1, n = 5, trim = 0.10)
  #> [1] -13.84  -8.71  -3.58   1.55   6.68
  seq_range(x1, n = 5, trim = 0.25)
  #> [1] -2.1735 -1.0594  0.0547  1.1687  2.2828
  seq_range(x1, n = 5, trim = 0.50)
  #> [1] -0.725 -0.268  0.189  0.647  1.104
  ```

- `expand = 0.1` is in some sense the opposite of `trim()`; it expands the range by 10%:

```
x2 <- c(0, 1)
seq_range(x2, n = 5)
#> [1] 0.00 0.25 0.50 0.75 1.00
seq_range(x2, n = 5, expand = 0.10)
#> [1] -0.050  0.225  0.500  0.775  1.050
seq_range(x2, n = 5, expand = 0.25)
#> [1] -0.125  0.188  0.500  0.812  1.125
seq_range(x2, n = 5, expand = 0.50)
#> [1] -0.250  0.125  0.500  0.875  1.250
```

Next let's try and visualize that model. We have two continuous predictors, so you can imagine the model like a 3D surface. We could display that using `geom_tile()`:

```
ggplot(grid, aes(x1, x2)) +
  geom_tile(aes(fill = pred)) +
  facet_wrap(~ model)
```

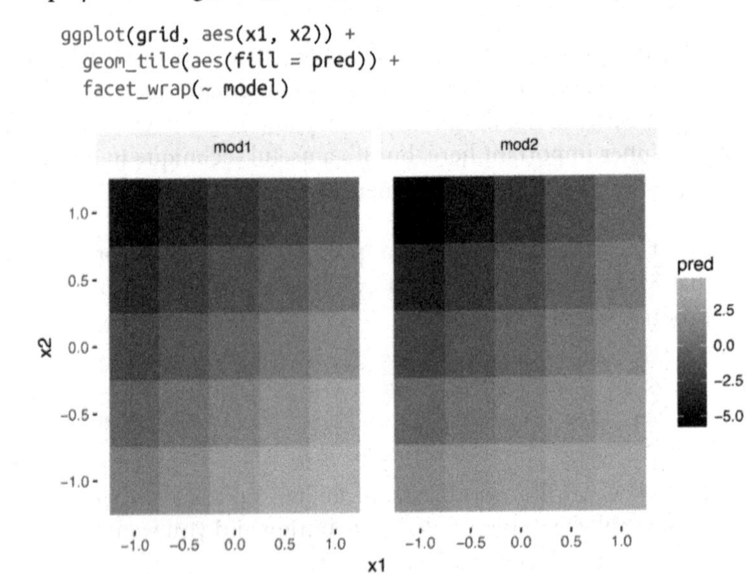

That doesn't suggest that the models are very different! But that's partly an illusion: our eyes and brains are not very good at accurately comparing shades of color. Instead of looking at the surface from the top, we could look at it from either side, showing multiple slices:

```
ggplot(grid, aes(x1, pred, color = x2, group = x2)) +
  geom_line() +
  facet_wrap(~ model)
ggplot(grid, aes(x2, pred, color = x1, group = x1)) +
```

```
  geom_line() +
  facet_wrap(~ model)
```

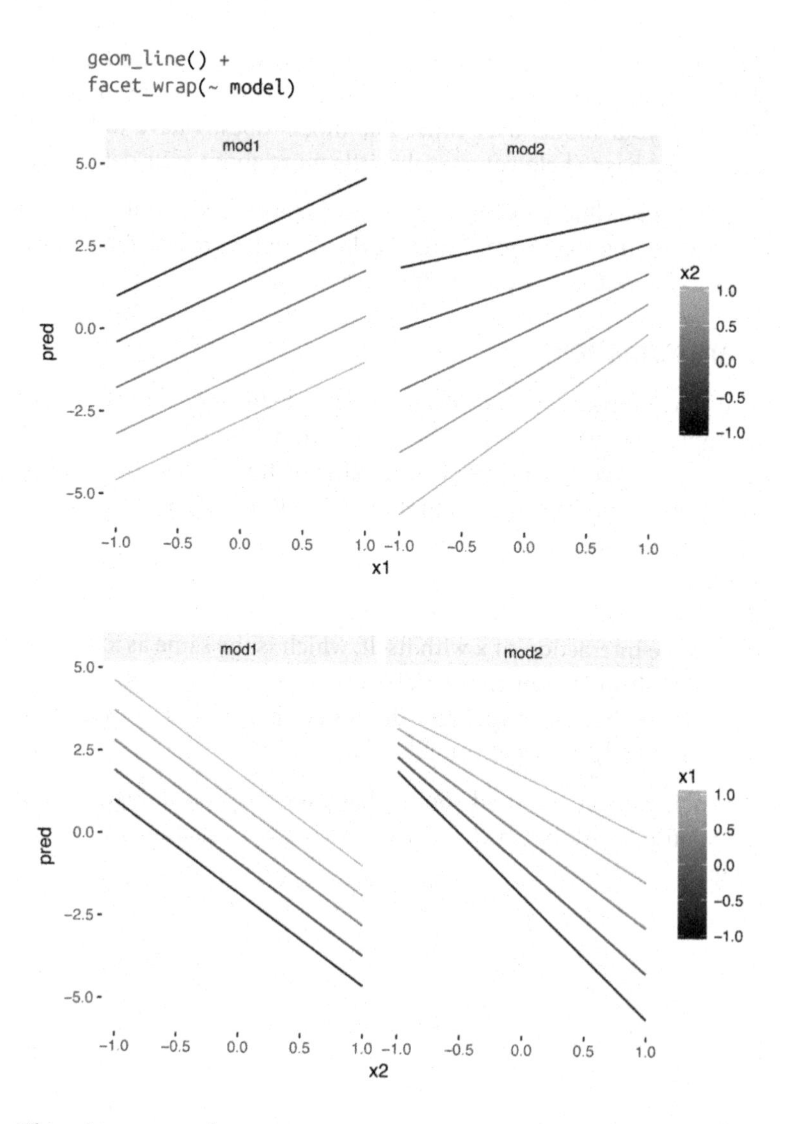

This shows you that interaction between two continuous variables works basically the same way as for a categorical and continuous variable. An interaction says that there's not a fixed offset: you need to consider both values of `x1` and `x2` simultaneously in order to predict `y`.

You can see that even with just two continuous variables, coming up with good visualizations are hard. But that's reasonable: you shouldn't expect it will be easy to understand how three or more variables simultaneously interact! But again, we're saved a little because we're using models for exploration, and you can gradually build up your model over time. The model doesn't have to be perfect, it just has to help you reveal a little more about your data.

I spent some time looking at the residuals to see if I could figure if `mod2` did better than `mod1`. I think it does, but it's pretty subtle. You'll have a chance to work on it in the exercises.

Transformations

You can also perform transformations inside the model formula. For example, `log(y) ~ sqrt(x1) + x2` is transformed to `y = a_1 + a_2 * x1 * sqrt(x) + a_3 * x2`. If your transformation involves +, *, ^, or -, you'll need to wrap it in `I()` so R doesn't treat it like part of the model specification. For example, `y ~ x + I(x ^ 2)` is translated to `y = a_1 + a_2 * x + a_3 * x^2`. If you forget the `I()` and specify `y ~ x ^ 2 + x`, R will compute `y ~ x * x + x`. `x * x` means the interaction of `x` with itself, which is the same as `x`. R automatically drops redundant variables so `x + x` becomes `x`, meaning that `y ~ x ^ 2 + x` specifies the function `y = a_1 + a_2 * x`. That's probably not what you intended!

Again, if you get confused about what your model is doing, you can always use `model_matrix()` to see exactly what equation `lm()` is fitting:

```
df <- tribble(
  ~y, ~x,
   1,  1,
   2,  2,
   3,  3
)
model_matrix(df, y ~ x^2 + x)
#> # A tibble: 3 × 2
#>   `(Intercept)`     x
#>           <dbl> <dbl>
#> 1             1     1
#> 2             1     2
#> 3             1     3
model_matrix(df, y ~ I(x^2) + x)
#> # A tibble: 3 × 3
#>   `(Intercept)` `I(x^2)`     x
#>           <dbl>    <dbl> <dbl>
#> 1             1        1     1
```

```
#> 2                  1         4     2
#> 3                  1         9     3
```

Transformations are useful because you can use them to approximate nonlinear functions. If you've taken a calculus class, you may have heard of Taylor's theorem, which says you can approximate any smooth function with an infinite sum of polynomials. That means you can use a linear function to get arbitrarily close to a smooth function by fitting an equation like `y = a_1 + a_2 * x + a_3 * x^2 + a_4 * x ^ 3`. Typing that sequence by hand is tedious, so R provides a helper function, `poly()`:

```
model_matrix(df, y ~ poly(x, 2))
#> # A tibble: 3 × 3
#>   `(Intercept)` `poly(x, 2)1` `poly(x, 2)2`
#>           <dbl>         <dbl>         <dbl>
#> 1             1     -7.07e-01         0.408
#> 2             1     -7.85e-17        -0.816
#> 3             1      7.07e-01         0.408
```

However there's one major problem with using `poly()`: outside the range of the data, polynomials rapidly shoot off to positive or negative infinity. One safer alternative is to use the natural spline, `splines::ns()`:

```
library(splines)
model_matrix(df, y ~ ns(x, 2))
#> # A tibble: 3 × 3
#>   `(Intercept)` `ns(x, 2)1` `ns(x, 2)2`
#>           <dbl>       <dbl>       <dbl>
#> 1             1       0.000       0.000
#> 2             1       0.566      -0.211
#> 3             1       0.344       0.771
```

Let's see what that looks like when we try and approximate a nonlinear function:

```
sim5 <- tibble(
  x = seq(0, 3.5 * pi, length = 50),
  y = 4 * sin(x) + rnorm(length(x))
)

ggplot(sim5, aes(x, y)) +
  geom_point()
```

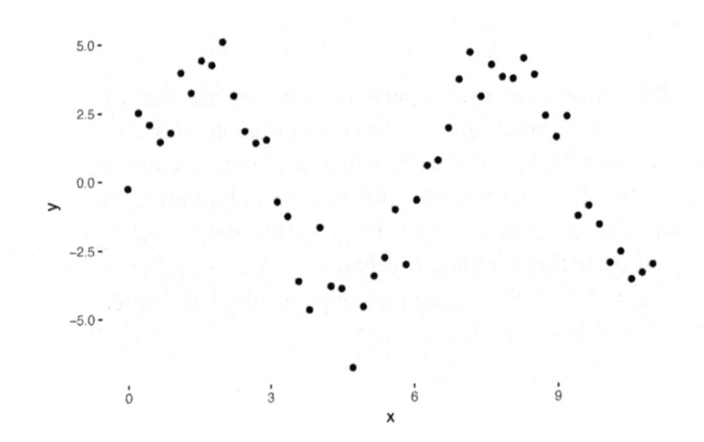

I'm going to fit five models to this data:

```
mod1 <- lm(y ~ ns(x, 1), data = sim5)
mod2 <- lm(y ~ ns(x, 2), data = sim5)
mod3 <- lm(y ~ ns(x, 3), data = sim5)
mod4 <- lm(y ~ ns(x, 4), data = sim5)
mod5 <- lm(y ~ ns(x, 5), data = sim5)

grid <- sim5 %>%
  data_grid(x = seq_range(x, n = 50, expand = 0.1)) %>%
  gather_predictions(mod1, mod2, mod3, mod4, mod5, .pred = "y")

ggplot(sim5, aes(x, y)) +
  geom_point() +
  geom_line(data = grid, color = "red") +
  facet_wrap(~ model)
```

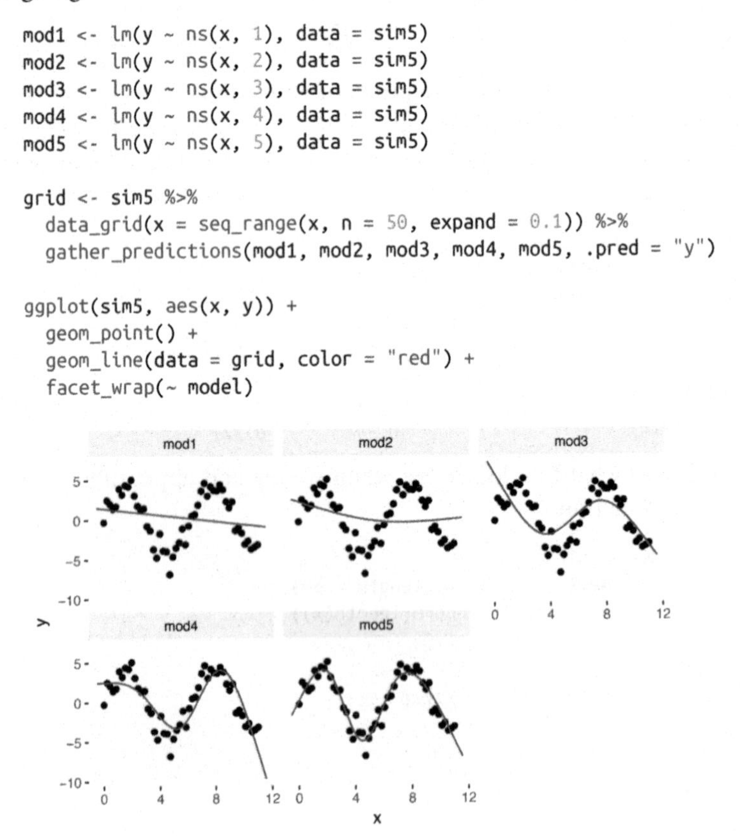

Notice that the extrapolation outside the range of the data is clearly bad. This is the downside to approximating a function with a polynomial. But this is a very real problem with every model: the model can never tell you if the behavior is true when you start extrapolating outside the range of the data that you have seen. You must rely on theory and science.

Exercises

1. What happens if you repeat the analysis of `sim2` using a model without an intercept? What happens to the model equation? What happens to the predictions?

2. Use `model_matrix()` to explore the equations generated for the models I fit to `sim3` and `sim4`. Why is `*` a good shorthand for interaction?

3. Using the basic principles, convert the formulas in the following two models into functions. (Hint: start by converting the categorical variable into 0-1 variables.)

   ```
   mod1 <- lm(y ~ x1 + x2, data = sim3)
   mod2 <- lm(y ~ x1 * x2, data = sim3)
   ```

4. For `sim4`, which of `mod1` and `mod2` is better? I think `mod2` does a slightly better job at removing patterns, but it's pretty subtle. Can you come up with a plot to support my claim?

Missing Values

Missing values obviously cannot convey any information about the relationship between the variables, so modeling functions will drop any rows that contain missing values. R's default behavior is to silently drop them, but `options(na.action = na.warn)` (run in the prerequisites), makes sure you get a warning:

```
df <- tribble(
  ~x, ~y,
  1, 2.2,
  2, NA,
  3, 3.5,
  4, 8.3,
  NA, 10
)
```

```
mod <- lm(y ~ x, data = df)
#> Warning: Dropping 2 rows with missing values
```

To suppress the warning, set `na.action = na.exclude`:

```
mod <- lm(y ~ x, data = df, na.action = na.exclude)
```

You can always see exactly how many observations were used with `nobs()`:

```
nobs(mod)
#> [1] 3
```

Other Model Families

This chapter has focused exclusively on the class of linear models, which assume a relationship of the form `y = a_1 * x1 + a_2 * x2 + ... + a_n * xn`. Linear models additionally assume that the residuals have a normal distribution, which we haven't talked about. There is a large set of model classes that extend the linear model in various interesting ways. Some of them are:

- *Generalized linear models*, e.g., `stats::glm()`. Linear models assume that the response is continuous and the error has a normal distribution. Generalized linear models extend linear models to include noncontinuous responses (e.g., binary data or counts). They work by defining a distance metric based on the statistical idea of likelihood.
- *Generalized additive models*, e.g., `mgcv::gam()`, extend generalized linear models to incorporate arbitrary smooth functions. That means you can write a formula like `y ~ s(x)`, which becomes an equation like `y = f(x)`, and let `gam()` estimate what that function is (subject to some smoothness constraints to make the problem tractable).
- *Penalized linear models*, e.g., `glmnet::glmnet()`, add a penalty term to the distance that penalizes complex models (as defined by the distance between the parameter vector and the origin). This tends to make models that generalize better to new datasets from the same population.
- *Robust linear models*, e.g., `MASS:rlm()`, tweak the distance to downweight points that are very far away. This makes them less sensitive to the presence of outliers, at the cost of being not quite as good when there are no outliers.

- *Trees*, e.g., `rpart::rpart()`, attack the problem in a completely different way than linear models. They fit a piece-wise constant model, splitting the data into progressively smaller and smaller pieces. Trees aren't terribly effective by themselves, but they are very powerful when used in aggregate by models like *random forests* (e.g., `randomForest::randomForest()`) or *gradient boosting machines* (e.g., `xgboost::xgboost`.)

These models all work similarly from a programming perspective. Once you've mastered linear models, you should find it easy to master the mechanics of these other model classes. Being a skilled modeler is a mixture of some good general principles and having a big toolbox of techniques. Now that you've learned some general tools and one useful class of models, you can go on and learn more classes from other sources.

CHAPTER 19

Model Building

Introduction

In the previous chapter you learned how linear models worked, and learned some basic tools for understanding what a model is telling you about your data. The previous chapter focused on simulated datasets to help you learn about how models work. This chapter will focus on real data, showing you how you can progressively build up a model to aid your understanding of the data.

We will take advantage of the fact that you can think about a model partitioning your data into patterns and residuals. We'll find patterns with visualization, then make them concrete and precise with a model. We'll then repeat the process, but replace the old response variable with the residuals from the model. The goal is to transition from implicit knowledge in the data and your head to explicit knowledge in a quantitative model. This makes it easier to apply to new domains, and easier for others to use.

For very large and complex datasets this will be a lot of work. There are certainly alternative approaches—a more machine learning approach is simply to focus on the predictive ability of the model. These approaches tend to produce black boxes: the model does a really good job at generating predictions, but you don't know why. This is a totally reasonable approach, but it does make it hard to apply your real-world knowledge to the model. That, in turn, makes it difficult to assess whether or not the model will continue to work in the long term, as fundamentals change. For most real models, I'd expect you to use some combination of this approach and a more classic automated approach.

It's a challenge to know when to stop. You need to figure out when your model is good enough, and when additional investment is unlikely to pay off. I particularly like this quote from reddit user Broseidon241:

> A long time ago in art class, my teacher told me "An artist needs to know when a piece is done. You can't tweak something into perfection—wrap it up. If you don't like it, do it over again. Otherwise begin something new." Later in life, I heard "A poor seamstress makes many mistakes. A good seamstress works hard to correct those mistakes. A great seamstress isn't afraid to throw out the garment and start over."
>
> —Broseidon241 (*https://www.reddit.com/r/datascience/comments/4irajq*)

Prerequisites

We'll use the same tools as in the previous chapter, but add in some real datasets: `diamonds` from **ggplot2**, and `flights` from **nycflights13**. We'll also need **lubridate** in order to work with the date/times in `flights`.

```
library(tidyverse)
library(modelr)
options(na.action = na.warn)

library(nycflights13)
library(lubridate)
```

Why Are Low-Quality Diamonds More Expensive?

In previous chapters we've seen a surprising relationship between the quality of diamonds and their price: low-quality diamonds (poor cuts, bad colors, and inferior clarity) have higher prices:

```
ggplot(diamonds, aes(cut, price)) + geom_boxplot()
ggplot(diamonds, aes(color, price)) + geom_boxplot()
ggplot(diamonds, aes(clarity, price)) + geom_boxplot()
```

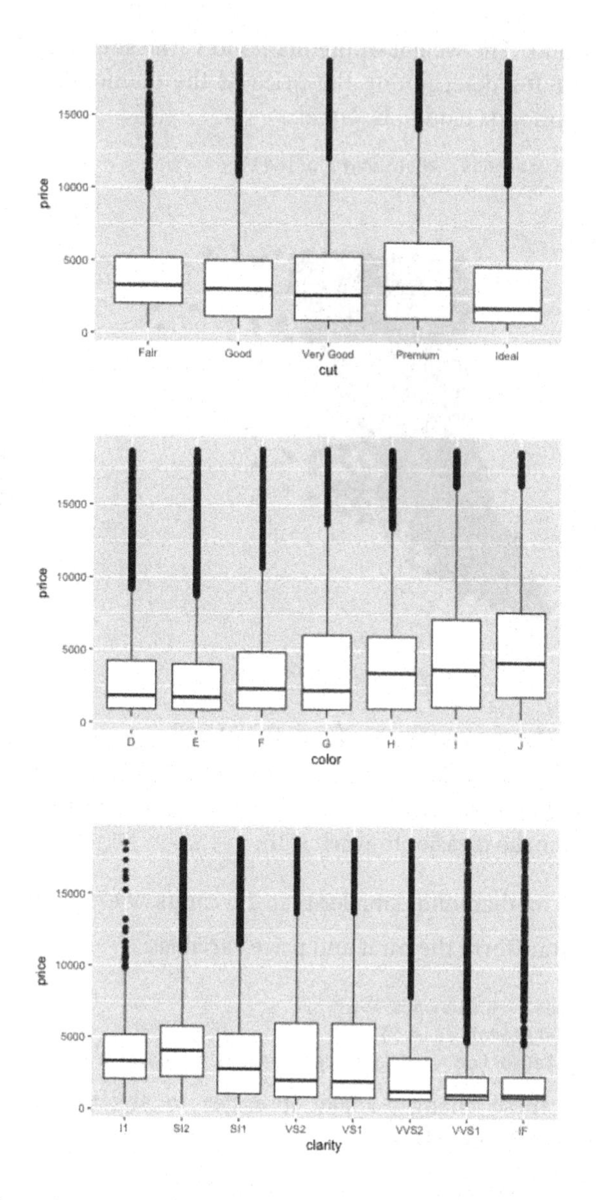

Note that the worst diamond color is J (slightly yellow), and the worst clarity is I1 (inclusions visible to the naked eye).

Price and Carat

It looks like lower-quality diamonds have higher prices because there is an important confounding variable: the weight (`carat`) of the diamond. The weight of the diamond is the single most important factor for determining the price of the diamond, and lower-quality diamonds tend to be larger:

```
ggplot(diamonds, aes(carat, price)) +
  geom_hex(bins = 50)
```

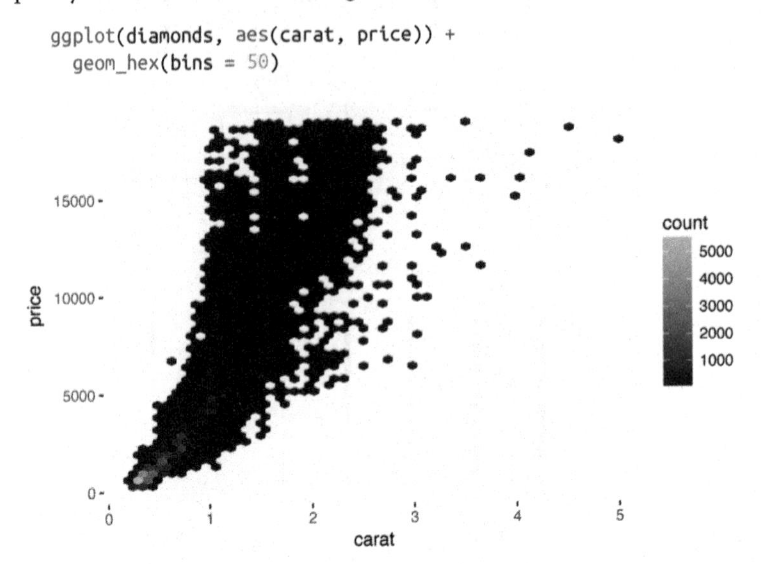

We can make it easier to see how the other attributes of a diamond affect its relative `price` by fitting a model to separate out the effect of `carat`. But first, let's make a couple of tweaks to the diamonds dataset to make it easier to work with:

1. Focus on diamonds smaller than 2.5 carats (99.7% of the data).
2. Log-transform the carat and price variables:

```
diamonds2 <- diamonds %>%
  filter(carat <= 2.5) %>%
  mutate(lprice = log2(price), lcarat = log2(carat))
```

Together, these changes make it easier to see the relationship between `carat` and `price`:

```
ggplot(diamonds2, aes(lcarat, lprice)) +
  geom_hex(bins = 50)
```

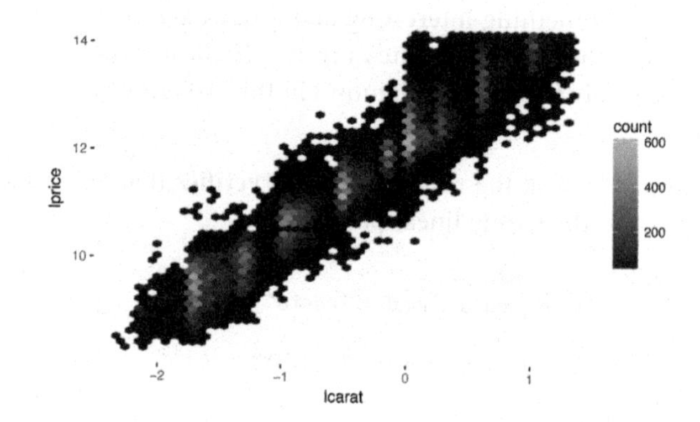

The log transformation is particularly useful here because it makes the pattern linear, and linear patterns are the easiest to work with. Let's take the next step and remove that strong linear pattern. We first make the pattern explicit by fitting a model:

```
mod_diamond <- lm(lprice ~ lcarat, data = diamonds2)
```

Then we look at what the model tells us about the data. Note that I back-transform the predictions, undoing the log transformation, so I can overlay the predictions on the raw data:

```
grid <- diamonds2 %>%
  data_grid(carat = seq_range(carat, 20)) %>%
  mutate(lcarat = log2(carat)) %>%
  add_predictions(mod_diamond, "lprice") %>%
  mutate(price = 2 ^ lprice)

ggplot(diamonds2, aes(carat, price)) +
  geom_hex(bins = 50) +
  geom_line(data = grid, color = "red", size = 1)
```

That tells us something interesting about our data. If we believe our model, then the large diamonds are much cheaper than expected. This is probably because no diamond in this dataset costs more than $19,000.

Now we can look at the residuals, which verifies that we've successfully removed the strong linear pattern:

```
diamonds2 <- diamonds2 %>%
  add_residuals(mod_diamond, "lresid")

ggplot(diamonds2, aes(lcarat, lresid)) +
  geom_hex(bins = 50)
```

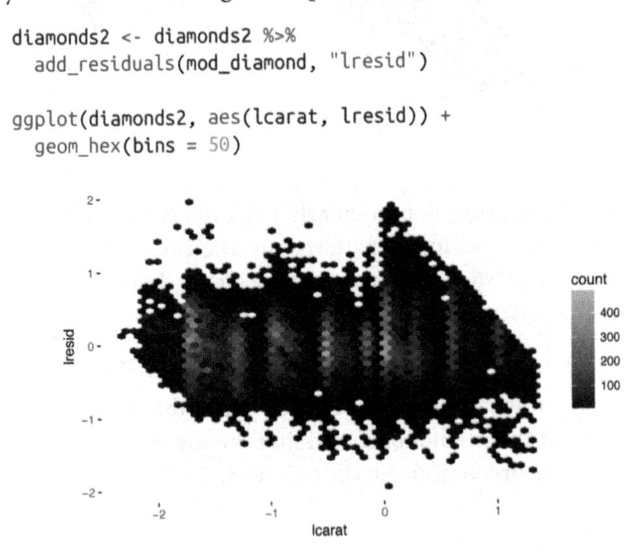

Importantly, we can now redo our motivating plots using those residuals instead of `price`:

```
ggplot(diamonds2, aes(cut, lresid)) + geom_boxplot()
ggplot(diamonds2, aes(color, lresid)) + geom_boxplot()
ggplot(diamonds2, aes(clarity, lresid)) + geom_boxplot()
```

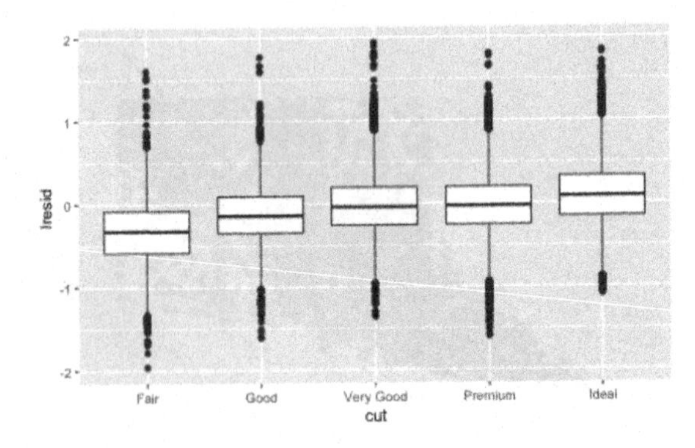

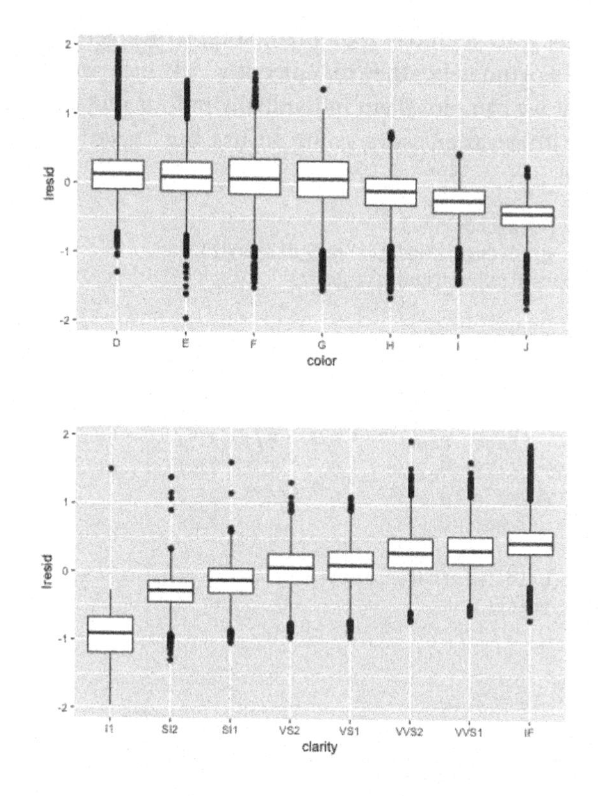

Now we see the relationship we expect: as the quality of the diamond increases, so to does its relative price. To interpret the y-axis, we need to think about what the residuals are telling us, and what scale they are on. A residual of –1 indicates that `lprice` was 1 unit lower than a prediction based solely on its weight. 2^{-1} is 1/2, so points with a value of –1 are half the expected price, and residuals with value 1 are twice the predicted price.

A More Complicated Model

If we wanted to, we could continue to build up our model, moving the effects we've observed into the model to make them explicit. For example, we could include `color`, `cut`, and `clarity` into the model so that we also make explicit the effect of these three categorical variables:

```
mod_diamond2 <- lm(
  lprice ~ lcarat + color + cut + clarity,
  data = diamonds2
)
```

This model now includes four predictors, so it's getting harder to visualize. Fortunately, they're currently all independent, which means that we can plot them individually in four plots. To make the process a little easier, we're going to use the `.model` argument to `data_grid`:

```
grid <- diamonds2 %>%
  data_grid(cut, .model = mod_diamond2) %>%
  add_predictions(mod_diamond2)
grid
#> # A tibble: 5 × 5
#>         cut lcarat color clarity  pred
#>       <ord>  <dbl> <chr>   <chr> <dbl>
#> 1      Fair -0.515     G     SI1  11.0
#> 2      Good -0.515     G     SI1  11.1
#> 3 Very Good -0.515     G     SI1  11.2
#> 4   Premium -0.515     G     SI1  11.2
#> 5     Ideal -0.515     G     SI1  11.2

ggplot(grid, aes(cut, pred)) +
  geom_point()
```


If the model needs variables that you haven't explicitly supplied, `data_grid()` will automatically fill them in with the "typical" value. For continuous variables, it uses the median, and for categorical variables, it uses the most common value (or values, if there's a tie):

```
diamonds2 <- diamonds2 %>%
  add_residuals(mod_diamond2, "lresid2")
```

```r
ggplot(diamonds2, aes(lcarat, lresid2)) +
  geom_hex(bins = 50)
```

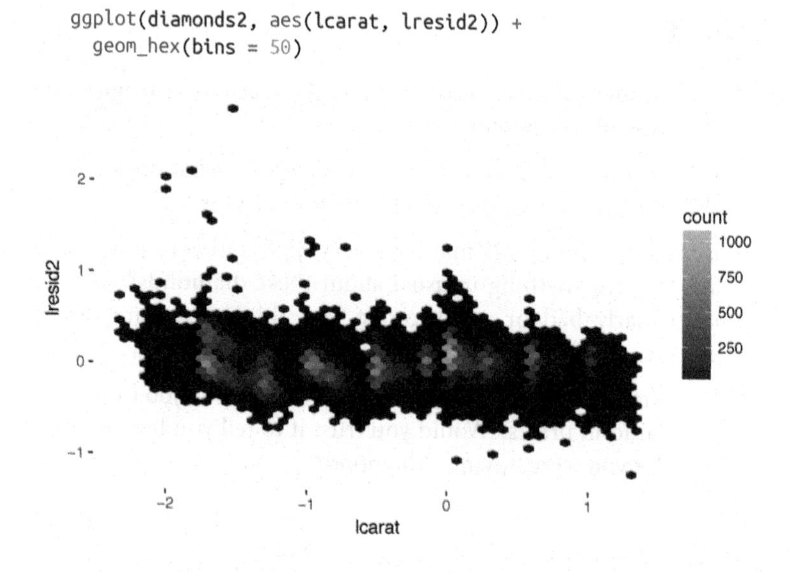

This plot indicates that there are some diamonds with quite large residuals—remember a residual of 2 indicates that the diamond is 4x the price that we expected. It's often useful to look at unusual values individually:

```r
diamonds2 %>%
  filter(abs(lresid2) > 1) %>%
  add_predictions(mod_diamond2) %>%
  mutate(pred = round(2 ^ pred)) %>%
  select(price, pred, carat:table, x:z) %>%
  arrange(price)
#> # A tibble: 16 × 11
#>   price  pred carat     cut color clarity depth table     x
#>   <int> <dbl> <dbl>   <ord> <ord>   <ord> <dbl> <dbl> <dbl>
#> 1  1013   264  0.25    Fair     F     SI2  54.4    64  4.30
#> 2  1186   284  0.25 Premium     G     SI2  59.0    60  5.33
#> 3  1186   284  0.25 Premium     G     SI2  58.8    60  5.33
#> 4  1262  2644  1.03    Fair     E      I1  78.2    54  5.72
#> 5  1415   639  0.35    Fair     G     VS2  65.9    54  5.57
#> 6  1415   639  0.35    Fair     G     VS2  65.9    54  5.57
#> # ... with 10 more rows, and 2 more variables: y <dbl>,
#> #   z <dbl>
```

Nothing really jumps out at me here, but it's probably worth spending time considering if this indicates a problem with our model, or if there are errors in the data. If there are mistakes in the data, this could be an opportunity to buy diamonds that have been priced low incorrectly.

Exercises

1. In the plot of `lcarat` versus `lprice`, there are some bright vertical strips. What do they represent?
2. If `log(price) = a_0 + a_1 * log(carat)`, what does that say about the relationship between `price` and `carat`?
3. Extract the diamonds that have very high and very low residuals. Is there anything unusual about these diamonds? Are they particularly bad or good, or do you think these are pricing errors?
4. Does the final model, `mod_diamonds2`, do a good job of predicting diamond prices? Would you trust it to tell you how much to spend if you were buying a diamond?

What Affects the Number of Daily Flights?

Let's work through a similar process for a dataset that seems even simpler at first glance: the number of flights that leave NYC per day. This is a really small dataset—only 365 rows and 2 columns—and we're not going to end up with a fully realized model, but as you'll see, the steps along the way will help us better understand the data. Let's get started by counting the number of flights per day and visualizing it with **ggplot2**:

```
daily <- flights %>%
  mutate(date = make_date(year, month, day)) %>%
  group_by(date) %>%
  summarize(n = n())
daily
#> # A tibble: 365 × 2
#>         date     n
#>       <date> <int>
#> 1 2013-01-01   842
#> 2 2013-01-02   943
#> 3 2013-01-03   914
#> 4 2013-01-04   915
#> 5 2013-01-05   720
#> 6 2013-01-06   832
#> # ... with 359 more rows

ggplot(daily, aes(date, n)) +
  geom_line()
```

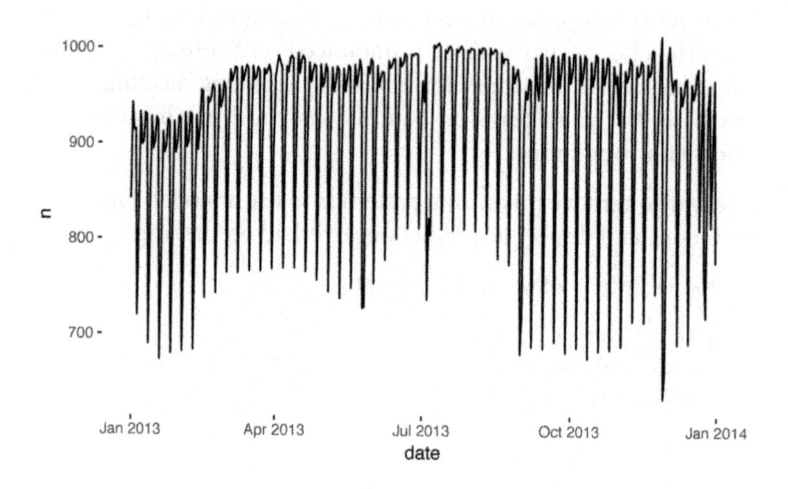

Day of Week

Understanding the long-term trend is challenging because there's a very strong day-of-week effect that dominates the subtler patterns. Let's start by looking at the distribution of flight numbers by day of week:

```
daily <- daily %>%
  mutate(wday = wday(date, label = TRUE))
ggplot(daily, aes(wday, n)) +
  geom_boxplot()
```

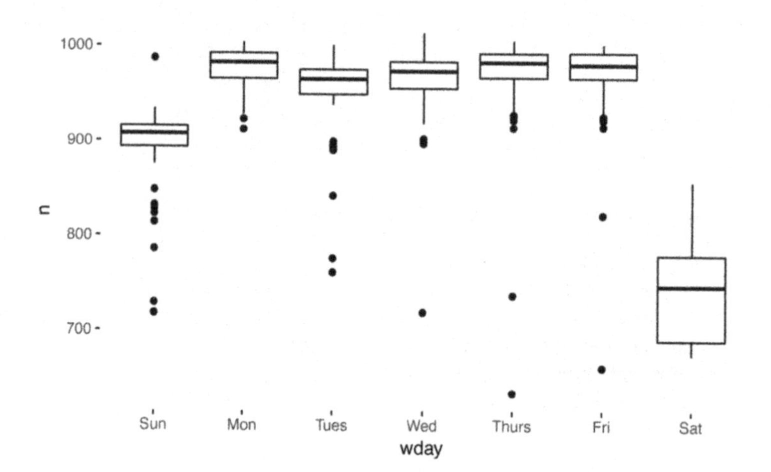

There are fewer flights on weekends because most travel is for business. The effect is particularly pronounced on Saturday: you might sometimes leave on Sunday for a Monday morning meeting, but it's very rare that you'd leave on Saturday as you'd much rather be at home with your family.

One way to remove this strong pattern is to use a model. First, we fit the model, and display its predictions overlaid on the original data:

```
mod <- lm(n ~ wday, data = daily)

grid <- daily %>%
  data_grid(wday) %>%
  add_predictions(mod, "n")

ggplot(daily, aes(wday, n)) +
  geom_boxplot() +
  geom_point(data = grid, color = "red", size = 4)
```

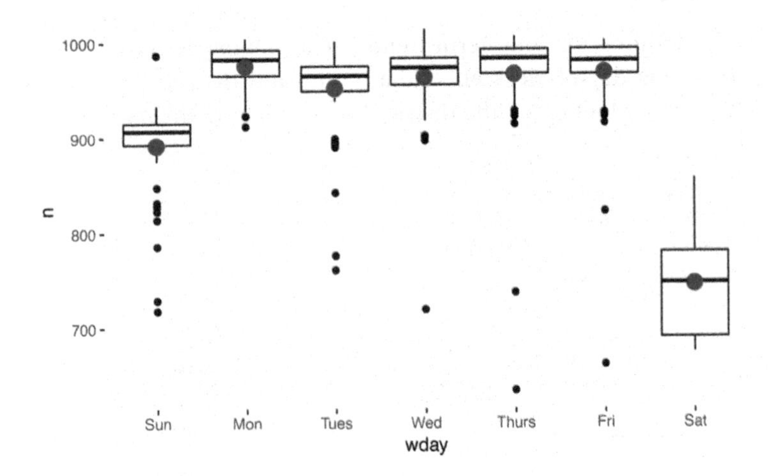

Next we compute and visualize the residuals:

```
daily <- daily %>%
  add_residuals(mod)
daily %>%
  ggplot(aes(date, resid)) +
  geom_ref_line(h = 0) +
  geom_line()
```

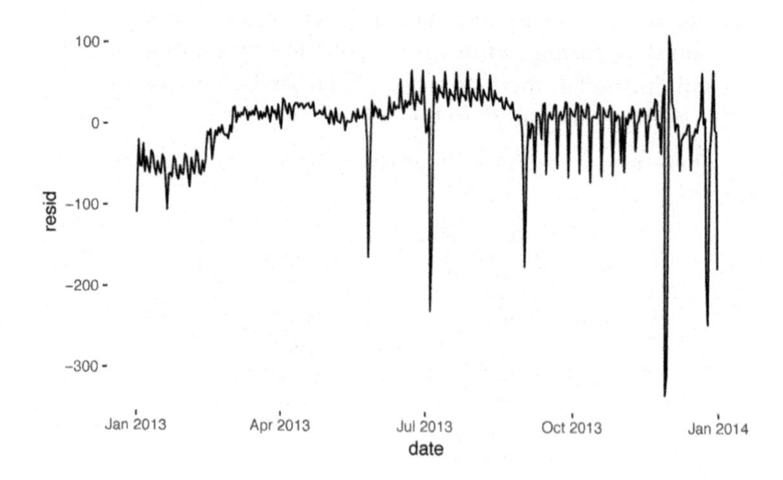

Note the change in the y-axis: now we are seeing the deviation from the expected number of flights, given the day of week. This plot is useful because now that we've removed much of the large day-of-week effect, we can see some of the subtler patterns that remain:

- Our model seems to fail starting in June: you can still see a strong regular pattern that our model hasn't captured. Drawing a plot with one line for each day of the week makes the cause easier to see:

```
ggplot(daily, aes(date, resid, color = wday)) +
  geom_ref_line(h = 0) +
  geom_line()
```

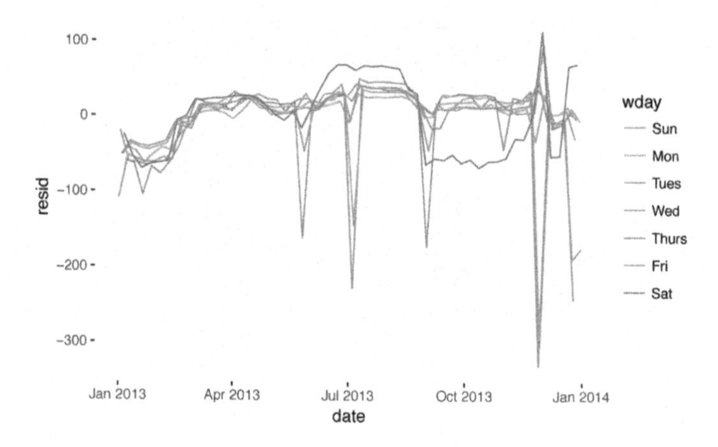

Our model fails to accurately predict the number of flights on Saturday: during summer there are more flights than we expect, and during fall there are fewer. We'll see how we can do better to capture this pattern in the next section.

- There are some days with far fewer flights than expected:

```
daily %>%
  filter(resid < -100)
#> # A tibble: 11 × 4
#>         date     n  wday resid
#>       <date> <int> <ord> <dbl>
#> 1 2013-01-01   842  Tues  -109
#> 2 2013-01-20   786   Sun  -105
#> 3 2013-05-26   729   Sun  -162
#> 4 2013-07-04   737 Thurs  -229
#> 5 2013-07-05   822   Fri  -145
#> 6 2013-09-01   718   Sun  -173
#> # ... with 5 more rows
```

  If you're familiar with American public holidays, you might spot New Year's Day, July 4th, Thanksgiving, and Christmas. There are some others that don't seem to correspond to public holidays. You'll work on those in one of the exercises.

- There seems to be some smoother long-term trend over the course of a year. We can highlight that trend with `geom_smooth()`:

```
daily %>%
  ggplot(aes(date, resid)) +
  geom_ref_line(h = 0) +
  geom_line(color = "grey50") +
  geom_smooth(se = FALSE, span = 0.20)
#> `geom_smooth()` using method = 'loess'
```

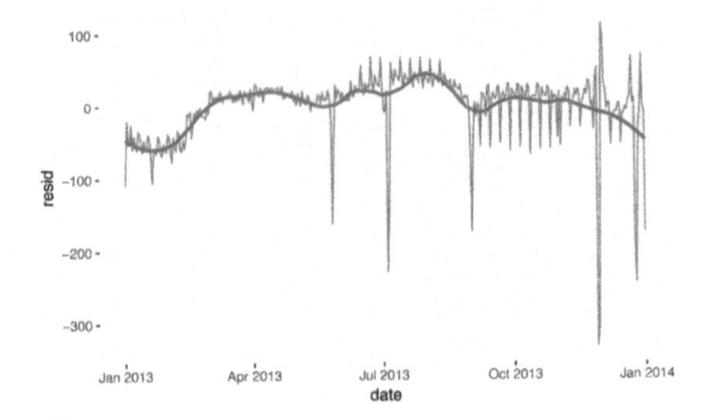

There are fewer flights in January (and December), and more in summer (May–Sep). We can't do much with this pattern quantitatively, because we only have a single year of data. But we can use our domain knowledge to brainstorm potential explanations.

Seasonal Saturday Effect

Let's first tackle our failure to accurately predict the number of flights on Saturday. A good place to start is to go back to the raw numbers, focusing on Saturdays:

```
daily %>%
  filter(wday == "Sat") %>%
  ggplot(aes(date, n)) +
    geom_point() +
    geom_line() +
    scale_x_date(
      NULL,
      date_breaks = "1 month",
      date_labels = "%b"
    )
```

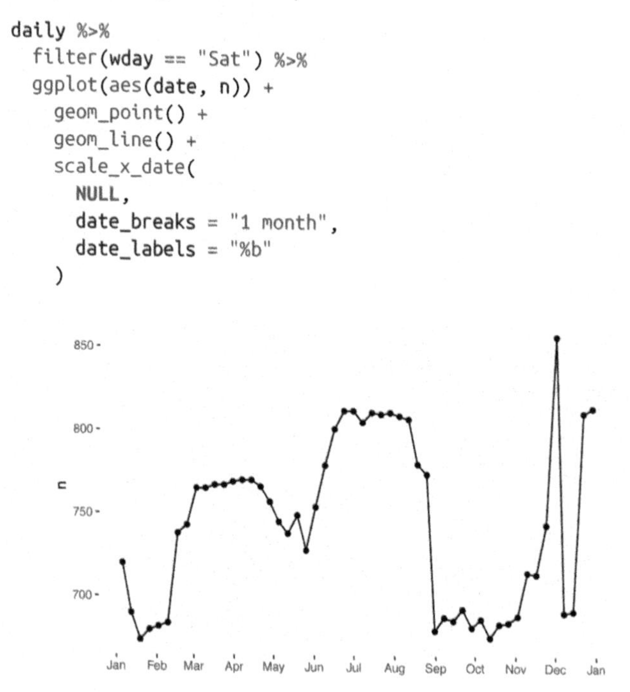

(I've used both points and lines to make it more clear what is data and what is interpolation.)

I suspect this pattern is caused by summer holidays: many people go on holiday in the summer, and people don't mind travelling on Saturdays for vacation. Looking at this plot, we might guess that summer holidays are from early June to late August. That seems to line up fairly well with the state's school terms (*http://on.nyc.gov/2gWAbBR*): summer break in 2013 was June 26–September 9.

Why are there more Saturday flights in the spring than the fall? I asked some American friends and they suggested that it's less common to plan family vacations during the fall because of the big Thanksgiving and Christmas holidays. We don't have the data to know for sure, but it seems like a plausible working hypothesis.

Let's create a "term" variable that roughly captures the three school terms, and check our work with a plot:

```
term <- function(date) {
  cut(date,
    breaks = ymd(20130101, 20130605, 20130825, 20140101),
    labels = c("spring", "summer", "fall")
  )
}

daily <- daily %>%
  mutate(term = term(date))

daily %>%
  filter(wday == "Sat") %>%
  ggplot(aes(date, n, color = term)) +
  geom_point(alpha = 1/3) +
  geom_line() +
  scale_x_date(
    NULL,
    date_breaks = "1 month",
    date_labels = "%b"
  )
```

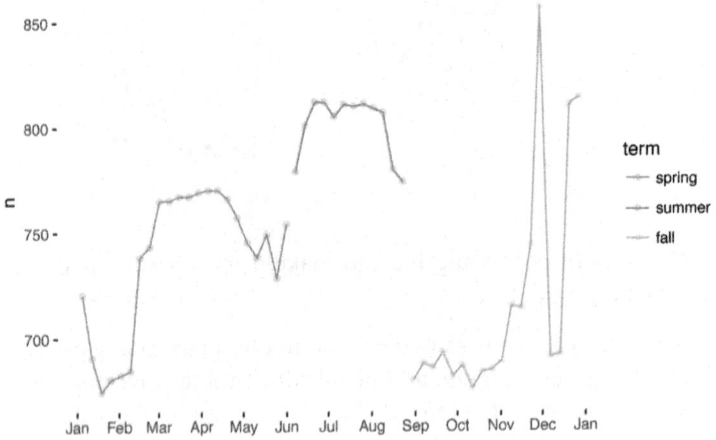

(I manually tweaked the dates to get nice breaks in the plot. Using a visualization to help you understand what your function is doing is a really powerful and general technique.)

It's useful to see how this new variable affects the other days of the week:

```
daily %>%
  ggplot(aes(wday, n, color = term)) +
    geom_boxplot()
```

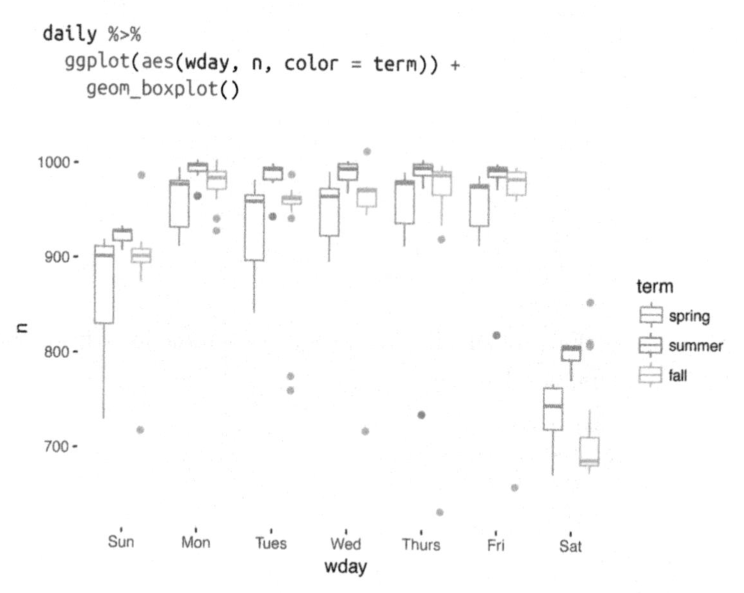

It looks like there is significant variation across the terms, so fitting a separate day-of-week effect for each term is reasonable. This improves our model, but not as much as we might hope:

```
mod1 <- lm(n ~ wday, data = daily)
mod2 <- lm(n ~ wday * term, data = daily)

daily %>%
  gather_residuals(without_term = mod1, with_term = mod2) %>%
  ggplot(aes(date, resid, color = model)) +
    geom_line(alpha = 0.75)
```

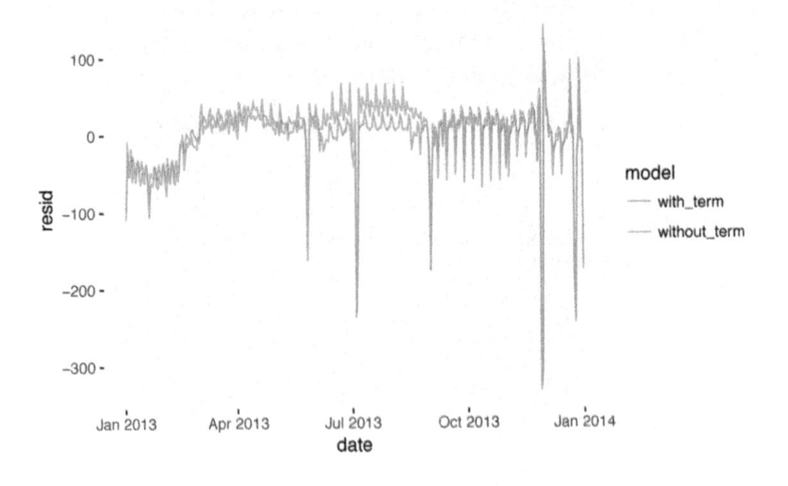

We can see the problem by overlaying the predictions from the model onto the raw data:

```
grid <- daily %>%
  data_grid(wday, term) %>%
  add_predictions(mod2, "n")

ggplot(daily, aes(wday, n)) +
  geom_boxplot() +
  geom_point(data = grid, color = "red") +
  facet_wrap(~ term)
```

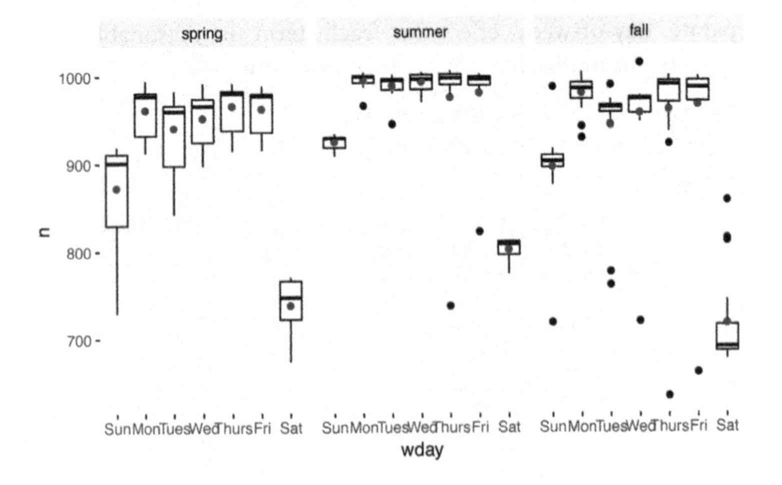

Our model is finding the *mean* effect, but we have a lot of big outliers, so the mean tends to be far away from the typical value. We can alleviate this problem by using a model that is robust to the effect of outliers: `MASS::rlm()`. This greatly reduces the impact of the outliers on our estimates, and gives a model that does a good job of removing the day-of-week pattern:

```
mod3 <- MASS::rlm(n ~ wday * term, data = daily)

daily %>%
  add_residuals(mod3, "resid") %>%
  ggplot(aes(date, resid)) +
  geom_hline(yintercept = 0, size = 2, color = "white") +
  geom_line()
```

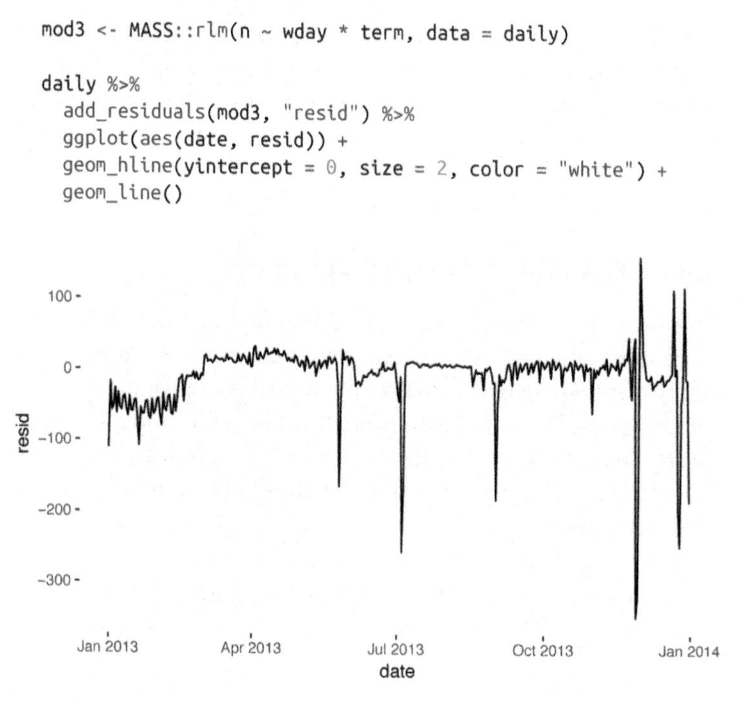

It's now much easier to see the long-term trend, and the positive and negative outliers.

Computed Variables

If you're experimenting with many models and many visualizations, it's a good idea to bundle the creation of variables up into a function so there's no chance of accidentally applying a different transformation in different places. For example, we could write:

```
compute_vars <- function(data) {
  data %>%
    mutate(
      term = term(date),
      wday = wday(date, label = TRUE)
    )
}
```

Another option is to put the transformations directly in the model formula:

```
wday2 <- function(x) wday(x, label = TRUE)
mod3 <- lm(n ~ wday2(date) * term(date), data = daily)
```

Either approach is reasonable. Making the transformed variable explicit is useful if you want to check your work, or use them in a visualization. But you can't easily use transformations (like splines) that return multiple columns. Including the transformations in the model function makes life a little easier when you're working with many different datasets because the model is self-contained.

Time of Year: An Alternative Approach

In the previous section we used our domain knowledge (how the US school term affects travel) to improve the model. An alternative to making our knowledge explicit in the model is to give the data more room to speak. We could use a more flexible model and allow that to capture the pattern we're interested in. A simple linear trend isn't adequate, so we could try using a natural spline to fit a smooth curve across the year:

```
library(splines)
mod <- MASS::rlm(n ~ wday * ns(date, 5), data = daily)

daily %>%
  data_grid(wday, date = seq_range(date, n = 13)) %>%
  add_predictions(mod) %>%
  ggplot(aes(date, pred, color = wday)) +
    geom_line() +
    geom_point()
```

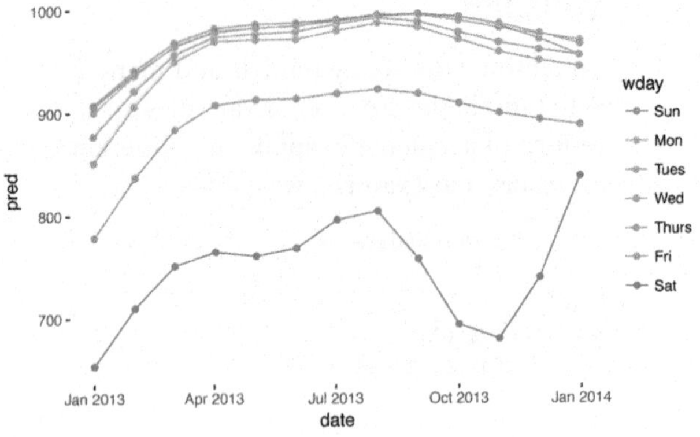

We see a strong pattern in the numbers of Saturday flights. This is reassuring, because we also saw that pattern in the raw data. It's a good sign when you get the same signal from different approaches.

Exercises

1. Use your Google sleuthing skills to brainstorm why there were fewer than expected flights on January 20, May 26, and September 1. (Hint: they all have the same explanation.) How would these days generalize to another year?
2. What do the three days with high positive residuals represent? How would these days generalize to another year?

```
daily %>%
  top_n(3, resid)
#> # A tibble: 3 × 5
#>         date     n  wday resid   term
#>       <date> <int> <ord> <dbl> <fctr>
#> 1 2013-11-30   857   Sat 112.4   fall
#> 2 2013-12-01   987   Sun  95.5   fall
#> 3 2013-12-28   814   Sat  69.4   fall
```

3. Create a new variable that splits the `wday` variable into terms, but only for Saturdays, i.e., it should have `Thurs`, `Fri`, but `Sat-summer`, `Sat-spring`, `Sat-fall`. How does this model compare with the model with every combination of `wday` and `term`?
4. Create a new `wday` variable that combines the day of week, term (for Saturdays), and public holidays. What do the residuals of that model look like?
5. What happens if you fit a day-of-week effect that varies by month (i.e., `n ~ wday * month`)? Why is this not very helpful?
6. What would you expect the model `n ~ wday + ns(date, 5)` to look like? Knowing what you know about the data, why would you expect it to be not particularly effective?
7. We hypothesized that people leaving on Sundays are more likely to be business travelers who need to be somewhere on Monday. Explore that hypothesis by seeing how it breaks down based on distance and time: if it's true, you'd expect to see more Sunday evening flights to places that are far away.

8. It's a little frustrating that Sunday and Saturday are on separate ends of the plot. Write a small function to set the levels of the factor so that the week starts on Monday.

Learning More About Models

We have only scratched the absolute surface of modeling, but you have hopefully gained some simple, but general-purpose tools that you can use to improve your own data analyses. It's OK to start simple! As you've seen, even very simple models can make a dramatic difference in your ability to tease out interactions between variables.

These modeling chapters are even more opinionated than the rest of the book. I approach modeling from a somewhat different perspective to most others, and there is relatively little space devoted to it. Modeling really deserves a book on its own, so I'd highly recommend that you read at least one of these three books:

- *Statistical Modeling: A Fresh Approach* (*http://bit.ly/statmodfresh*) by Danny Kaplan. This book provides a gentle introduction to modeling, where you build your intuition, mathematical tools, and R skills in parallel. The book replaces a traditional "introduction to statistics" course, providing a curriculum that is up-to-date and relevant to data science.
- *An Introduction to Statistical Learning* (*http://bit.ly/introstatlearn*) by Gareth James, Daniela Witten, Trevor Hastie, and Robert Tibshirani (available online for free). This book presents a family of modern modeling techniques collectively known as statistical learning. For an even deeper understanding of the math behind the models, read the classic *Elements of Statistical Learning* (*http://stanford.io/1ycOXbo*) by Trevor Hastie, Robert Tibshirani, and Jerome Friedman (also available online for free).
- *Applied Predictive Modeling* (*http://appliedpredictivemodeling.com*) by Max Kuhn and Kjell Johnson. This book is a companion to the **caret** package and provides practical tools for dealing with real-life predictive modeling challenges.

CHAPTER 20

Many Models with purrr and broom

Introduction

In this chapter you're going to learn three powerful ideas that help you to work with large numbers of models with ease:

- Using many simple models to better understand complex datasets.
- Using list-columns to store arbitrary data structures in a data frame. For example, this will allow you to have a column that contains linear models.
- Using the **broom** package, by David Robinson, to turn models into tidy data. This is a powerful technique for working with large numbers of models because once you have tidy data, you can apply all of the techniques that you've learned about earlier in the book.

We'll start by diving into a motivating example using data about life expectancy around the world. It's a small dataset but it illustrates how important modeling can be for improving your visualizations. We'll use a large number of simple models to partition out some of the strongest signals so we can see the subtler signals that remain. We'll also see how model summaries can help us pick out outliers and unusual trends.

The following sections will dive into more detail about the individual techniques:

- In "gapminder" on page 398, you'll see a motivating example that puts list-columns to use to fit per-county models to world economic data.
- In "List-Columns" on page 402, you'll learn more about the list-column data structure, and why it's valid to put lists in data frames.
- In "Creating List-Columns" on page 411, you'll learn the three main ways in which you'll create list-columns.
- In "Simplifying List-Columns" on page 416 you'll learn how to convert list-columns back to regular atomic vectors (or sets of atomic vectors) so you can work with them more easily.
- In "Making Tidy Data with broom" on page 419, you'll learn about the full set of tools provided by broom, and see how they can be applied to other types of data structure.

This chapter is somewhat aspirational: if this book is your first introduction to R, this chapter is likely to be a struggle. It requires you to have deeply internalized ideas about modeling, data structures, and iteration. So don't worry if you don't get it—just put this chapter aside for a few months, and come back when you want to stretch your brain.

Prerequisites

Working with many models requires many of the packages of the tidyverse (for data exploration, wrangling, and programming) and **modelr** to facilitate modeling.

```
library(modelr)
library(tidyverse)
```

gapminder

To motivate the power of many simple models, we're going to look into the "gapminder" data. This data was popularized by Hans Rosling, a Swedish doctor and statistician. If you've never heard of him, stop reading this chapter right now and go watch one of his videos! He is a fantastic data presenter and illustrates how you can use data to present a compelling story. A good place to start is this short video (*https://youtu.be/jbkSRLYSojo*) filmed in conjunction with the BBC.

The gapminder data summarizes the progression of countries over time, looking at statistics like life expectancy and GDP. The data is easy to access in R, thanks to Jenny Bryan, who created the **gapminder** package:

```
library(gapminder)
gapminder
#> # A tibble: 1,704 × 6
#>       country continent  year lifeExp      pop gdpPercap
#>        <fctr>    <fctr> <int>   <dbl>    <int>     <dbl>
#> 1 Afghanistan      Asia  1952    28.8  8425333       779
#> 2 Afghanistan      Asia  1957    30.3  9240934       821
#> 3 Afghanistan      Asia  1962    32.0 10267083       853
#> 4 Afghanistan      Asia  1967    34.0 11537966       836
#> 5 Afghanistan      Asia  1972    36.1 13079460       740
#> 6 Afghanistan      Asia  1977    38.4 14880372       786
#> # ... with 1,698 more rows
```

In this case study, we're going to focus on just three variables to answer the question "How does life expectancy (`lifeExp`) change over time (`year`) for each country (`country`)?" A good place to start is with a plot:

```
gapminder %>%
  ggplot(aes(year, lifeExp, group = country)) +
    geom_line(alpha = 1/3)
```

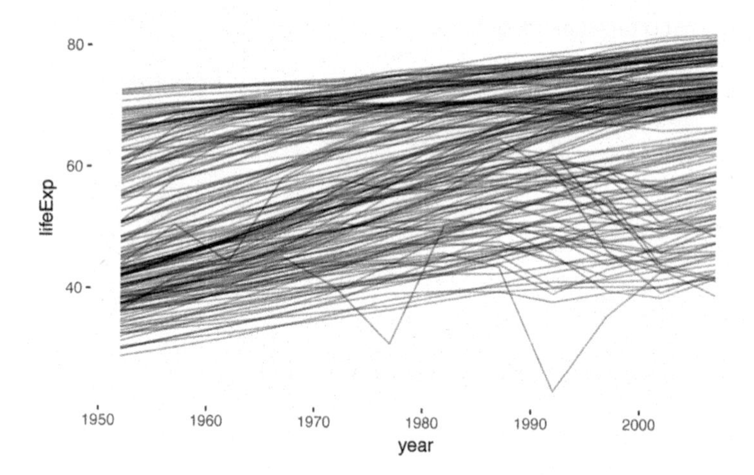

This is a small dataset: it only has ~1,700 observations and 3 variables. But it's still hard to see what's going on! Overall, it looks like life expectancy has been steadily improving. However, if you look closely, you might notice some countries that don't follow this pattern. How can we make those countries easier to see?

One way is to use the same approach as in the last chapter: there's a strong signal (overall linear growth) that makes it hard to see subtler trends. We'll tease these factors apart by fitting a model with a linear trend. The model captures steady growth over time, and the residuals will show what's left.

You already know how to do that if we had a single country:

```
nz <- filter(gapminder, country == "New Zealand")
nz %>%
  ggplot(aes(year, lifeExp)) +
  geom_line() +
  ggtitle("Full data = ")

nz_mod <- lm(lifeExp ~ year, data = nz)
nz %>%
  add_predictions(nz_mod) %>%
  ggplot(aes(year, pred)) +
  geom_line() +
  ggtitle("Linear trend + ")

nz %>%
  add_residuals(nz_mod) %>%
  ggplot(aes(year, resid)) +
  geom_hline(yintercept = 0, color = "white", size = 3) +
  geom_line() +
  ggtitle("Remaining pattern")
```

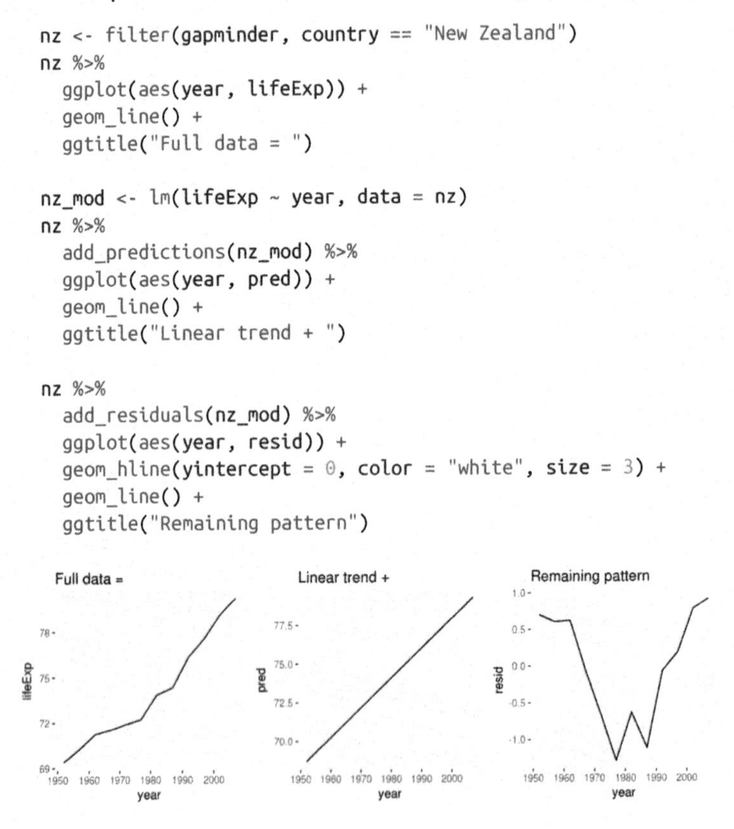

How can we easily fit that model to every country?

Nested Data

You could imagine copying and pasting that code multiple times; but you've already learned a better way! Extract out the common code with a function and repeat using a map function from **purrr**. This problem is structured a little differently to what you've seen before. Instead of repeating an action for each variable, we want to repeat an action for each country, a subset of rows. To do that, we need a new data structure: the *nested data frame*. To create a nested data frame we start with a grouped data frame, and "nest" it:

```
by_country <- gapminder %>%
  group_by(country, continent) %>%
  nest()

by_country
#> # A tibble: 142 × 3
#>       country continent               data
#>        <fctr>    <fctr>             <list>
#> 1 Afghanistan      Asia <tibble [12 × 4]>
#> 2     Albania    Europe <tibble [12 × 4]>
#> 3     Algeria    Africa <tibble [12 × 4]>
#> 4      Angola    Africa <tibble [12 × 4]>
#> 5   Argentina  Americas <tibble [12 × 4]>
#> 6   Australia   Oceania <tibble [12 × 4]>
#> # ... with 136 more rows
```

(I'm cheating a little by grouping on both `continent` and `country`. Given `country`, `continent` is fixed, so this doesn't add any more groups, but it's an easy way to carry an extra variable along for the ride.)

This creates a data frame that has one row per group (per country), and a rather unusual column: `data`. `data` is a list of data frames (or tibbles, to be precise). This seems like a crazy idea: we have a data frame with a column that is a list of other data frames! I'll explain shortly why I think this is a good idea.

The `data` column is a little tricky to look at because it's a moderately complicated list, and we're still working on good tools to explore these objects. Unfortunately using `str()` is not recommended as it will often produce very long output. But if you pluck out a single element from the `data` column you'll see that it contains all the data for that country (in this case, Afghanistan):

```
by_country$data[[1]]
#> # A tibble: 12 × 4
#>    year lifeExp      pop gdpPercap
#>   <int>   <dbl>    <int>     <dbl>
#> 1  1952    28.8  8425333       779
#> 2  1957    30.3  9240934       821
#> 3  1962    32.0 10267083       853
```

```
#> 4  1967     34.0 11537966         836
#> 5  1972     36.1 13079460         740
#> 6  1977     38.4 14880372         786
#> # ... with 6 more rows
```

Note the difference between a standard grouped data frame and a nested data frame: in a grouped data frame, each row is an observation; in a nested data frame, each row is a group. Another way to think about a nested dataset is we now have a meta-observation: a row that represents the complete time course for a country, rather than a single point in time.

List-Columns

Now that we have our nested data frame, we're in a good position to fit some models. We have a model-fitting function:

```
country_model <- function(df) {
  lm(lifeExp ~ year, data = df)
}
```

And we want to apply it to every data frame. The data frames are in a list, so we can use `purrr::map()` to apply `country_model` to each element:

```
models <- map(by_country$data, country_model)
```

However, rather than leaving the list of models as a free-floating object, I think it's better to store it as a column in the `by_country` data frame. Storing related objects in columns is a key part of the value of data frames, and why I think list-columns are such a good idea. In the course of working with these countries, we are going to have lots of lists where we have one element per country. So why not store them all together in one data frame?

In other words, instead of creating a new object in the global environment, we're going to create a new variable in the `by_country` data frame. That's a job for `dplyr::mutate()`:

```
by_country <- by_country %>%
  mutate(model = map(data, country_model))
by_country
#> # A tibble: 142 × 4
#>         country continent                data     model
#>          <fctr>    <fctr>              <list>    <list>
#> 1 Afghanistan      Asia <tibble [12 × 4]> <S3: lm>
#> 2     Albania    Europe <tibble [12 × 4]> <S3: lm>
#> 3     Algeria    Africa <tibble [12 × 4]> <S3: lm>
```

```
#> 4      Angola    Africa <tibble [12 × 4]> <S3: lm>
#> 5   Argentina  Americas <tibble [12 × 4]> <S3: lm>
#> 6   Australia   Oceania <tibble [12 × 4]> <S3: lm>
#> # ... with 136 more rows
```

This has a big advantage: because all the related objects are stored together, you don't need to manually keep them in sync when you filter or arrange. The semantics of the data frame takes care of that for you:

```
by_country %>%
  filter(continent == "Europe")
#> # A tibble: 30 × 4
#>                   country continent                 data    model
#>                    <fctr>    <fctr>               <list>   <list>
#> 1                 Albania    Europe <tibble [12 × 4]> <S3: lm>
#> 2                 Austria    Europe <tibble [12 × 4]> <S3: lm>
#> 3                 Belgium    Europe <tibble [12 × 4]> <S3: lm>
#> 4 Bosnia and Herzegovina    Europe <tibble [12 × 4]> <S3: lm>
#> 5                Bulgaria    Europe <tibble [12 × 4]> <S3: lm>
#> 6                 Croatia    Europe <tibble [12 × 4]> <S3: lm>
#> # ... with 24 more rows
by_country %>%
  arrange(continent, country)
#> # A tibble: 142 × 4
#>         country continent                 data    model
#>          <fctr>    <fctr>               <list>   <list>
#> 1       Algeria    Africa <tibble [12 × 4]> <S3: lm>
#> 2        Angola    Africa <tibble [12 × 4]> <S3: lm>
#> 3         Benin    Africa <tibble [12 × 4]> <S3: lm>
#> 4      Botswana    Africa <tibble [12 × 4]> <S3: lm>
#> 5  Burkina Faso    Africa <tibble [12 × 4]> <S3: lm>
#> 6       Burundi    Africa <tibble [12 × 4]> <S3: lm>
#> # ... with 136 more rows
```

If your list of data frames and list of models were separate objects, you have to remember that whenever you reorder or subset one vector, you need to reorder or subset all the others in order to keep them in sync. If you forget, your code will continue to work, but it will give the wrong answer!

Unnesting

Previously we computed the residuals of a single model with a single dataset. Now we have 142 data frames and 142 models. To compute the residuals, we need to call `add_residuals()` with each model-data pair:

```
by_country <- by_country %>%
  mutate(
    resids = map2(data, model, add_residuals)
  )
by_country
#> # A tibble: 142 × 5
#>       country continent                data    model
#>        <fctr>    <fctr>              <list>   <list>
#> 1 Afghanistan      Asia <tibble [12 × 4]> <S3: lm>
#> 2     Albania    Europe <tibble [12 × 4]> <S3: lm>
#> 3     Algeria    Africa <tibble [12 × 4]> <S3: lm>
#> 4      Angola    Africa <tibble [12 × 4]> <S3: lm>
#> 5   Argentina  Americas <tibble [12 × 4]> <S3: lm>
#> 6   Australia   Oceania <tibble [12 × 4]> <S3: lm>
#> # ... with 136 more rows, and 1 more variable:
#> #   resids <list>
```

But how can you plot a list of data frames? Instead of struggling to answer that question, let's turn the list of data frames back into a regular data frame. Previously we used `nest()` to turn a regular data frame into a nested data frame, and now we do the opposite with `unnest()`:

```
resids <- unnest(by_country, resids)
resids
#> # A tibble: 1,704 × 7
#>       country continent  year lifeExp      pop gdpPercap
#>        <fctr>    <fctr> <int>   <dbl>    <int>     <dbl>
#> 1 Afghanistan      Asia  1952    28.8  8425333       779
#> 2 Afghanistan      Asia  1957    30.3  9240934       821
#> 3 Afghanistan      Asia  1962    32.0 10267083       853
#> 4 Afghanistan      Asia  1967    34.0 11537966       836
#> 5 Afghanistan      Asia  1972    36.1 13079460       740
#> 6 Afghanistan      Asia  1977    38.4 14880372       786
#> # ... with 1,698 more rows, and 1 more variable: resid <dbl>
```

Note that each regular column is repeated once for each row in the nested column.

Now that we have regular data frame, we can plot the residuals:

```
resids %>%
  ggplot(aes(year, resid)) +
    geom_line(aes(group = country), alpha = 1 / 3) +
    geom_smooth(se = FALSE)
#> `geom_smooth()` using method = 'gam'
```

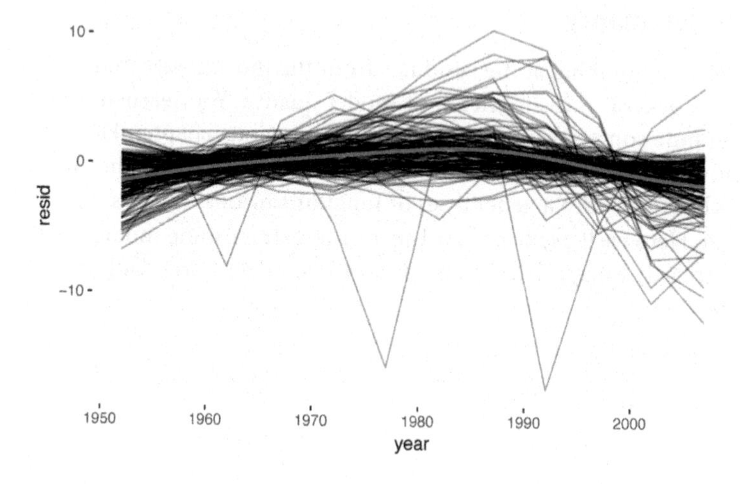

Faceting by continent is particularly revealing:

```
resids %>%
  ggplot(aes(year, resid, group = country)) +
    geom_line(alpha = 1 / 3) +
    facet_wrap(~continent)
```

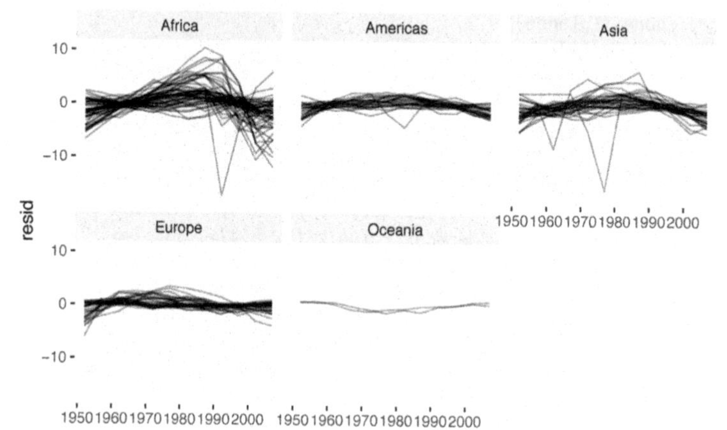

It looks like we've missed some mild pattern. There's also something interesting going on in Africa: we see some very large residuals, which suggests our model isn't fitting so well there. We'll explore that more in the next section, attacking it from a slightly different angle.

Model Quality

Instead of looking at the residuals from the model, we could look at some general measurements of model quality. You learned how to compute some specific measures in the previous chapter. Here we'll show a different approach using the **broom** package. The **broom** package provides a general set of functions to turn models into tidy data. Here we'll use `broom::glance()` to extract some model quality metrics. If we apply it to a model, we get a data frame with a single row:

```
broom::glance(nz_mod)
#>   r.squared adj.r.squared sigma statistic  p.value df logLik
#>   AIC  BIC
#> 1     0.954         0.949 0.804       205 5.41e-08  2  -13.3
#>   32.6 34.1
#>   deviance df.residual
#> 1     6.47          10
```

We can use `mutate()` and `unnest()` to create a data frame with a row for each country:

```
by_country %>%
  mutate(glance = map(model, broom::glance)) %>%
  unnest(glance)
#> # A tibble: 142 × 16
#>       country continent              data    model
#>        <fctr>    <fctr>            <list>   <list>
#> 1 Afghanistan      Asia <tibble [12 × 4]> <S3: lm>
#> 2     Albania    Europe <tibble [12 × 4]> <S3: lm>
#> 3     Algeria    Africa <tibble [12 × 4]> <S3: lm>
#> 4      Angola    Africa <tibble [12 × 4]> <S3: lm>
#> 5   Argentina  Americas <tibble [12 × 4]> <S3: lm>
#> 6   Australia   Oceania <tibble [12 × 4]> <S3: lm>
#> # ... with 136 more rows, and 12 more variables:
#> #   resids <list>, r.squared <dbl>, adj.r.squared <dbl>,
#> #   sigma <dbl>, statistic <dbl>, p.value <dbl>, df <int>,
#> #   logLik <dbl>, AIC <dbl>, BIC <dbl>, deviance <dbl>,
#> #   df.residual <int>
```

This isn't quite the output we want, because it still includes all the list-columns. This is default behavior when `unnest()` works on single-row data frames. To suppress these columns we use `.drop = TRUE`:

```
glance <- by_country %>%
  mutate(glance = map(model, broom::glance)) %>%
  unnest(glance, .drop = TRUE)
glance
```

```
#> # A tibble: 142 × 13
#>       country continent r.squared adj.r.squared sigma
#>        <fctr>    <fctr>     <dbl>         <dbl> <dbl>
#> 1 Afghanistan      Asia     0.948         0.942 1.223
#> 2     Albania    Europe     0.911         0.902 1.983
#> 3     Algeria    Africa     0.985         0.984 1.323
#> 4      Angola    Africa     0.888         0.877 1.407
#> 5   Argentina  Americas     0.996         0.995 0.292
#> 6   Australia   Oceania     0.980         0.978 0.621
#> # ... with 136 more rows, and 8 more variables:
#> #   statistic <dbl>, p.value <dbl>, df <int>, logLik <dbl>,
#> #   AIC <dbl>, BIC <dbl>, deviance <dbl>, df.residual <int>
```

(Pay attention to the variables that aren't printed: there's a lot of useful stuff there.)

With this data frame in hand, we can start to look for models that don't fit well:

```
glance %>%
  arrange(r.squared)
#> # A tibble: 142 × 13
#>     country continent r.squared adj.r.squared sigma
#>      <fctr>    <fctr>     <dbl>         <dbl> <dbl>
#> 1    Rwanda    Africa    0.0172      -0.08112  6.56
#> 2  Botswana    Africa    0.0340      -0.06257  6.11
#> 3  Zimbabwe    Africa    0.0562      -0.03814  7.21
#> 4    Zambia    Africa    0.0598      -0.03418  4.53
#> 5 Swaziland    Africa    0.0682      -0.02497  6.64
#> 6   Lesotho    Africa    0.0849      -0.00666  5.93
#> # ... with 136 more rows, and 8 more variables:
#> #   statistic <dbl>, p.value <dbl>, df <int>, logLik <dbl>,
#> #   AIC <dbl>, BIC <dbl>, deviance <dbl>, df.residual <int>
```

The worst models all appear to be in Africa. Let's double-check that with a plot. Here we have a relatively small number of observations and a discrete variable, so `geom_jitter()` is effective:

```
glance %>%
  ggplot(aes(continent, r.squared)) +
    geom_jitter(width = 0.5)
```

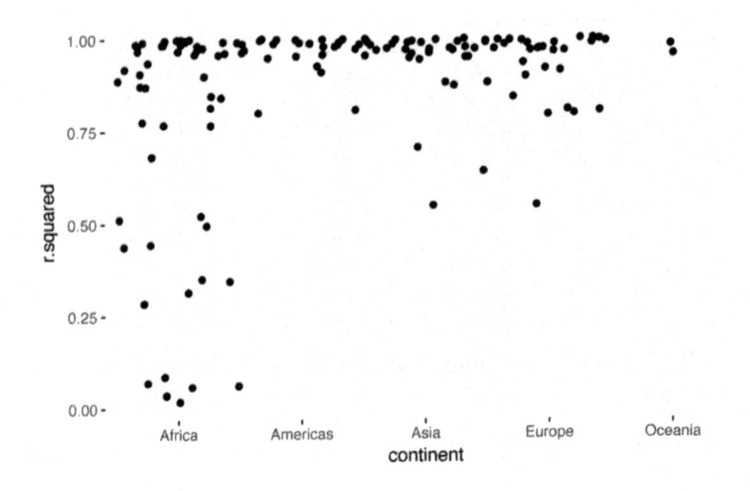

We could pull out the countries with particularly bad R^2 and plot the data:

```
bad_fit <- filter(glance, r.squared < 0.25)

gapminder %>%
  semi_join(bad_fit, by = "country") %>%
  ggplot(aes(year, lifeExp, color = country)) +
    geom_line()
```

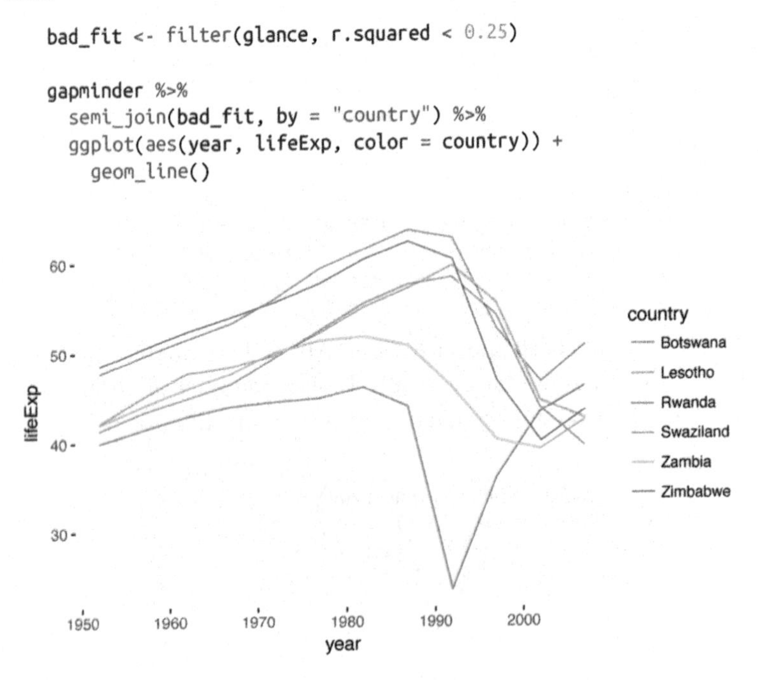

We see two main effects here: the tragedies of the HIV/AIDS epidemic and the Rwandan genocide.

Exercises

1. A linear trend seems to be slightly too simple for the overall trend. Can you do better with a quadratic polynomial? How can you interpret the coefficients of the quadratic? (Hint: you might want to transform `year` so that it has mean zero.)

2. Explore other methods for visualizing the distribution of R^2 per continent. You might want to try the **ggbeeswarm** package, which provides similar methods for avoiding overlaps as jitter, but uses deterministic methods.

3. To create the last plot (showing the data for the countries with the worst model fits), we needed two steps: we created a data frame with one row per country and then semi-joined it to the original dataset. It's possible avoid this join if we use `unnest()` instead of `unnest(.drop = TRUE)`. How?

List-Columns

Now that you've seen a basic workflow for managing many models, let's dive back into some of the details. In this section, we'll explore the list-column data structure in a little more detail. It's only recently that I've really appreciated the idea of the list-column. List-columns are implicit in the definition of the data frame: a data frame is a named list of equal length vectors. A list is a vector, so it's always been legitimate to use a list as a column of a data frame. However, base R doesn't make it easy to create list-columns, and `data.frame()` treats a list as a list of columns:

```
data.frame(x = list(1:3, 3:5))
#>   x.1.3 x.3.5
#> 1     1     3
#> 2     2     4
#> 3     3     5
```

You can prevent `data.frame()` from doing this with `I()`, but the result doesn't print particularly well:

```
data.frame(
  x = I(list(1:3, 3:5)),
  y = c("1, 2", "3, 4, 5")
)
#>          x       y
```

```
#> 1 1, 2, 3    1, 2
#> 2 3, 4, 5 3, 4, 5
```

Tibble alleviates this problem by being lazier (`tibble()` doesn't modify its inputs) and by providing a better print method:

```
tibble(
  x = list(1:3, 3:5),
  y = c("1, 2", "3, 4, 5")
)
#> # A tibble: 2 × 2
#>          x       y
#>     <list>   <chr>
#> 1 <int [3]>    1, 2
#> 2 <int [3]> 3, 4, 5
```

It's even easier with `tribble()` as it can automatically work out that you need a list:

```
tribble(
   ~x, ~y,
  1:3, "1, 2",
  3:5, "3, 4, 5"
)
#> # A tibble: 2 × 2
#>          x       y
#>     <list>   <chr>
#> 1 <int [3]>    1, 2
#> 2 <int [3]> 3, 4, 5
```

List-columns are often most useful as an intermediate data structure. They're hard to work with directly, because most R functions work with atomic vectors or data frames, but the advantage of keeping related items together in a data frame is worth a little hassle.

Generally there are three parts of an effective list-column pipeline:

1. You create the list-column using one of `nest()`, `summarize()` + `list()`, or `mutate()` + a map function, as described in "Creating List-Columns" on page 411.
2. You create other intermediate list-columns by transforming existing list columns with `map()`, `map2()`, or `pmap()`. For example, in the previous case study, we created a list-column of models by transforming a list-column of data frames.
3. You simplify the list-column back down to a data frame or atomic vector, as described in "Simplifying List-Columns" on page 416.

Creating List-Columns

Typically, you won't create list-columns with `tibble()`. Instead, you'll create them from regular columns, using one of three methods:

1. With `tidyr::nest()` to convert a grouped data frame into a nested data frame where you have list-column of data frames.
2. With `mutate()` and vectorized functions that return a list.
3. With `summarize()` and summary functions that return multiple results.

Alternatively, you might create them from a named list, using `tibble::enframe()`.

Generally, when creating list-columns, you should make sure they're homogeneous: each element should contain the same type of thing. There are no checks to make sure this is true, but if you use **purrr** and remember what you've learned about type-stable functions, you should find it happens naturally.

With Nesting

`nest()` creates a nested data frame, which is a data frame with a list-column of data frames. In a nested data frame each row is a meta-observation: the other columns give variables that define the observation (like country and continent earlier), and the list-column of data frames gives the individual observations that make up the meta-observation.

There are two ways to use `nest()`. So far you've seen how to use it with a grouped data frame. When applied to a grouped data frame, `nest()` keeps the grouping columns as is, and bundles everything else into the list-column:

```
gapminder %>%
  group_by(country, continent) %>%
  nest()
#> # A tibble: 142 × 3
#>       country continent              data
#>        <fctr>    <fctr>            <list>
#> 1 Afghanistan      Asia <tibble [12 × 4]>
#> 2     Albania    Europe <tibble [12 × 4]>
#> 3     Algeria    Africa <tibble [12 × 4]>
```

```
#> 4        Angola    Africa <tibble [12 × 4]>
#> 5   Argentina  Americas <tibble [12 × 4]>
#> 6   Australia   Oceania <tibble [12 × 4]>
#> # ... with 136 more rows
```

You can also use it on an ungrouped data frame, specifying which columns you want to nest:

```
gapminder %>%
  nest(year:gdpPercap)
#> # A tibble: 142 × 3
#>        country continent                data
#>         <fctr>    <fctr>              <list>
#> 1 Afghanistan      Asia <tibble [12 × 4]>
#> 2      Albania    Europe <tibble [12 × 4]>
#> 3      Algeria    Africa <tibble [12 × 4]>
#> 4       Angola    Africa <tibble [12 × 4]>
#> 5   Argentina  Americas <tibble [12 × 4]>
#> 6   Australia   Oceania <tibble [12 × 4]>
#> # ... with 136 more rows
```

From Vectorized Functions

Some useful functions take an atomic vector and return a list. For example, in Chapter 11 you learned about `stringr::str_split()`, which takes a character vector and returns a list of character vectors. If you use that inside mutate, you'll get a list-column:

```
df <- tribble(
  ~x1,
  "a,b,c",
  "d,e,f,g"
)

df %>%
  mutate(x2 = stringr::str_split(x1, ","))
#> # A tibble: 2 × 2
#>          x1        x2
#>       <chr>    <list>
#> 1   a,b,c <chr [3]>
#> 2 d,e,f,g <chr [4]>
```

`unnest()` knows how to handle these lists of vectors:

```
df %>%
  mutate(x2 = stringr::str_split(x1, ",")) %>%
  unnest()
#> # A tibble: 7 × 2
#>          x1    x2
#>       <chr> <chr>
#> 1   a,b,c     a
```

```
#> 2   a,b,c     b
#> 3   a,b,c     c
#> 4 d,e,f,g     d
#> 5 d,e,f,g     e
#> 6 d,e,f,g     f
#> # ... with 1 more rows
```

(If you find yourself using this pattern a lot, make sure to check out `tidyr:separate_rows()`, which is a wrapper around this common pattern).

Another example of this pattern is using the `map()`, `map2()`, `pmap()` functions from **purrr**. For example, we could take the final example from "Invoking Different Functions" on page 334 and rewrite it to use `mutate()`:

```
sim <- tribble(
  ~f,      ~params,
  "runif", list(min = -1, max = -1),
  "rnorm", list(sd = 5),
  "rpois", list(lambda = 10)
)

sim %>%
  mutate(sims = invoke_map(f, params, n = 10))
#> # A tibble: 3 × 3
#>       f     params        sims
#>   <chr>     <list>      <list>
#> 1 runif <list [2]> <dbl [10]>
#> 2 rnorm <list [1]> <dbl [10]>
#> 3 rpois <list [1]> <int [10]>
```

Note that technically `sim` isn't homogeneous because it contains both double and integer vectors. However, this is unlikely to cause many problems since integers and doubles are both numeric vectors.

From Multivalued Summaries

One restriction of `summarize()` is that it only works with summary functions that return a single value. That means that you can't use it with functions like `quantile()` that return a vector of arbitrary length:

```
mtcars %>%
  group_by(cyl) %>%
  summarize(q = quantile(mpg))
#> Error in eval(expr, envir, enclos): expecting a single value
```

You can however, wrap the result in a list! This obeys the contract of `summarize()`, because each summary is now a list (a vector) of length 1:

```
mtcars %>%
  group_by(cyl) %>%
  summarize(q = list(quantile(mpg)))
#> # A tibble: 3 × 2
#>     cyl        q
#>   <dbl>   <list>
#> 1     4 <dbl [5]>
#> 2     6 <dbl [5]>
#> 3     8 <dbl [5]>
```

To make useful results with `unnest()`, you'll also need to capture the probabilities:

```
probs <- c(0.01, 0.25, 0.5, 0.75, 0.99)
mtcars %>%
  group_by(cyl) %>%
  summarize(p = list(probs), q = list(quantile(mpg, probs))) %>%
  unnest()
#> # A tibble: 15 × 3
#>     cyl     p     q
#>   <dbl> <dbl> <dbl>
#> 1     4  0.01  21.4
#> 2     4  0.25  22.8
#> 3     4  0.50  26.0
#> 4     4  0.75  30.4
#> 5     4  0.99  33.8
#> 6     6  0.01  17.8
#> # ... with 9 more rows
```

From a Named List

Data frames with list-columns provide a solution to a common problem: what do you do if you want to iterate over both the contents of a list and its elements? Instead of trying to jam everything into one object, it's often easier to make a data frame: one column can contain the elements, and one column can contain the list. An easy way to create such a data frame from a list is `tibble::enframe()`:

```
x <- list(
  a = 1:5,
  b = 3:4,
  c = 5:6
)
```

```
df <- enframe(x)
df
#> # A tibble: 3 × 2
#>     name       value
#>    <chr>      <list>
#> 1      a <int [5]>
#> 2      b <int [2]>
#> 3      c <int [2]>
```

The advantage of this structure is that it generalizes in a straightforward way—names are useful if you have a character vector of metadata, but don't help if you have other types of data, or multiple vectors.

Now if you want to iterate over names and values in parallel, you can use `map2()`:

```
df %>%
  mutate(
    smry = map2_chr(
      name,
      value,
      ~ stringr::str_c(.x, ": ", .y[1])
    )
  )
#> # A tibble: 3 × 3
#>     name       value  smry
#>    <chr>      <list> <chr>
#> 1      a <int [5]>  a: 1
#> 2      b <int [2]>  b: 3
#> 3      c <int [2]>  c: 5
```

Exercises

1. List all the functions that you can think of that take an atomic vector and return a list.
2. Brainstorm useful summary functions that, like `quantile()`, return multiple values.
3. What's missing in the following data frame? How does `quantile()` return that missing piece? Why isn't that helpful here?

```
mtcars %>%
  group_by(cyl) %>%
  summarize(q = list(quantile(mpg))) %>%
  unnest()
#> # A tibble: 15 × 2
#>      cyl     q
#>    <dbl> <dbl>
```

```
#> 1      4  21.4
#> 2      4  22.8
#> 3      4  26.0
#> 4      4  30.4
#> 5      4  33.9
#> 6      6  17.8
#> # ... with 9 more rows
```

4. What does this code do? Why might might it be useful?

```
mtcars %>%
  group_by(cyl) %>%
  summarize_each(funs(list))
```

Simplifying List-Columns

To apply the techniques of data manipulation and visualization you've learned in this book, you'll need to simplify the list-column back to a regular column (an atomic vector), or set of columns. The technique you'll use to collapse back down to a simpler structure depends on whether you want a single value per element, or multiple values:

- If you want a single value, use `mutate()` with `map_lgl()`, `map_int()`, `map_dbl()`, and `map_chr()` to create an atomic vector.
- If you want many values, use `unnest()` to convert list-columns back to regular columns, repeating the rows as many times as necessary.

These are described in more detail in the following sections.

List to Vector

If you can reduce your list column to an atomic vector then it will be a regular column. For example, you can always summarize an object with it's type and length, so this code will work regardless of what sort of list-column you have:

```
df <- tribble(
  ~x,
  letters[1:5],
  1:3,
  runif(5)
)
```

```
df %>% mutate(
  type = map_chr(x, typeof),
  length = map_int(x, length)
)
#> # A tibble: 3 × 3
#>             x      type length
#>        <list>     <chr>  <int>
#> 1 <chr [5]> character      5
#> 2 <int [3]>   integer      3
#> 3 <dbl [5]>    double      5
```

This is the same basic information that you get from the default `tbl` `print` method, but now you can use it for filtering. This is a useful technique if you have a heterogeneous list, and want to filter out the parts that aren't working for you.

Don't forget about the `map_*()` shortcuts—you can use `map_chr(x, "apple")` to extract the string stored in `apple` for each element of `x`. This is useful for pulling apart nested lists into regular columns. Use the `.null` argument to provide a value to use if the element is missing (instead of returning `NULL`):

```
df <- tribble(
  ~x,
  list(a = 1, b = 2),
  list(a = 2, c = 4)
)
df %>% mutate(
  a = map_dbl(x, "a"),
  b = map_dbl(x, "b", .null = NA_real_)
)
#> # A tibble: 2 × 3
#>              x     a     b
#>         <list> <dbl> <dbl>
#> 1 <list [2]>     1     2
#> 2 <list [2]>     2    NA
```

Unnesting

`unnest()` works by repeating the regular columns once for each element of the list-column. For example, in the following very simple example we repeat the first row four times (because there the first element of `y` has length four), and the second row once:

```
tibble(x = 1:2, y = list(1:4, 1)) %>% unnest(y)
#> # A tibble: 5 × 2
#>       x     y
#>   <int> <dbl>
```

```
#> 1       1     1
#> 2       1     2
#> 3       1     3
#> 4       1     4
#> 5       2     1
```

This means that you can't simultaneously unnest two columns that contain a different number of elements:

```
# Ok, because y and z have the same number of elements in
# every row
df1 <- tribble(
  ~x, ~y,           ~z,
   1, c("a", "b"), 1:2,
   2, "c",           3
)
df1
#> # A tibble: 2 × 3
#>         x         y           z
#>     <dbl>    <list>      <list>
#> 1       1 <chr [2]> <int [2]>
#> 2       2 <chr [1]> <dbl [1]>
df1 %>% unnest(y, z)
#> # A tibble: 3 × 3
#>         x     y     z
#>     <dbl> <chr> <dbl>
#> 1       1     a     1
#> 2       1     b     2
#> 3       2     c     3

# Doesn't work because y and z have different number of elements
df2 <- tribble(
  ~x, ~y,           ~z,
   1, "a",         1:2,
   2, c("b", "c"),   3
)
df2
#> # A tibble: 2 × 3
#>         x         y           z
#>     <dbl>    <list>      <list>
#> 1       1 <chr [1]> <int [2]>
#> 2       2 <chr [2]> <dbl [1]>
df2 %>% unnest(y, z)
#> Error: All nested columns must have
#> the same number of elements.
```

The same principle applies when unnesting list-columns of data frames. You can unnest multiple list-columns as long as all the data frames in each row have the same number of rows.

Exercises

1. Why might the `lengths()` function be useful for creating atomic vector columns from list-columns?
2. List the most common types of vector found in a data frame. What makes lists different?

Making Tidy Data with broom

The **broom** package provides three general tools for turning models into tidy data frames:

- `broom::glance(model)` returns a row for each model. Each column gives a model summary: either a measure of model quality, or complexity, or a combination of the two.
- `broom::tidy(model)` returns a row for each coefficient in the model. Each column gives information about the estimate or its variability.
- `broom::augment(model, data)` returns a row for each row in `data`, adding extra values like residuals, and influence statistics.

Broom works with a wide variety of models produced by the most popular modelling packages. See *https://github.com/tidyverse/broom* for a list of currently supported models.

PART V
Communicate

So far, you've learned the tools to get your data into R, tidy it into a form convenient for analysis, and then understand your data through transformation, visualization, and modeling. However, it doesn't matter how great your analysis is unless you can explain it to others: you need to *communicate* your results.

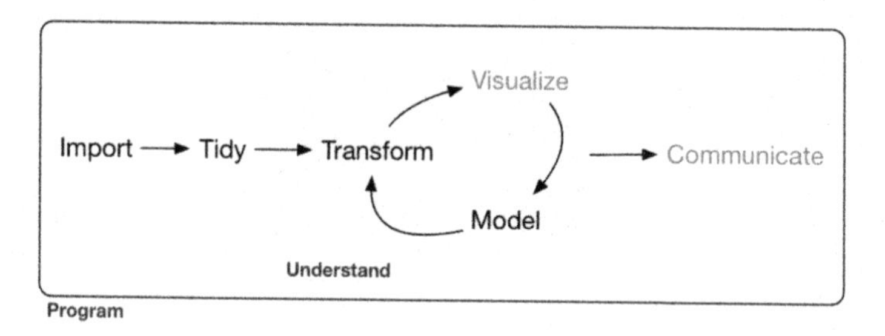

Communication is the theme of the following four chapters:

- In Chapter 21, you will learn about R Markdown, a tool for integrating prose, code, and results. You can use R Markdown in notebook mode for analyst-to-analyst communication, and in report mode for analyst-to-decision-maker communication. Thanks to the power of R Markdown formats, you can even use the same document for both purposes.

- In Chapter 22, you will learn how to take your exploratory graphics and turn them into expository graphics, graphics that help the newcomer to your analysis understand what's going on as quickly and easily as possible.
- In Chapter 23, you'll learn a little about the many other varieties of outputs you can produce using R Markdown, including dashboards, websites, and books.
- We'll finish up with Chapter 24, where you'll learn about the "analysis notebook" and how to systematically record your successes and failures so that you can learn from them.

Unfortunately, these chapters focus mostly on the technical mechanics of communication, not the really hard problems of communicating your thoughts to other humans. However, there are lot of other great books about communication, which we'll point you to at the end of each chapter.

CHAPTER 21

R Markdown

Introduction

R Markdown provides a unified authoring framework for data science, combining your code, its results, and your prose commentary. R Markdown documents are fully reproducible and support dozens of output formats, like PDFs, Word files, slideshows, and more.

R Markdown files are designed to be used in three ways:

- For communicating to decision makers, who want to focus on the conclusions, not the code behind the analysis.
- For collaborating with other data scientists (including future you!), who are interested in both your conclusions, and how you reached them (i.e., the code).
- As an environment in which to *do* data science, as a modern day lab notebook where you can capture not only what you did, but also what you were thinking.

R Markdown integrates a number of R packages and external tools. This means that help is, by and large, not available through `?`. Instead, as you work through this chapter, and use R Markdown in the future, keep these resources close to hand:

- R Markdown Cheat Sheet: available in the RStudio IDE under *Help → Cheatsheets → R Markdown Cheat Sheet*

- R Markdown Reference Guide: available in the RStudio IDE under *Help → Cheatsheets → R Markdown Reference Guide*

Both cheatsheets are also available at *http://rstudio.com/cheatsheets*.

Prerequisites

You need the **rmarkdown** package, but you don't need to explicitly install it or load it, as RStudio automatically does both when needed.

R Markdown Basics

This is an R Markdown file, a plain-text file that has the extension *.Rmd*:

````
---
title: "Diamond sizes"
date: 2016-08-25
output: html_document
---

```{r setup, include = FALSE}
library(ggplot2)
library(dplyr)

smaller <- diamonds %>%
  filter(carat <= 2.5)
```

We have data about `r nrow(diamonds)` diamonds. Only
`r nrow(diamonds) - nrow(smaller)` are larger than
2.5 carats. The distribution of the remainder is shown
below:

```{r, echo = FALSE}
smaller %>%
  ggplot(aes(carat)) +
  geom_freqpoly(binwidth = 0.01)
```
````

It contains three important types of content:

1. An (optional) *YAML header* surrounded by `---`s.
2. *Chunks* of R code surrounded by ```` ``` ````.
3. Text mixed with simple text formatting like `# heading` and `_italics_`.

When you open an *.Rmd*, you get a notebook interface where code and output are interleaved. You can run each code chunk by clicking the Run icon (it looks like a play button at the top of the chunk), or by pressing Cmd/Ctrl-Shift-Enter. RStudio executes the code and displays the results inline with the code:

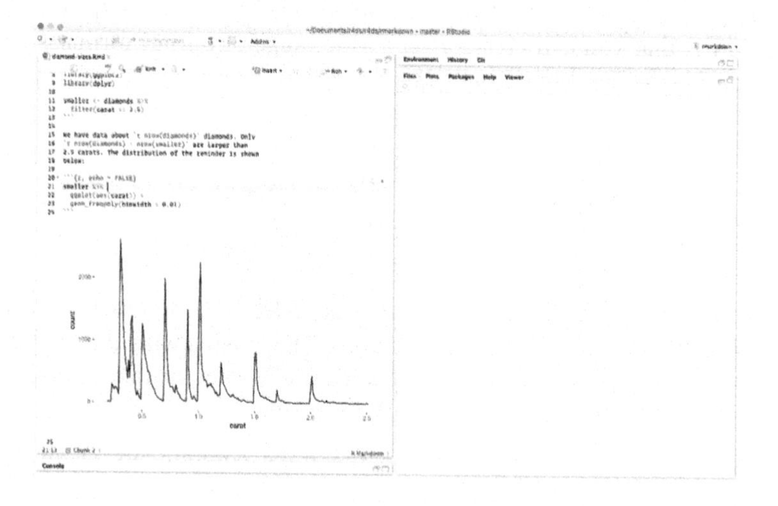

To produce a complete report containing all text, code, and results, click "Knit" or press Cmd/Ctrl-Shift-K. You can also do this programmatically with `rmarkdown::render("1-example.Rmd")`. This will display the report in the viewer pane, and create a self-contained HTML file that you can share with others.

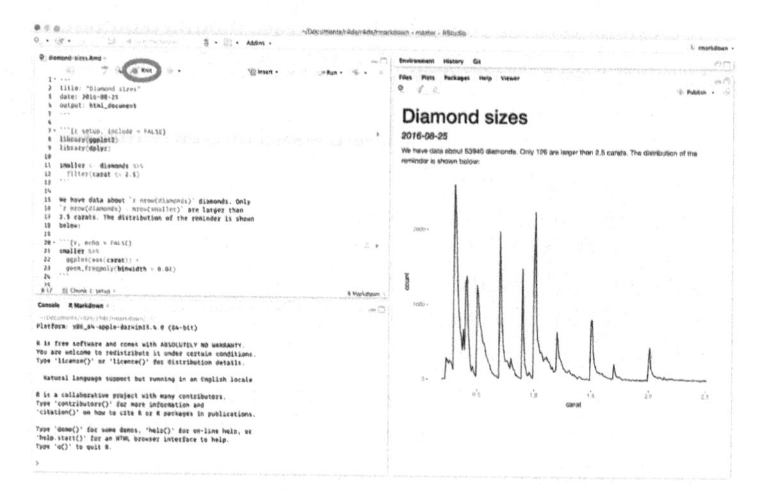

When you *knit* the document R Markdown sends the *.Rmd* file to **knitr** (*http://yihui.name/knitr/*), which executes all of the code chunks and creates a new Markdown (*.md*) document that includes the code and its output. The Markdown file generated by **knitr** is then processed by **pandoc** (*http://pandoc.org/*), which is responsible for creating the finished file. The advantage of this two-step workflow is that you can create a very wide range of output formats, as you'll learn about in Chapter 23.

To get started with your own *.Rmd* file, select *File → New File → R Markdown...* in the menu bar. RStudio will launch a wizard that you can use to pre-populate your file with useful content that reminds you how the key features of R Markdown work.

The following sections dive into the three components of an R Markdown document in more detail: the Markdown text, the code chunks, and the YAML header.

Exercises

1. Create a new notebook using *File → New File → R Notebook*. Read the instructions. Practice running the chunks. Verify that you can modify the code, rerun it, and see modified output.
2. Create a new R Markdown document with *File → New File → R Markdown...* Knit it by clicking the appropriate button. Knit it by using the appropriate keyboard shortcut. Verify that you can modify the input and see the output update.
3. Compare and contrast the R Notebook and R Markdown files you created earlier. How are the outputs similar? How are they different? How are the inputs similar? How are they different? What happens if you copy the YAML header from one to the other?
4. Create one new R Markdown document for each of the three built-in formats: HTML, PDF, and Word. Knit each of the three documents. How does the output differ? How does the input

differ? (You may need to install LaTeX in order to build the PDF output—RStudio will prompt you if this is necessary.)

Text Formatting with Markdown

Prose in *.Rmd* files is written in Markdown, a lightweight set of conventions for formatting plain-text files. Markdown is designed to be easy to read and easy to write. It is also very easy to learn. The following guide shows how to use Pandoc's Markdown, a slightly extended version of Markdown that R Markdown understands:

```
Text formatting
------------------------------------------------------------

*italic*  or _italic_
**bold**   __bold__
`code`
superscript^2^ and subscript~2~

Headings
------------------------------------------------------------

# 1st Level Header

## 2nd Level Header

### 3rd Level Header

Lists
------------------------------------------------------------

*   Bulleted list item 1

*   Item 2

    * Item 2a

    * Item 2b

1.  Numbered list item 1

1.  Item 2. The numbers are incremented automatically in
the output.

Links and images
------------------------------------------------------------

<http://example.com>
```

```
[linked phrase](http://example.com)

![optional caption text](path/to/img.png)

Tables
------------------------------------------------------------

First Header  | Second Header
------------- | -------------
Content Cell  | Content Cell
Content Cell  | Content Cell
```

The best way to learn these is simply to try them out. It will take a few days, but soon they will become second nature, and you won't need to think about them. If you forget, you can get to a handy reference sheet with *Help → Markdown Quick Reference*.

Exercises

1. Practice what you've learned by creating a brief CV. The title should be your name, and you should include headings for (at least) education or employment. Each of the sections should include a bulleted list of jobs/degrees. Highlight the year in bold.

2. Using the R Markdown quick reference, figure out how to:

   a. Add a footnote.

   b. Add a horizontal rule.

   c. Add a block quote.

3. Copy and paste the contents of *diamond-sizes.Rmd* from *https://github.com/hadley/r4ds/tree/master/rmarkdown* into a local R Markdown document. Check that you can run it, then add text after the frequency polygon that describes its most striking features.

Code Chunks

To run code inside an R Markdown document, you need to insert a chunk. There are three ways to do so:

1. The keyboard shortcut Cmd/Ctrl-Alt-I
2. The "Insert" button icon in the editor toolbar
3. By manually typing the chunk delimiters ` ```{r} ` and ` ``` `

Obviously, I'd recommend you learn the keyboard shortcut. It will save you a lot of time in the long run!

You can continue to run the code using the keyboard shortcut that by now (I hope!) you know and love: Cmd/Ctrl-Enter. However, chunks get a new keyboard shortcut: Cmd/Ctrl-Shift-Enter, which runs all the code in the chunk. Think of a chunk like a function. A chunk should be relatively self-contained, and focused around a single task.

The following sections describe the chunk header, which consists of ` ```{r `, followed by an optional chunk name, followed by comma-separated options, followed by `}`. Next comes your R code and the chunk end is indicated by a final ` ``` `.

Chunk Name

Chunks can be given an optional name: ` ```{r by-name} `. This has three advantages:

- You can more easily navigate to specific chunks using the drop-down code navigator in the bottom-left of the script editor:

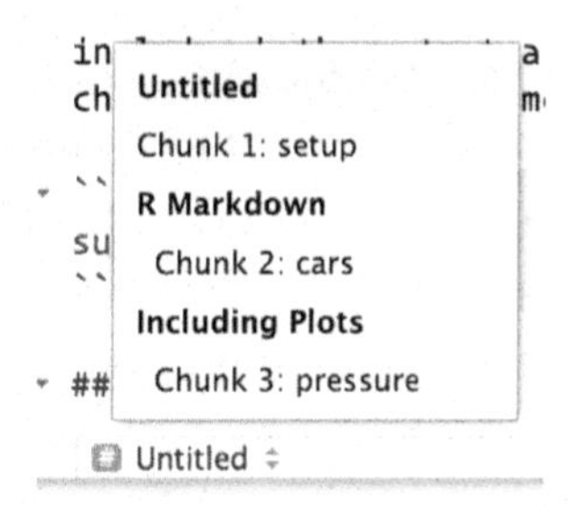

- Graphics produced by the chunks will have useful names that make them easier to use elsewhere. More on that in "Other Important Options" on page 467.

- You can set up networks of cached chunks to avoid re-performing expensive computations on every run. More on that in a bit.

There is one chunk name that imbues special behavior: `setup`. When you're in a notebook mode, the chunk named `setup` will be run automatically once, before any other code is run.

Chunk Options

Chunk output can be customized with *options*, arguments supplied to the chunk header. **knitr** provides almost 60 options that you can use to customize your code chunks. Here we'll cover the most important chunk options that you'll use frequently. You can see the full list at *http://yihui.name/knitr/options/*.

The most important set of options controls if your code block is executed and what results are inserted in the finished report:

- `eval = FALSE` prevents code from being evaluated. (And obviously if the code is not run, no results will be generated.) This is useful for displaying example code, or for disabling a large block of code without commenting each line.
- `include = FALSE` runs the code, but doesn't show the code or results in the final document. Use this for setup code that you don't want cluttering your report.
- `echo = FALSE` prevents code, but not the results from appearing in the finished file. Use this when writing reports aimed at people who don't want to see the underlying R code.
- `message = FALSE` or `warning = FALSE` prevents messages or warnings from appearing in the finished file.
- `results = 'hide'` hides printed output; `fig.show = 'hide'` hides plots.
- `error = TRUE` causes the render to continue even if code returns an error. This is rarely something you'll want to include in the final version of your report, but can be very useful if you need to debug exactly what is going on inside your *.Rmd*. It's also useful if you're teaching R and want to deliberately include an error. The default, `error = FALSE`, causes knitting to fail if there is a single error in the document.

The following table summarizes which types of output each option suppressess:

Option	Run code	Show code	Output	Plots	Messages	Warnings
`eval = FALSE`	X		X	X	X	X
`include = FALSE`		X	X	X	X	X
`echo = FALSE`		X				
`results = "hide"`			X			
`fig.show = "hide"`				X		
`message = FALSE`					X	
`warning = FALSE`						X

Table

By default, R Markdown prints data frames and matrices as you'd see them in the console:

```
mtcars[1:5, 1:10]
#>                    mpg cyl disp  hp drat   wt qsec vs am gear
#> Mazda RX4         21.0   6  160 110 3.90 2.62 16.5  0  1    4
#> Mazda RX4 Wag     21.0   6  160 110 3.90 2.88 17.0  0  1    4
#> Datsun 710        22.8   4  108  93 3.85 2.32 18.6  1  1    4
#> Hornet 4 Drive    21.4   6  258 110 3.08 3.21 19.4  1  0    3
#> Hornet Sportabout 18.7   8  360 175 3.15 3.44 17.0  0  0    3
```

If you prefer that data be displayed with additional formatting you can use the `knitr::kable` function. The following code generates Table 21-1:

```
knitr::kable(
  mtcars[1:5, ],
  caption = "A knitr kable."
)
```

Table 21-1. A knitr kable

	mpg	cyl	disp	hp	drat	wt	qsec	vs	am	gear	carb
Mazda RX4	21.0	6	160	110	3.90	2.62	16.5	0	1	4	4
Mazda RX4 Wag	21.0	6	160	110	3.90	2.88	17.0	0	1	4	4
Datsun 710	22.8	4	108	93	3.85	2.32	18.6	1	1	4	1
Hornet 4 Drive	21.4	6	258	110	3.08	3.21	19.4	1	0	3	1
Hornet Sportabout	18.7	8	360	175	3.15	3.44	17.0	0	0	3	2

Read the documentation for ?knitr::kable to see the other ways in which you can customize the table. For even deeper customization, consider the **xtable**, **stargazer**, **pander**, **tables**, and **ascii** packages. Each provides a set of tools for returning formatted tables from R code.

There is also a rich set of options for controlling how figures are embedded. You'll learn about these in "Saving Your Plots" on page 464.

Caching

Normally, each knit of a document starts from a completely clean slate. This is great for reproducibility, because it ensures that you've captured every important computation in code. However, it can be painful if you have some computations that take a long time. The solution is cache = TRUE. When set, this will save the output of the chunk to a specially named file on disk. On subsequent runs, **knitr** will check to see if the code has changed, and if it hasn't, it will reuse the cached results.

The caching system must be used with care, because by default it is based on the code only, not its dependencies. For example, here the processed_data chunk depends on the raw_data chunk:

````
```{r raw_data}
rawdata <- readr::read_csv("a_very_large_file.csv")
```

```{r processed_data, cached = TRUE}
processed_data <- rawdata %>%
  filter(!is.na(import_var)) %>%
  mutate(new_variable = complicated_transformation(x, y, z))
```
````

Caching the processed_data chunk means that it will get rerun if the **dplyr** pipeline is changed, but it won't get rerun if the read_csv() call changes. You can avoid that problem with the dependson chunk option:

````
```{r processed_data, cached = TRUE, dependson = "raw_data"}
processed_data <- rawdata %>%
  filter(!is.na(import_var)) %>%
  mutate(new_variable = complicated_transformation(x, y, z))
```
````

dependson should contain a character vector of *every* chunk that the cached chunk depends on. **knitr** will update the results for the cached chunk whenever it detects that one of its dependencies has changed.

Note that the chunks won't update if *a_very_large_file.csv* changes, because **knitr** caching only tracks changes within the *.Rmd* file. If you want to also track changes to that file you can use the cache.extra option. This is an arbitrary R expression that will invalidate the cache whenever it changes. A good function to use is file.info(): it returns a bunch of information about the file including when it was last modified. Then you can write:

```
```{r raw_data, cache.extra = file.info("a_very_large_file.csv")}
rawdata <- readr::read_csv("a_very_large_file.csv")
```
```

As your caching strategies get progressively more complicated, it's a good idea to regularly clear out all your caches with knitr::clean_cache().

I've used the advice of David Robinson (*http://bit.ly/DavidRobinsonTwitter*) to name these chunks: each chunk is named after the primary object that it creates. This makes it easier to understand the dependson specification.

Global Options

As you work more with **knitr**, you will discover that some of the default chunk options don't fit your needs, and want to change them. You can do that by calling knitr::opts_chunk$set() in a code chunk. For example, when writing books and tutorials I set:

```
knitr::opts_chunk$set(
  comment = "#>",
  collapse = TRUE
)
```

This uses my preferred comment formatting, and ensures that the code and output are kept closely entwined. On the other hand, if you were preparing a report, you might set:

```
knitr::opts_chunk$set(
  echo = FALSE
)
```

That will hide the code by default, only showing the chunks you deliberately choose to show (with `echo = TRUE`). You might consider setting `message = FALSE` and `warning = FALSE`, but that would make it harder to debug problems because you wouldn't see any messages in the final document.

Inline Code

There is one other way to embed R code into an R Markdown document: directly into the text, with: `` `r ` ``. This can be very useful if you mention properties of your data in the text. For example, in the example document I used at the start of the chapter I had:

> We have data about `` `r nrow(diamonds)` `` diamonds. Only `` `r nrow(diamonds) - nrow(smaller)` `` are larger than 2.5 carats. The distribution of the remainder is shown below:

When the report is knit, the results of these computations are inserted into the text:

> We have data about 53940 diamonds. Only 126 are larger than 2.5 carats. The distribution of the remainder is shown below:

When inserting numbers into text, `format()` is your friend. It allows you to set the number of `digits` so you don't print to a ridiculous degree of accuracy, and a `big.mark` to make numbers easier to read. I'll often combine these into a helper function:

```
comma <- function(x) format(x, digits = 2, big.mark = ",")
comma(3452345)
#> [1] "3,452,345"
comma(.12358124331)
#> [1] "0.12"
```

Exercises

1. Add a section that explores how diamond sizes vary by cut, color, and clarity. Assume you're writing a report for someone who doesn't know R, and instead of setting `echo = FALSE` on each chunk, set a global option.

2. Download *diamond-sizes.Rmd* from *https://github.com/hadley/r4ds/tree/master/rmarkdown*. Add a section that describes the largest 20 diamonds, including a table that displays their most important attributes.

3. Modify *diamonds-sizes.Rmd* to use `comma()` to produce nicely formatted output. Also include the percentage of diamonds that are larger than 2.5 carats.
4. Set up a network of chunks where `d` depends on `c` and `b`, and both `b` and `c` depend on `a`. Have each chunk print `lubridate::now()`, set `cache = TRUE`, then verify your understanding of caching.

Troubleshooting

Troubleshooting R Markdown documents can be challenging because you are no longer in an interactive R environment, and you will need to learn some new tricks. The first thing you should always try is to re-create the problem in an interactive session. Restart R, then "Run all chunks" (either from the the Code menu, under the Run region, or with the keyboard shortcut Ctrl-Alt-R). If you're lucky, that will re-create the problem, and you can figure out what's going on interactively.

If that doesn't help, there must be something different between your interactive environment and the R Markdown environment. You're going to need to systematically explore the options. The most common difference is the working directory: the working directory of an R Markdown document is the directory in which it lives. Check that the working directory is what you expect by including `getwd()` in a chunk.

Next, brainstorm all of the things that might cause the bug. You'll need to systematically check that they're the same in your R session and your R Markdown session. The easiest way to do that is to set `error = TRUE` on the chunk causing the problem, then use `print()` and `str()` to check that settings are as you expect.

YAML Header

You can control many other "whole document" settings by tweaking the parameters of the YAML header. You might wonder what YAML stands for: it's "yet another markup language," which is designed for representing hierarchical data in a way that's easy for humans to read and write. R Markdown uses it to control many details of the output. Here we'll discuss two: document parameters and bibliographies.

Parameters

R Markdown documents can include one or more parameters whose values can be set when you render the report. Parameters are useful when you want to re-render the same report with distinct values for various key inputs. For example, you might be producing sales reports per branch, exam results by student, or demographic summaries by country. To declare one or more parameters, use the `params` field.

This example use a `my_class` parameter to determine which class of cars to display:

```
---
output: html_document
params:
  my_class: "suv"
---

```{r setup, include = FALSE}
library(ggplot2)
library(dplyr)

class <- mpg %>% filter(class == params$my_class)
```

Fuel economy for `r params$my_class`s

```{r, message = FALSE}
ggplot(class, aes(displ, hwy)) +
  geom_point() +
  geom_smooth(se = FALSE)
```
```

As you can see, parameters are available within the code chunks as a read-only list named `params`.

You can write atomic vectors directly into the YAML header. You can also run arbitrary R expressions by prefacing the parameter value with `!r`. This is a good way to specify date/time parameters:

```
params:
  start: !r lubridate::ymd("2015-01-01")
  snapshot: !r lubridate::ymd_hms("2015-01-01 12:30:00")
```

In RStudio, you can click the "Knit with Parameters" option in the Knit drop-down menu to set parameters, render, and preview the report in a single user-friendly step. You can customize the dialog by setting other options in the header. See *http://bit.ly/ParamReports* for more details.

Alternatively, if you need to produce many such parameterized reports, you can call `rmarkdown::render()` with a list of `params`:

```
rmarkdown::render(
  "fuel-economy.Rmd",
  params = list(my_class = "suv")
)
```

This is particularly powerful in conjunction with `purrr:pwalk()`. The following example creates a report for each value of `class` found in `mpg`. First we create a data frame that has one row for each class, giving the `filename` of report and the `params` it should be given:

```
reports <- tibble(
  class = unique(mpg$class),
  filename = stringr::str_c("fuel-economy-", class, ".html"),
  params = purrr::map(class, ~ list(my_class = .))
)
reports
#> # A tibble: 7 × 3
#>     class                   filename     params
#>     <chr>                      <chr>     <list>
#> 1 compact fuel-economy-compact.html <list [1]>
#> 2 midsize fuel-economy-midsize.html <list [1]>
#> 3     suv     fuel-economy-suv.html <list [1]>
#> 4 2seater fuel-economy-2seater.html <list [1]>
#> 5 minivan fuel-economy-minivan.html <list [1]>
#> 6  pickup  fuel-economy-pickup.html <list [1]>
#> # ... with 1 more rows
```

Then we match the column names to the argument names of `render()`, and use **purrr**'s *parallel* walk to call `render()` once for each row:

```
reports %>%
  select(output_file = filename, params) %>%
  purrr::pwalk(rmarkdown::render, input = "fuel-economy.Rmd")
```

Bibliographies and Citations

Pandoc can automatically generate citations and a bibliography in a number of styles. To use this feature, specify a bibliography file using the `bibliography` field in your file's header. The field should contain a path from the directory that contains your *.Rmd* file to the file that contains the bibliography file:

```
bibliography: rmarkdown.bib
```

You can use many common bibliography formats including BibLaTeX, BibTeX, endnote, and medline.

To create a citation within your *.Rmd* file, use a key composed of "@" *and the citation identifier* from the bibliography file. Then place the citation in square brackets. Here are some examples:

```
Separate multiple citations with a `;`:
Blah blah [@smith04; @doe99].

You can add arbitrary comments inside the square brackets:
Blah blah [see @doe99, pp. 33-35; also @smith04, ch. 1].

Remove the square brackets to create an in-text citation:
@smith04 says blah, or @smith04 [p. 33] says blah.

Add a `-` before the citation to suppress the author's name:
Smith says blah [-@smith04].
```

When R Markdown renders your file, it will build and append a bibliography to the end of your document. The bibliography will contain each of the cited references from your bibliography file, but it will not contain a section heading. As a result it is common practice to end your file with a section header for the bibliography, such as `# References` or `# Bibliography`.

You can change the style of your citations and bibliography by referencing a CSL (citation style language) file to the `csl` field:

```
bibliography: rmarkdown.bib
csl: apa.csl
```

As with the bibliography field, your CSL file should contain a path to the file. Here I assume that the CSL file is in the same directory as the *.Rmd* file. A good place to find CSL style files for common bibliography styles is *http://github.com/citation-style-language/styles*.

Learning More

R Markdown is still relatively young, and is still growing rapidly. The best place to stay on top of innovations is the official R Markdown website: *http://rmarkdown.rstudio.com*.

There are two important topics that we haven't covered here: collaboration, and the details of accurately communicating your ideas to other humans. Collaboration is a vital part of modern data science, and you can make your life much easier by using version control tools, like Git and GitHub. We recommend two free resources that will teach you about Git:

- "Happy Git with R": a user-friendly introduction to Git and GitHub from R users, by Jenny Bryan. The book is freely available online (*http://happygitwithr.com*).
- The "Git and GitHub" chapter of *R Packages*, by Hadley. You can also read it for free online: *http://r-pkgs.had.co.nz/git.html*.

I have also not talked about what you should actually write in order to clearly communicate the results of your analysis. To improve your writing, I highly recommend reading either *Style: Lessons in Clarity and Grace* by Joseph M. Williams and Joseph Bizup, or *The Sense of Structure: Writing from the Reader's Perspective* by George Gopen. Both books will help you understand the structure of sentences and paragraphs, and give you the tools to make your writing more clear. (These books are rather expensive if purchased new, but they're used by many English classes so there are plenty of cheap secondhand copies). George Gopen also has a number of short articles on writing (*http://georgegopen.com/articles/litigation/*). They are aimed at lawyers, but almost everything applies to data scientists too.

CHAPTER 22

Graphics for Communication with ggplot2

Introduction

In Chapter 5, you learned how to use plots as tools for *exploration*. When you make exploratory plots, you know—even before looking —which variables the plot will display. You made each plot for a purpose, could quickly look at it, and then move on to the next plot. In the course of most analyses, you'll produce tens or hundreds of plots, most of which are immediately thrown away.

Now that you understand your data, you need to *communicate* your understanding to others. Your audience will likely not share your background knowledge and will not be deeply invested in the data. To help others quickly build up a good mental model of the data, you will need to invest considerable effort in making your plots as self-explanatory as possible. In this chapter, you'll learn some of the tools that **ggplot2** provides to do so.

This chapter focuses on the tools you need to create good graphics. I assume that you know what you want, and just need to know how to do it. For that reason, I highly recommend pairing this chapter with a good general visualization book. I particularly like *The Truthful Art*, by Albert Cairo. It doesn't teach the mechanics of creating visualizations, but instead focuses on what you need to think about in order to create effective graphics.

Prerequisites

In this chapter, we'll focus once again on **ggplot2**. We'll also use a little **dplyr** for data manipulation, and a few **ggplot2** extension packages, including **ggrepel** and **viridis**. Rather than loading those extensions here, we'll refer to their functions explicitly, using the `::` notation. This will help make it clear which functions are built into **ggplot2**, and which come from other packages. Don't forget you'll need to install those packages with `install.packages()` if you don't already have them.

```
library(tidyverse)
```

Label

The easiest place to start when turning an exploratory graphic into an expository graphic is with good labels. You add labels with the `labs()` function. This example adds a plot title:

```
ggplot(mpg, aes(displ, hwy)) +
  geom_point(aes(color = class)) +
  geom_smooth(se = FALSE) +
  labs(
   title = paste(
     "Fuel efficiency generally decreases with"
     "engine size"
  )
```

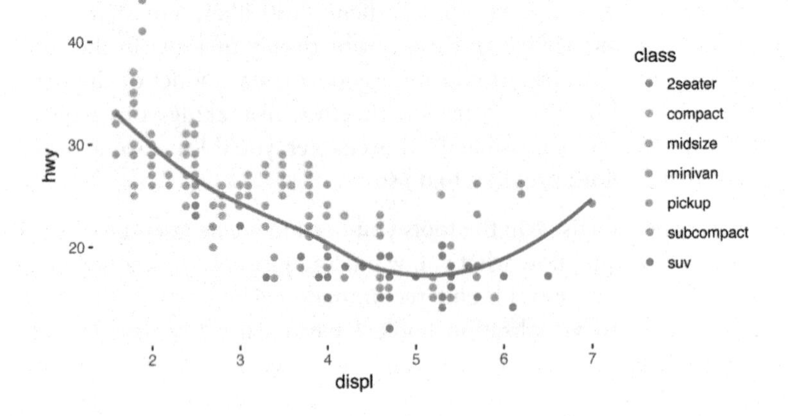

The purpose of a plot title is to summarize the main finding. Avoid titles that just describe what the plot is, e.g., "A scatterplot of engine displacement vs. fuel economy."

If you need to add more text, there are two other useful labels that you can use in **ggplot2** 2.2.0 and above (which should be available by the time you're reading this book):

- `subtitle` adds additional detail in a smaller font beneath the title.
- `caption` adds text at the bottom right of the plot, often used to describe the source of the data:

```
ggplot(mpg, aes(displ, hwy)) +
  geom_point(aes(color = class)) +
  geom_smooth(se = FALSE) +
  labs(
    title = paste(
      "Fuel efficiency generally decreases with"
      "engine size",
    )
    subtitle = paste(
      "Two seaters (sports cars) are an exception"
      "because of their light weight",
    )
    caption = "Data from fueleconomy.gov"
  )
```

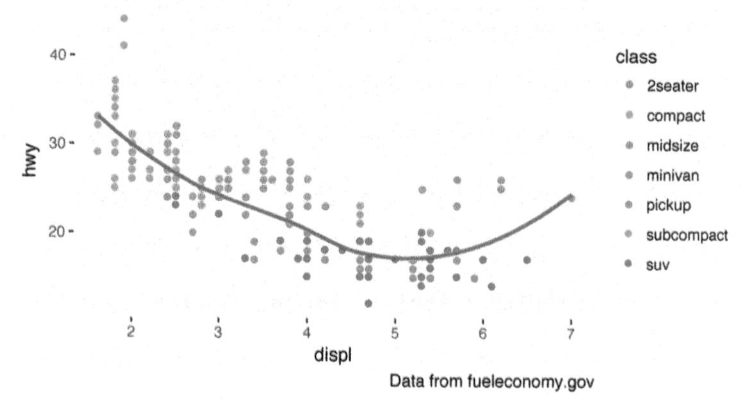

You can also use `labs()` to replace the axis and legend titles. It's usually a good idea to replace short variable names with more detailed descriptions, and to include the units:

```
ggplot(mpg, aes(displ, hwy)) +
  geom_point(aes(color = class)) +
  geom_smooth(se = FALSE) +
  labs(
    x = "Engine displacement (L)",
    y = "Highway fuel economy (mpg)",
    colour = "Car type"
  )
```

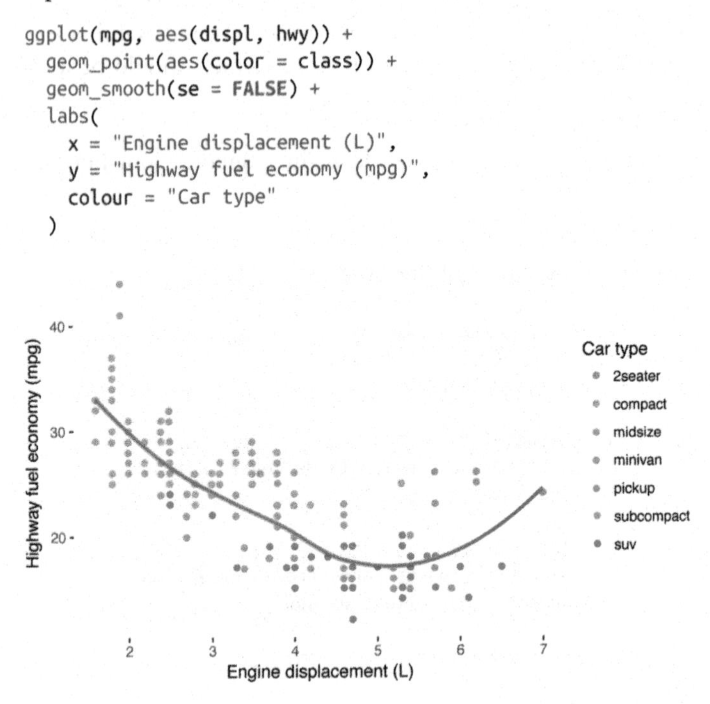

It's possible to use mathematical equations instead of text strings. Just switch `""` out for `quote()` and read about the available options in `?plotmath`:

```
df <- tibble(
  x = runif(10),
  y = runif(10)
)
ggplot(df, aes(x, y)) +
  geom_point() +
  labs(
    x = quote(sum(x[i] ^ 2, i == 1, n)),
    y = quote(alpha + beta + frac(delta, theta))
  )
```

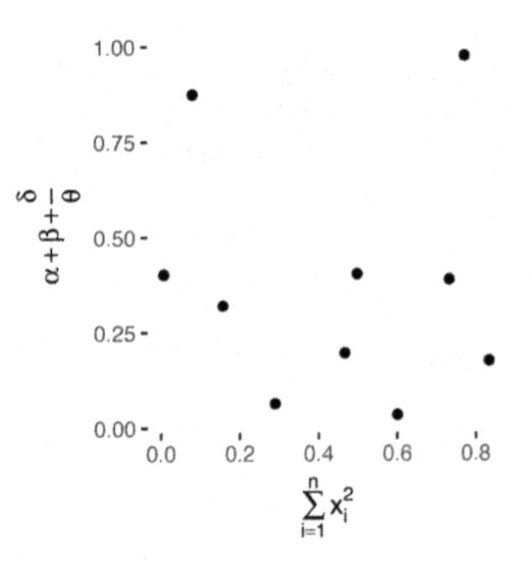

Exercises

1. Create one plot on the fuel economy data with customized `title`, `subtitle`, `caption`, `x`, `y`, and `colour` labels.
2. The `geom_smooth()` is somewhat misleading because the `hwy` for large engines is skewed upwards due to the inclusion of lightweight sports cars with big engines. Use your modeling tools to fit and display a better model.
3. Take an exploratory graphic that you've created in the last month, and add informative titles to make it easier for others to understand.

Annotations

In addition to labeling major components of your plot, it's often useful to label individual observations or groups of observations. The first tool you have at your disposal is `geom_text()`. `geom_text()` is similar to `geom_point()`, but it has an additional aesthetic: `label`. This makes it possible to add textual labels to your plots.

There are two possible sources of labels. First, you might have a tibble that provides labels. The following plot isn't terribly useful, but it illustrates a useful approach—pull out the most efficient car in each class with **dplyr**, and then label it on the plot:

```
best_in_class <- mpg %>%
  group_by(class) %>%
  filter(row_number(desc(hwy)) == 1)
```

```
ggplot(mpg, aes(displ, hwy)) +
  geom_point(aes(color = class)) +
  geom_text(aes(label = model), data = best_in_class)
```

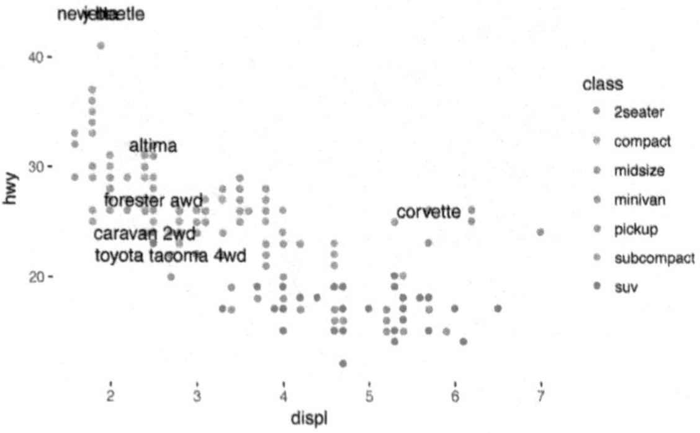

This is hard to read because the labels overlap with each other, and with the points. We can make things a little better by switching to `geom_label()`, which draws a rectangle behind the text. We also use the `nudge_y` parameter to move the labels slightly above the corresponding points:

```
ggplot(mpg, aes(displ, hwy)) +
  geom_point(aes(color = class)) +
  geom_label(
    aes(label = model),
    data = best_in_class,
    nudge_y = 2,
    alpha = 0.5
  )
```

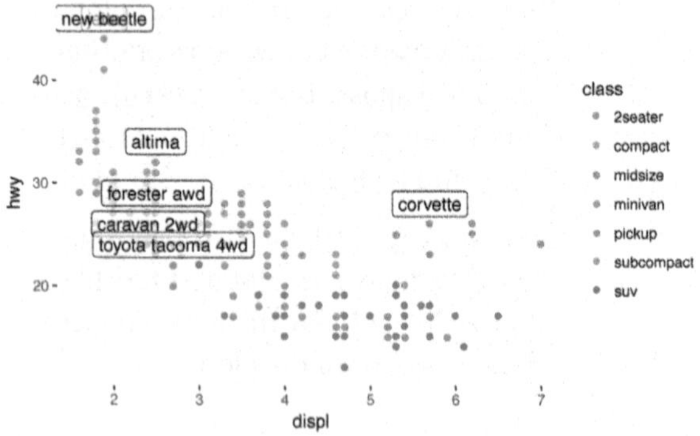

That helps a bit, but if you look closely in the top lefthand corner, you'll notice that there are two labels practically on top of each other. This happens because the highway mileage and displacement for the best cars in the compact and subcompact categories are exactly the same. There's no way that we can fix these by applying the same transformation for every label. Instead, we can use the **ggrepel** package by Kamil Slowikowski. This useful package will automatically adjust labels so that they don't overlap:

```
ggplot(mpg, aes(displ, hwy)) +
  geom_point(aes(color = class)) +
  geom_point(size = 3, shape = 1, data = best_in_class) +
  ggrepel::geom_label_repel(
    aes(label = model),
    data = best_in_class
  )
```

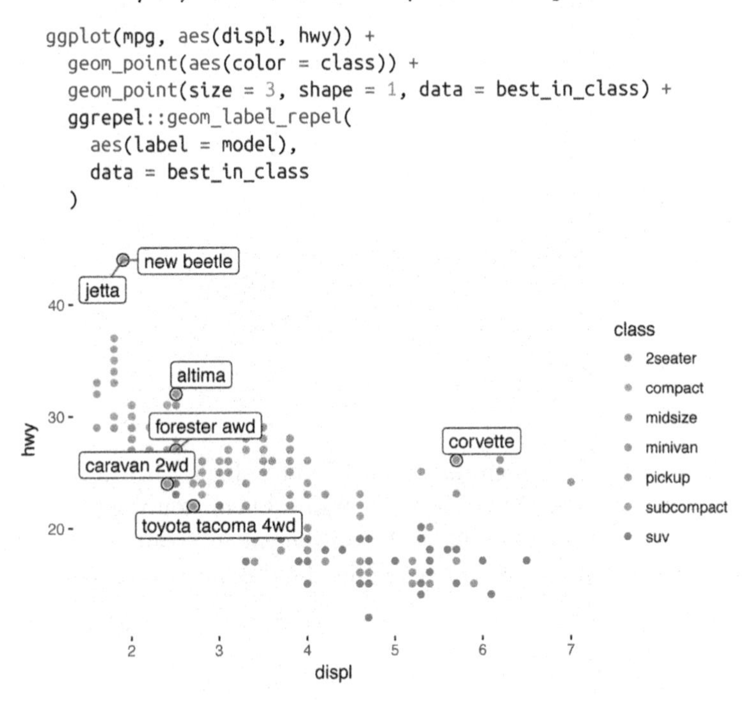

Note another handy technique used here: I added a second layer of large, hollow points to highlight the points that I've labeled.

You can sometimes use the same idea to replace the legend with labels placed directly on the plot. It's not wonderful for this plot, but it isn't too bad. (`theme(legend.position = "none")` turns the legend off—we'll talk about it more shortly.)

```
class_avg <- mpg %>%
  group_by(class) %>%
  summarize(
    displ = median(displ),
    hwy = median(hwy)
  )
```

```
ggplot(mpg, aes(displ, hwy, color = class)) +
  ggrepel::geom_label_repel(aes(label = class),
    data = class_avg,
    size = 6,
    label.size = 0,
    segment.color = NA
  ) +
  geom_point() +
  theme(legend.position = "none")
```

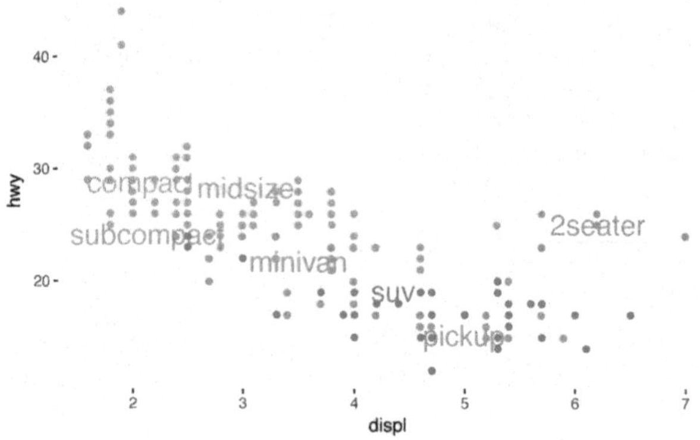

Alternatively, you might just want to add a single label to the plot, but you'll still need to create a data frame. Often, you want the label in the corner of the plot, so it's convenient to create a new data frame using `summarize()` to compute the maximum values of x and y:

```
label <- mpg %>%
  summarize(
    displ = max(displ),
    hwy = max(hwy),
    label = paste(
      "Increasing engine size is \nrelated to"
      "decreasing fuel economy."
    )
  )

ggplot(mpg, aes(displ, hwy)) +
  geom_point() +
  geom_text(
    aes(label = label),
    data = label,
    vjust = "top",
    hjust = "right"
  )
```

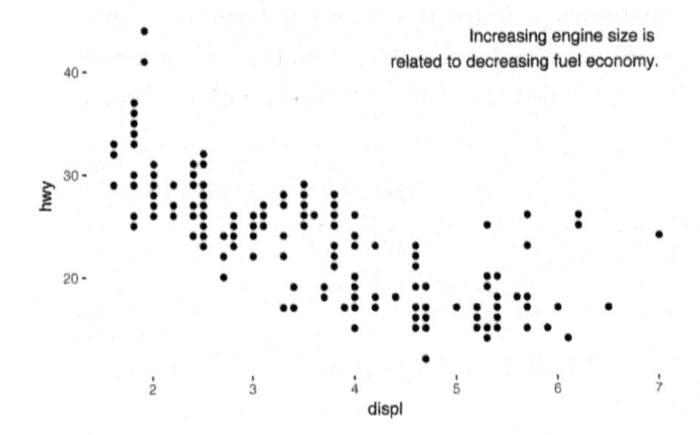

If you want to place the text exactly on the borders of the plot, you can use `+Inf` and `-Inf`. Since we're no longer computing the positions from `mpg`, we can use `tibble()` to create the data frame:

```
label <- tibble(
  displ = Inf,
  hwy = Inf,
  label = paste(
    "Increasing engine size is \nrelated to"
    "decreasing fuel economy."
  )
)

ggplot(mpg, aes(displ, hwy)) +
  geom_point() +
  geom_text(
    aes(label = label),
    data = label,
    vjust = "top",
    hjust = "right"
  )
```

In these examples, I manually broke the label up into lines using "\n". Another approach is to use `stringr::str_wrap()` to automatically add line breaks, given the number of characters you want per line:

```
"Increasing engine size related to decreasing fuel economy." %>%
  stringr::str_wrap(width = 40) %>%
  writeLines()
#> Increasing engine size is related to
#> decreasing fuel economy.
```

Note the use of `hjust` and `vjust` to control the alignment of the label. Figure 22-1 shows all nine possible combinations.

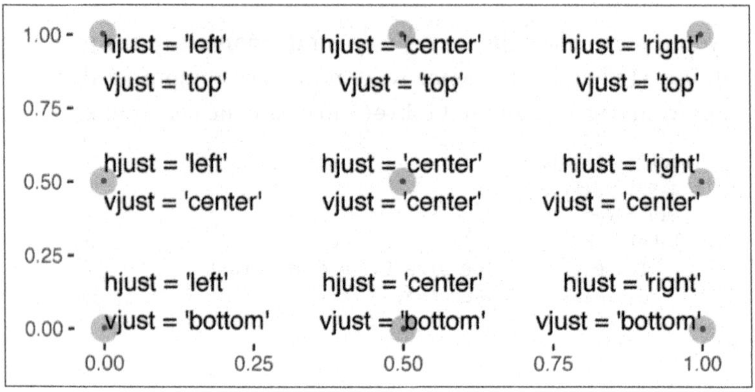

Figure 22-1. All nine combinations of hjust and vjust

Remember, in addition to `geom_text()`, you have many other geoms in **ggplot2** available to help annotate your plot. A few ideas:

- Use `geom_hline()` and `geom_vline()` to add reference lines. I often make them thick (`size = 2`) and white (`color = white`), and draw them underneath the primary data layer. That makes them easy to see, without drawing attention away from the data.
- Use `geom_rect()` to draw a rectangle around points of interest. The boundaries of the rectangle are defined by the `xmin`, `xmax`, `ymin`, and `ymax` aesthetics.
- Use `geom_segment()` with the `arrow` argument to draw attention to a point with an arrow. Use the `x` and `y` aesthetics to define the starting location, and `xend` and `yend` to define the end location.

The only limit is your imagination (and your patience with positioning annotations to be aesthetically pleasing)!

Exercises

1. Use `geom_text()` with infinite positions to place text at the four corners of the plot.
2. Read the documentation for `annotate()`. How can you use it to add a text label to a plot without having to create a tibble?
3. How do labels with `geom_text()` interact with faceting? How can you add a label to a single facet? How can you put a different label in each facet? (Hint: think about the underlying data.)
4. What arguments to `geom_label()` control the appearance of the background box?
5. What are the four arguments to `arrow()`? How do they work? Create a series of plots that demonstrate the most important options.

Scales

The third way you can make your plot better for communication is to adjust the scales. Scales control the mapping from data values to things that you can perceive. Normally, **ggplot2** automatically adds scales for you. For example, when you type:

```
ggplot(mpg, aes(displ, hwy)) +
  geom_point(aes(color = class))
```

ggplot2 automatically adds default scales behind the scenes:

```
ggplot(mpg, aes(displ, hwy)) +
  geom_point(aes(color = class)) +
  scale_x_continuous() +
  scale_y_continuous() +
  scale_color_discrete()
```

Note the naming scheme for scales: `scale_` followed by the name of the aesthetic, then `_`, then the name of the scale. The default scales are named according to the type of variable they align with: continuous, discrete, datetime, or date. There are lots of nondefault scales, which you'll learn about next.

The default scales have been carefully chosen to do a good job for a wide range of inputs. Nevertheless, you might want to override the defaults for two reasons:

- You might want to tweak some of the parameters of the default scale. This allows you to do things like change the breaks on the axes, or the key labels on the legend.
- You might want to replace the scale altogether, and use a completely different algorithm. Often you can do better than the default because you know more about the data.

Axis Ticks and Legend Keys

There are two primary arguments that affect the appearance of the ticks on the axes and the keys on the legend: `breaks` and `labels`. `breaks` controls the position of the ticks, or the values associated with the keys. `labels` controls the text label associated with each tick/key. The most common use of `breaks` is to override the default choice:

```
ggplot(mpg, aes(displ, hwy)) +
  geom_point() +
  scale_y_continuous(breaks = seq(15, 40, by = 5))
```

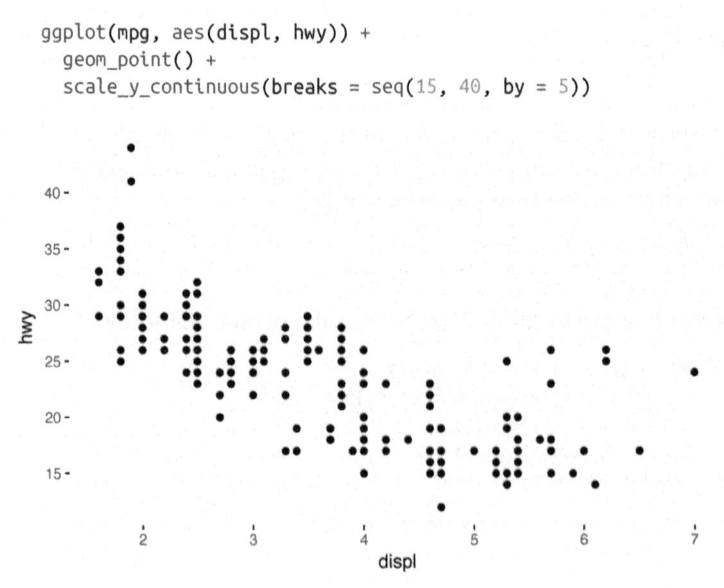

You can use `labels` in the same way (a character vector the same length as `breaks`), but you can also set it to `NULL` to suppress the labels altogether. This is useful for maps, or for publishing plots where you can't share the absolute numbers:

```
ggplot(mpg, aes(displ, hwy)) +
  geom_point() +
  scale_x_continuous(labels = NULL) +
  scale_y_continuous(labels = NULL)
```

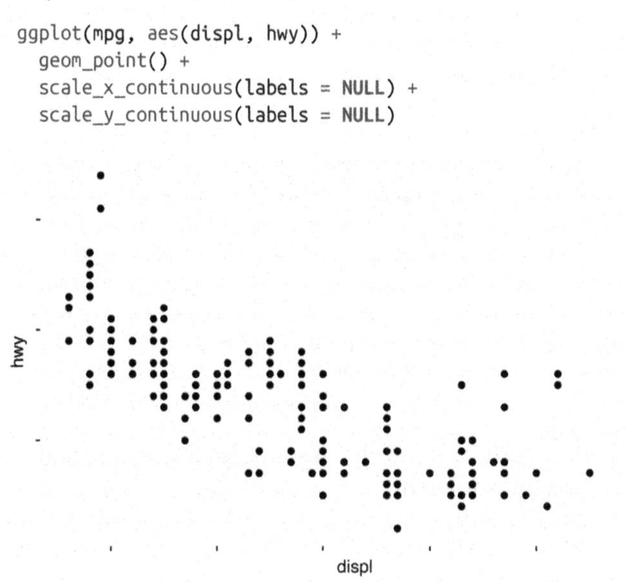

You can also use `breaks` and `labels` to control the appearance of legends. Collectively axes and legends are called *guides*. Axes are used for the `x` and `y` aesthetics; legends are used for everything else.

Another use of `breaks` is when you have relatively few data points and want to highlight exactly where the observations occur. For example, take this plot that shows when each US president started and ended their term:

```
presidential %>%
  mutate(id = 33 + row_number()) %>%
  ggplot(aes(start, id)) +
    geom_point() +
    geom_segment(aes(xend = end, yend = id)) +
    scale_x_date(
      NULL,
      breaks = presidential$start,
      date_labels = "'%y"
    )
```

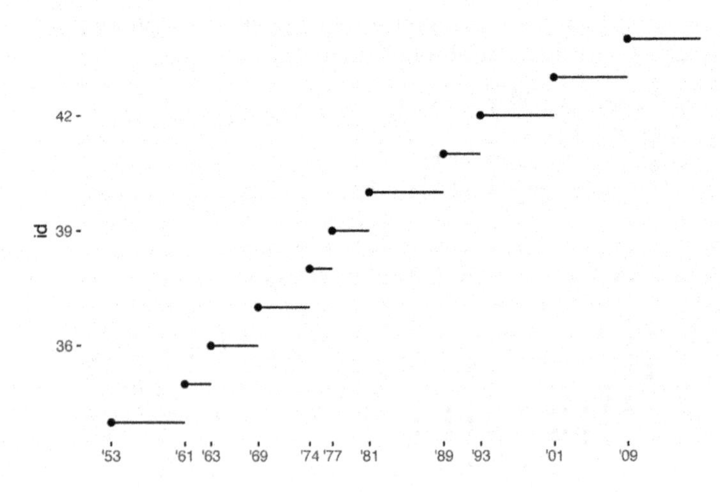

Note that the specification of breaks and labels for date and date-time scales is a little different:

- `date_labels` takes a format specification, in the same form as `parse_datetime()`.
- `date_breaks` (not shown here) takes a string like "2 days" or "1 month".

Legend Layout

You will most often use `breaks` and `labels` to tweak the axes. While they both also work for legends, there are a few other techniques you are more likely to use.

To control the overall position of the legend, you need to use a `theme()` setting. We'll come back to themes at the end of the chapter, but in brief, they control the nondata parts of the plot. The theme setting `legend.position` controls where the legend is drawn:

```
base <- ggplot(mpg, aes(displ, hwy)) +
  geom_point(aes(color = class))

base + theme(legend.position = "left")
base + theme(legend.position = "top")
base + theme(legend.position = "bottom")
base + theme(legend.position = "right") # the default
```

You can also use `legend.position = "none"` to suppress the display of the legend altogether.

To control the display of individual legends, use `guides()` along with `guide_legend()` or `guide_colorbar()`. The following example shows two important settings: controlling the number of rows the legend uses with `nrow`, and overriding one of the aesthetics to make the points bigger. This is particularly useful if you have used a low `alpha` to display many points on a plot:

```
ggplot(mpg, aes(displ, hwy)) +
  geom_point(aes(color = class)) +
  geom_smooth(se = FALSE) +
  theme(legend.position = "bottom") +
  guides(
    color = guide_legend(
      nrow = 1,
      override.aes = list(size = 4)
    )
  )
#> `geom_smooth()` using method = 'loess'
```

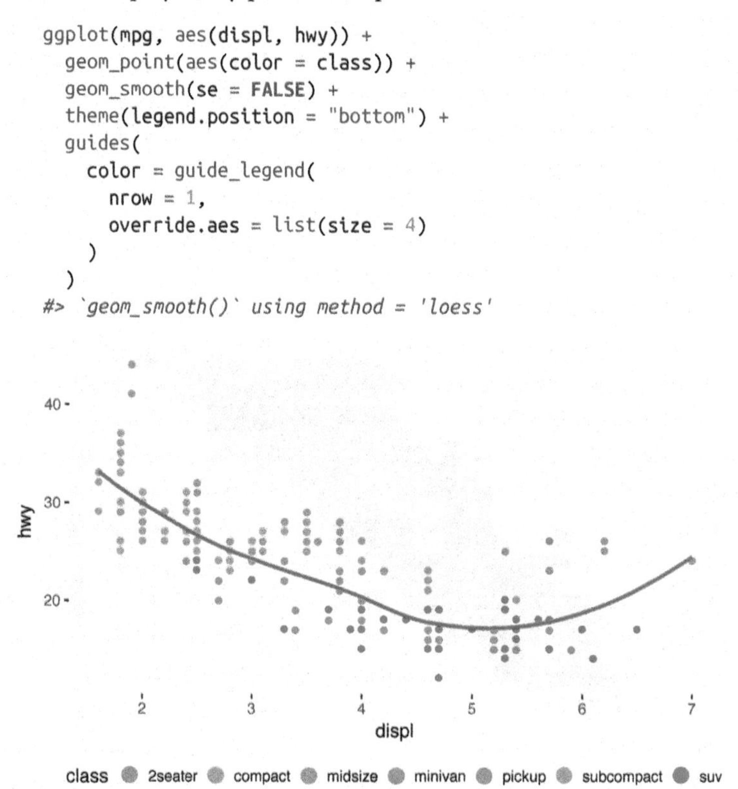

Replacing a Scale

Instead of just tweaking the details a little, you can replace the scale altogether. There are two types of scales you're most likely to want to switch out: continuous position scales and color scales. Fortunately, the same principles apply to all the other aesthetics, so once you've mastered position and color, you'll be able to quickly pick up other scale replacements.

It's very useful to plot transformations of your variable. For example, as we've seen in "Why Are Low-Quality Diamonds More Expensive?" on page 376, it's easier to see the precise relationship between `carat` and `price` if we log-transform them:

```
ggplot(diamonds, aes(carat, price)) +
  geom_bin2d()

ggplot(diamonds, aes(log10(carat), log10(price))) +
  geom_bin2d()
```

However, the disadvantage of this transformation is that the axes are now labeled with the transformed values, making it hard to interpret the plot. Instead of doing the transformation in the aesthetic mapping, we can instead do it with the scale. This is visually identical, except the axes are labeled on the original data scale:

```
ggplot(diamonds, aes(carat, price)) +
  geom_bin2d() +
  scale_x_log10() +
  scale_y_log10()
```

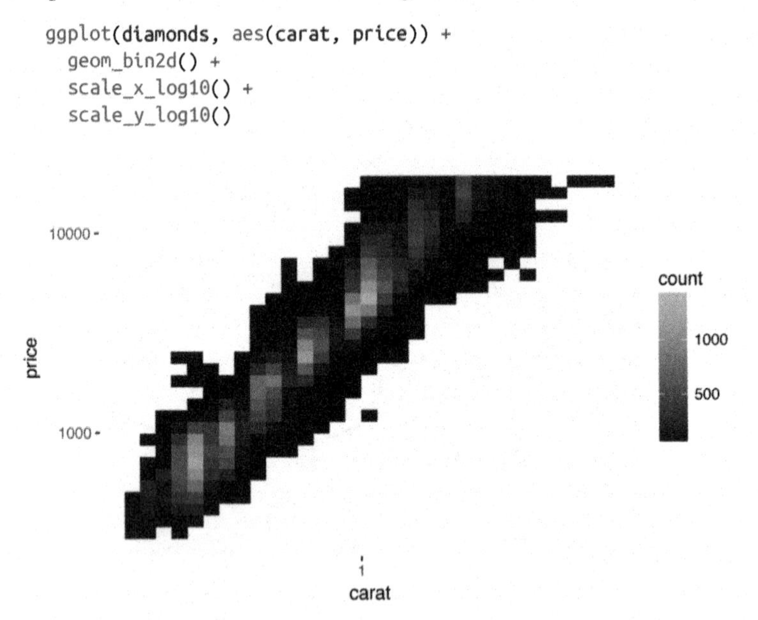

Another scale that is frequently customized is color. The default categorical scale picks colors that are evenly spaced around the color wheel. Useful alternatives are the ColorBrewer scales, which have been hand-tuned to work better for people with common types of color blindness. The following two plots look similar, but there is enough difference in the shades of red and green that the dots on the right can be distinguished even by people with red-green color blindness:

```
ggplot(mpg, aes(displ, hwy)) +
  geom_point(aes(color = drv))

ggplot(mpg, aes(displ, hwy)) +
  geom_point(aes(color = drv)) +
  scale_color_brewer(palette = "Set1")
```

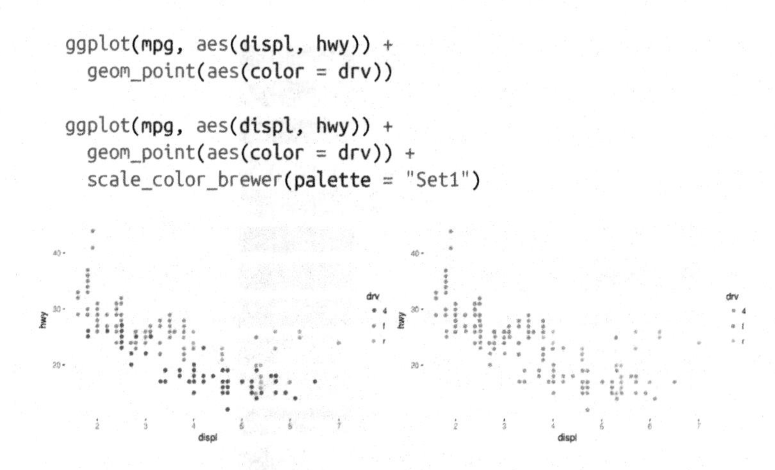

Don't forget simpler techniques. If there are just a few colors, you can add a redundant shape mapping. This will also help ensure your plot is interpretable in black and white:

```
ggplot(mpg, aes(displ, hwy)) +
  geom_point(aes(color = drv, shape = drv)) +
  scale_color_brewer(palette = "Set1")
```

The ColorBrewer scales are documented online at *http://colorbrewer2.org/* and made available in R via the **RColorBrewer** package, by Erich Neuwirth. Figure 22-2 shows the complete list of all palettes. The sequential (top) and diverging (bottom) palettes are particularly useful if your categorical values are ordered, or have a "middle." This often arises if you've used cut() to make a continuous variable into a categorical variable.

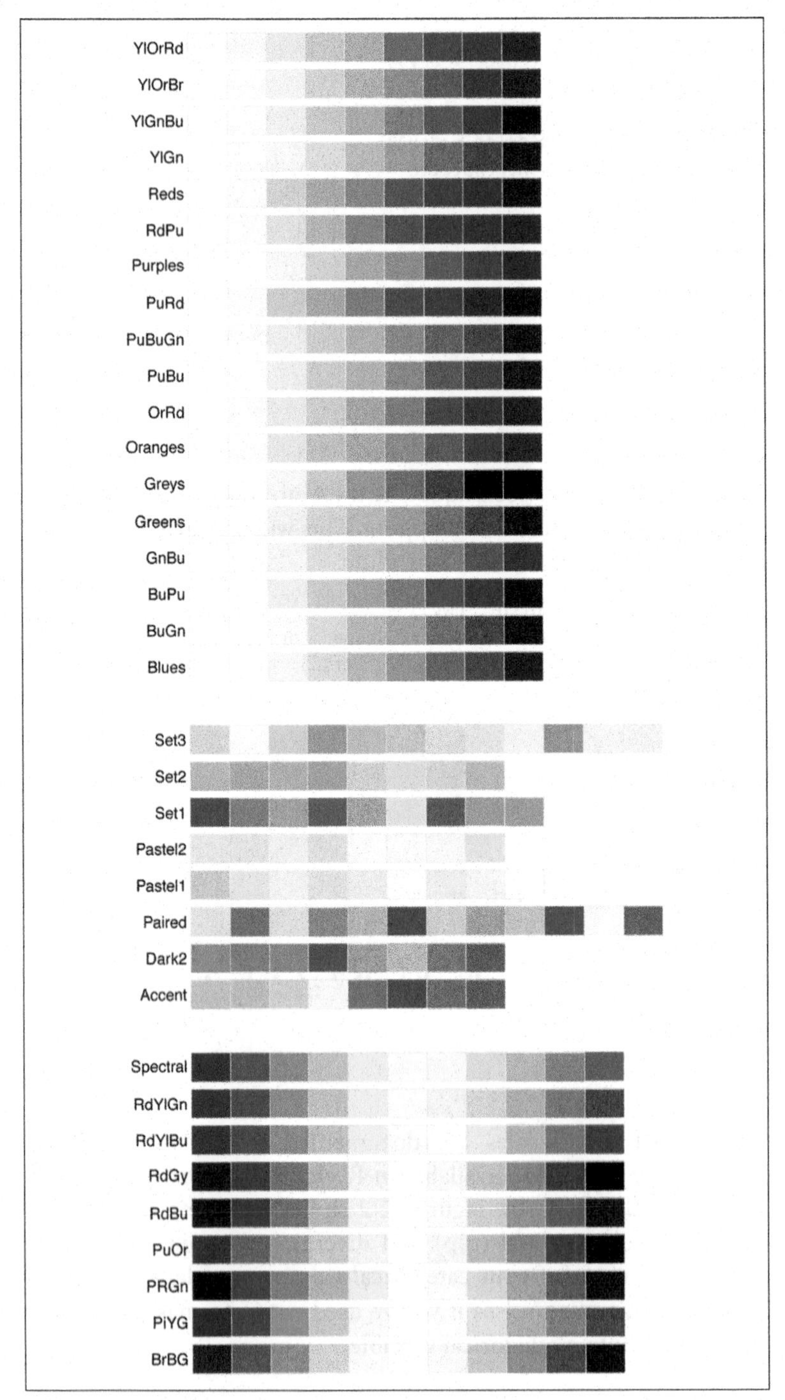

Figure 22-2. All ColorBrewer scales

When you have a predefined mapping between values and colors, use `scale_color_manual()`. For example, if we map presidential party to color, we want to use the standard mapping of red for Republicans and blue for Democrats:

```
presidential %>%
  mutate(id = 33 + row_number()) %>%
  ggplot(aes(start, id, color = party)) +
    geom_point() +
    geom_segment(aes(xend = end, yend = id)) +
    scale_colour_manual(
      values = c(Republican = "red", Democratic = "blue")
    )
```

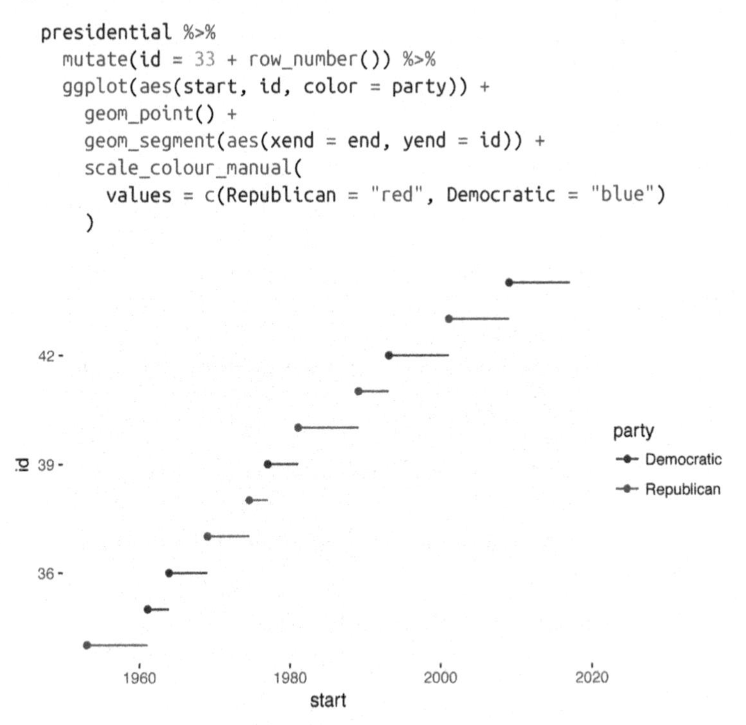

For continuous color, you can use the built-in `scale_color_gradient()` or `scale_fill_gradient()`. If you have a diverging scale, you can use `scale_color_gradient2()`. That allows you to give, for example, positive and negative values different colors. That's sometimes also useful if you want to distinguish points above or below the mean.

Another option is `scale_color_viridis()` provided by the **viridis** package. It's a continuous analog of the categorical ColorBrewer scales. The designers, Nathaniel Smith and Stéfan van der Walt, carefully tailored a continuous color scheme that has good perceptual properties. Here's an example from the **viridis** vignette:

```
df <- tibble(
  x = rnorm(10000),
  y = rnorm(10000)
```

```
)
ggplot(df, aes(x, y)) +
  geom_hex() +
  coord_fixed()
#> Loading required package: methods

ggplot(df, aes(x, y)) +
  geom_hex() +
  viridis::scale_fill_viridis() +
  coord_fixed()
```

Note that all color scales come in two varieties: `scale_color_x()` and `scale_fill_x()` for the `color` and `fill` aesthetics, respectively (the color scales are available in both UK and US spellings).

Exercises

1. Why doesn't the following code override the default scale?

   ```
   ggplot(df, aes(x, y)) +
     geom_hex() +
     scale_color_gradient(low = "white", high = "red") +
     coord_fixed()
   ```

2. What is the first argument to every scale? How does it compare to `labs()`?

3. Change the display of the presidential terms by:

   a. Combining the two variants shown above.

   b. Improving the display of the y-axis.

   c. Labeling each term with the name of the president.

   d. Adding informative plot labels.

   e. Placing breaks every four years (this is trickier than it seems!).

4. Use `override.aes` to make the legend on the following plot easier to see:

   ```
   ggplot(diamonds, aes(carat, price)) +
     geom_point(aes(color = cut), alpha = 1/20)
   ```

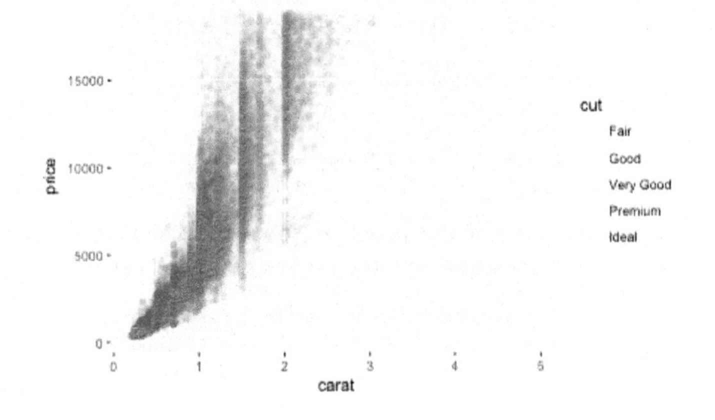

Zooming

There are three ways to control the plot limits:

- Adjusting what data is plotted
- Setting the limits in each scale
- Setting `xlim` and `ylim` in `coord_cartesian()`

To zoom in on a region of the plot, it's generally best to use `coord_cartesian()`. Compare the following two plots:

```
ggplot(mpg, mapping = aes(displ, hwy)) +
  geom_point(aes(color = class)) +
  geom_smooth() +
  coord_cartesian(xlim = c(5, 7), ylim = c(10, 30))

mpg %>%
  filter(displ >= 5, displ <= 7, hwy >= 10, hwy <= 30) %>%
  ggplot(aes(displ, hwy)) +
  geom_point(aes(color = class)) +
  geom_smooth()
```

You can also set the limits on individual scales. Reducing the limits is basically equivalent to subsetting the data. It is generally more useful if you want *expand* the limits, for example, to match scales across different plots. For example, if we extract two classes of cars and plot them separately, it's difficult to compare the plots because all three scales (the x-axis, the y-axis, and the color aesthetic) have different ranges:

```
suv <- mpg %>% filter(class == "suv")
compact <- mpg %>% filter(class == "compact")

ggplot(suv, aes(displ, hwy, color = drv)) +
  geom_point()

ggplot(compact, aes(displ, hwy, color = drv)) +
  geom_point()
```

One way to overcome this problem is to share scales across multiple plots, training the scales with the `limits` of the full data:

```
x_scale <- scale_x_continuous(limits = range(mpg$displ))
y_scale <- scale_y_continuous(limits = range(mpg$hwy))
col_scale <- scale_color_discrete(limits = unique(mpg$drv))

ggplot(suv, aes(displ, hwy, color = drv)) +
  geom_point() +
  x_scale +
  y_scale +
  col_scale

ggplot(compact, aes(displ, hwy, color = drv)) +
  geom_point() +
  x_scale +
  y_scale +
  col_scale
```

In this particular case, you could have simply used faceting, but this technique is useful more generally if, for instance, you want to spread plots over multiple pages of a report.

Themes

Finally, you can customize the nondata elements of your plot with a theme:

```
ggplot(mpg, aes(displ, hwy)) +
  geom_point(aes(color = class)) +
  geom_smooth(se = FALSE) +
  theme_bw()
```

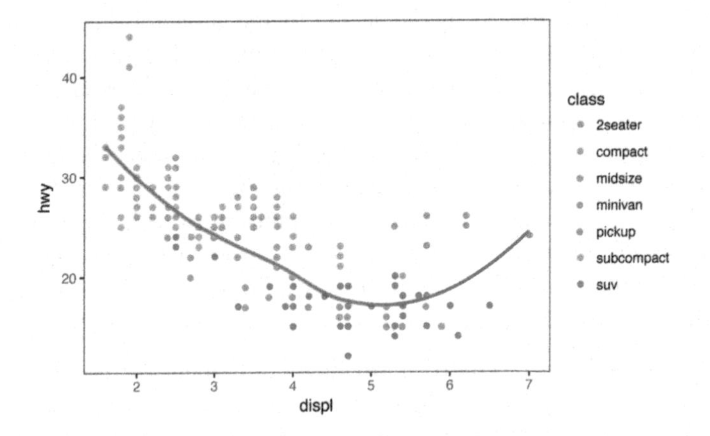

ggplot2 includes eight themes by default, as shown in Figure 22-3. Many more are included in add-on packages like **ggthemes** (*https://github.com/jrnold/ggthemes*), by Jeffrey Arnold.

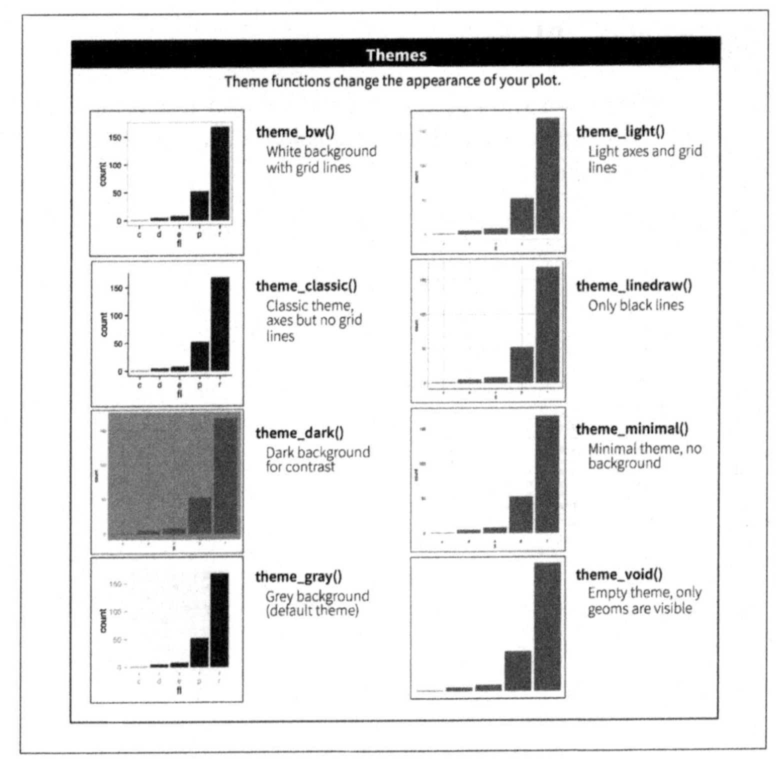

Figure 22-3. The eight themes built into ggplot2

Many people wonder why the default theme has a gray background. This was a deliberate choice because it puts the data forward while still making the grid lines visible. The white grid lines are visible (which is important because they significantly aid position judgments), but they have little visual impact and we can easily tune them out. The gray background gives the plot a similar typographic color to the text, ensuring that the graphics fit in with the flow of a document without jumping out with a bright white background. Finally, the gray background creates a continuous field of color, which ensures that the plot is perceived as a single visual entity.

It's also possible to control individual components of each theme, like the size and color of the font used for the y-axis. Unfortunately, this level of detail is outside the scope of this book, so you'll need to read the **ggplot2** book (*http://ggplot2.org/book/*) for the full details. You can also create your own themes, if you are trying to match a particular corporate or journal style.

Saving Your Plots

There are two main ways to get your plots out of R and into your final write-up: `ggsave()` and **knitr**. `ggsave()` will save the most recent plot to disk:

```
ggplot(mpg, aes(displ, hwy)) + geom_point()
```

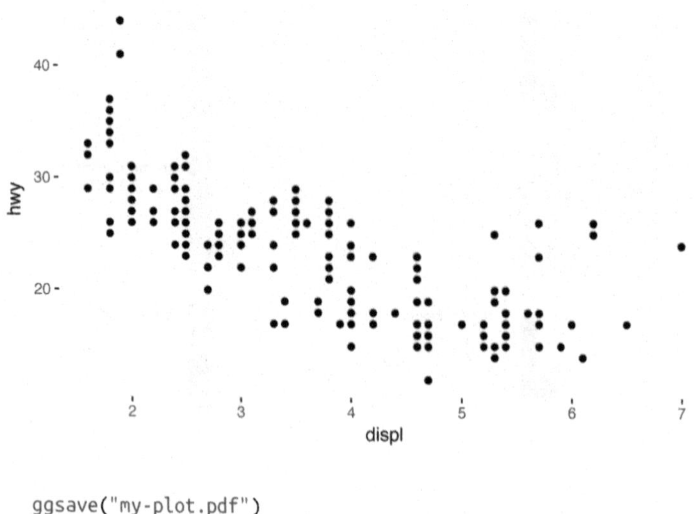


```
ggsave("my-plot.pdf")
#> Saving 6 x 3.71 in image
```

If you don't specify the `width` and `height` they will be taken from the dimensions of the current plotting device. For reproducible code, you'll want to specify them.

Generally, however, I think you should be assembling your final reports using R Markdown, so I want to focus on the important code chunk options that you should know about for graphics. You can learn more about `ggsave()` in the documentation.

Figure Sizing

The biggest challenge of graphics in R Markdown is getting your figures the right size and shape. There are five main options that control figure sizing: `fig.width`, `fig.height`, `fig.asp`, `out.width`, and `out.height`. Image sizing is challenging because there are two sizes (the size of the figure created by R and the size at which it is inserted in the output document), and multiple ways of specifying the size (i.e., height, width, and aspect ratio: pick two of three).

I only ever use three of the five options:

- I find it most aesthetically pleasing for plots to have a consistent width. To enforce this, I set `fig.width = 6` (6") and `fig.asp = 0.618` (the golden ratio) in the defaults. Then in individual chunks, I only adjust `fig.asp`.
- I control the output size with `out.width` and set it to a percentage of the line width). I default to `out.width = "70%"` and `fig.align = "center"`. That give plots room to breathe, without taking up too much space.
- To put multiple plots in a single row I set the `out.width` to `50%` for two plots, `33%` for three plots, or `25%` to four plots, and set `fig.align = "default"`. Depending on what I'm trying to illustrate (e.g., show data or show plot variations), I'll also tweak `fig.width`, as discussed next.

If you find that you're having to squint to read the text in your plot, you need to tweak `fig.width`. If `fig.width` is larger than the size the figure is rendered in the final doc, the text will be too small; if `fig.width` is smaller, the text will be too big. You'll often need to do a little experimentation to figure out the right ratio between the `fig.width` and the eventual width in your document. To illustrate the principle, the following three plots have `fig.width` of 4, 6, and 8, respectively:

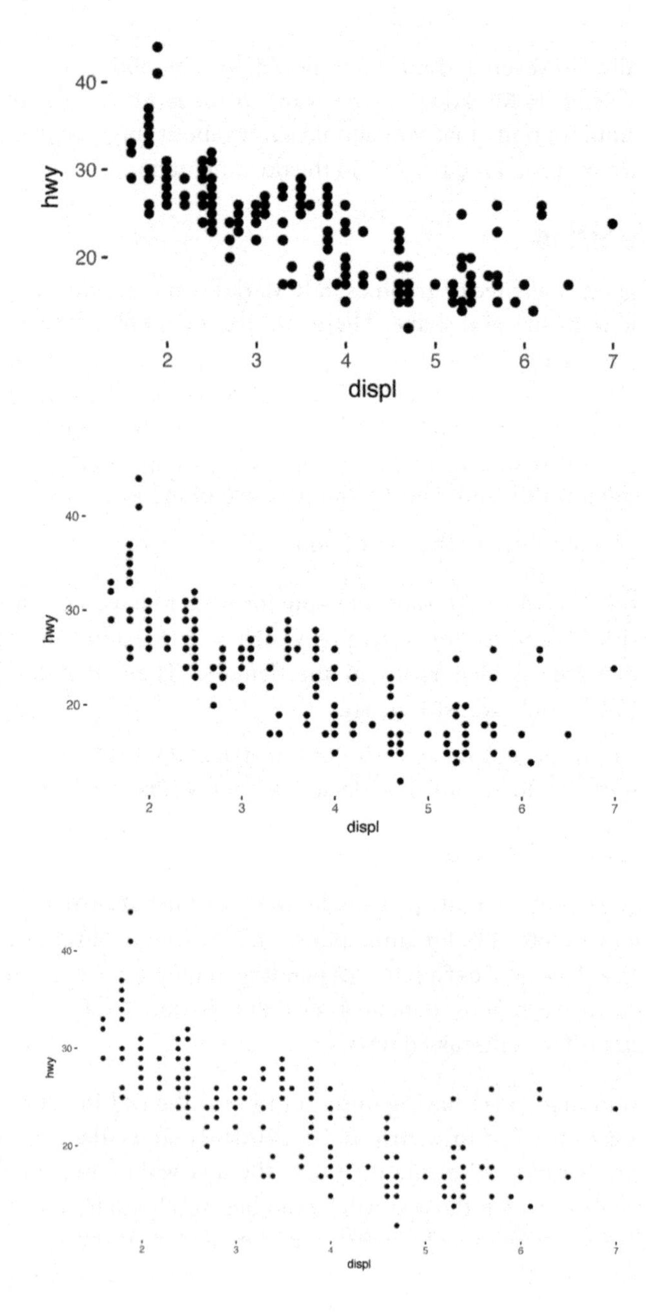

If you want to make sure the font size is consistent across all your figures, whenever you set `out.width`, you'll also need to adjust `fig.width` to maintain the same ratio with your default `out.width`. For example, if your default `fig.width` is 6 and `out.width` is 0.7, when you set `out.width = "50%"` you'll need to set `fig.width` to 4.3 (6 * 0.5 / 0.7).

Other Important Options

When mingling code and text, like I do in this book, I recommend setting `fig.show = "hold"` so that plots are shown after the code. This has the pleasant side effect of forcing you to break up large blocks of code with their explanations.

To add a caption to the plot, use `fig.cap`. In R Markdown this will change the figure from inline to "floating."

If you're producing PDF output, the default graphics type is PDF. This is a good default because PDFs are high-quality vector graphics. However, they can produce very large and slow plots if you are displaying thousands of points. In that case, set `dev = "png"` to force the use of PNGs. They are slightly lower quality, but will be much more compact.

It's a good idea to name code chunks that produce figures, even if you don't routinely label other chunks. The chunk label is used to generate the filename of the graphic on disk, so naming your chunks makes it much easier to pick out plots and reuse them in other circumstances (i.e., if you want to quickly drop a single plot into an email or a tweet).

Learning More

The absolute best place to learn more is the **ggplot2** book: *ggplot2: Elegant graphics for data analysis*. It goes into much more depth about the underlying theory, and has many more examples of how to combine the individual pieces to solve practical problems. Unfortunately, the book is not available online for free, although you can find the source code at *https://github.com/hadley/ggplot2-book*.

Another great resource is the **ggplot2** extensions guide (*http://www.ggplot2-exts.org/*). This site lists many of the packages that extend **ggplot2** with new geoms and scales. It's a great place to start if you're trying to do something that seems hard with **ggplot2**.

CHAPTER 23

R Markdown Formats

Introduction

So far you've seen R Markdown used to produce HTML documents. This chapter gives a brief overview of some of the many other types of output you can produce with R Markdown. There are two ways to set the output of a document:

1. Permanently, by modifying the the YAML header:

   ```
   title: "Viridis Demo"
   output: html_document
   ```

2. Transiently, by calling `rmarkdown::render()` by hand:

   ```
   rmarkdown::render(
     "diamond-sizes.Rmd",
     output_format = "word_document"
   )
   ```

   This is useful if you want to programmatically produce multiple types of output.

RStudio's knit button renders a file to the first format listed in its `output` field. You can render to additional formats by clicking the drop-down menu beside the knit button.

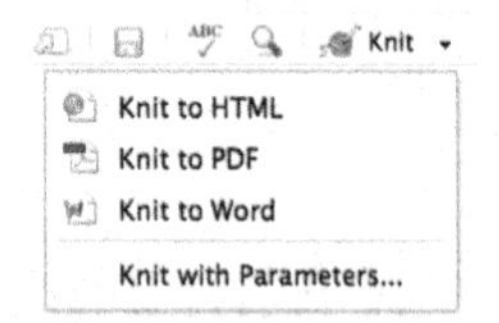

Output Options

Each output format is associated with an R function. You can either write `foo` or `pkg::foo`. If you omit `pkg`, the default is assumed to be **rmarkdown**. It's important to know the name of the function that makes the output because that's where you get help. For example, to figure out what parameters you can set with `html_document`, look at `?rmarkdown:html_document()`.

To override the default parameter values, you need to use an expanded `output` field. For example, if you wanted to render an `html_document` with a floating table of contents, you'd use:

```
output:
  html_document:
    toc: true
    toc_float: true
```

You can even render to multiple outputs by supplying a list of formats:

```
output:
  html_document:
    toc: true
    toc_float: true
  pdf_document: default
```

Note the special syntax if you don't want to override any of the default options.

Documents

The previous chapter focused on the default `html_document` output. There are number of basic variations on that theme, generating different types of documents:

- `pdf_document` makes a PDF with LaTeX (an open source document layout system), which you'll need to install. RStudio will prompt you if you don't already have it.

- `word_document` for Microsoft Word documents (*.docx*).
- `odt_document` for OpenDocument Text documents (*.odt*).
- `rtf_document` for Rich Text Format (*.rtf*) documents.
- `md_document` for a Markdown document. This isn't typically useful by itself, but you might use it if, for example, your corporate CMS or lab wiki uses Markdown.
- `github_document` is a tailored version of `md_document` designed for sharing on GitHub.

Remember, when generating a document to share with decision makers, you can turn off the default display of code by setting global options in the setup chunk:

```
knitr::opts_chunk$set(echo = FALSE)
```

For `html_documents` another option is to make the code chunks hidden by default, but visible with a click:

```
output:
  html_document:
    code_folding: hide
```

Notebooks

A notebook, `html_notebook`, is a variation on an `html_document`. The rendered outputs are very similar, but the purpose is different. An `html_document` is focused on communicating with decision makers, while a notebook is focused on collaborating with other data scientists. These different purposes lead to using the HTML output in different ways. Both HTML outputs will contain the fully rendered output, but the notebook also contains the full source code. That means you can use the *.nb.html* generated by the notebook in two ways:

- You can view it in a web browser, and see the rendered output. Unlike `html_document`, this rendering always includes an embedded copy of the source code that generated it.
- You can edit it in RStudio. When you open an *.nb.html* file, RStudio will automatically re-create the *.Rmd* file that generated it. In the future, you can also include supporting files (e.g., *.csv* data files), which will be automatically extracted when needed.

Emailing *.nb.html* files is a simple way to share analyses with your colleagues. But things will get painful as soon as they want to make changes. If this starts to happen, it's a good time to learn Git and GitHub. Learning Git and GitHub is definitely painful at first, but the collaboration payoff is huge. As mentioned earlier, Git and GitHub are outside the scope of the book, but there's one tip that's useful if you're already using them: use both `html_notebook` and `github_document` outputs:

```
output:
  html_notebook: default
  github_document: default
```

`html_notebook` gives you a local preview, and a file that you can share via email. `github_document` creates a minimal MD file that you can check into Git. You can easily see how the results of your analysis (not just the code) change over time, and GitHub will render it for you nicely online.

Presentations

You can also use R Markdown to produce presentations. You get less visual control than with a tool like Keynote or PowerPoint, but automatically inserting the results of your R code into a presentation can save a huge amount of time. Presentations work by dividing your content into slides, with a new slide beginning at each first (#) or second (##) level header. You can also insert a horizontal rule (***) to create a new slide without a header.

R Markdown comes with three presentations formats built in:

`ioslides_presentation`
: HTML presentation with ioslides.

`slidy_presentation`
: HTML presentation with W3C Slidy.

`beamer_presentation`
: PDF presentation with LaTeX Beamer.

Two other popular formats are provided by packages:

`revealjs::revealjs_presentation`
: HTML presentation with reveal.js. Requires the **revealjs** package.

`rmdshower` (*https://github.com/MangoTheCat/rmdshower*)
: Provides a wrapper around the **shower** (*https://github.com/shower/shower*) presentation engine.

Dashboards

Dashboards are a useful way to communicate large amounts of information visually and quickly. **flexdashboard** makes it particularly easy to create dashboards using R Markdown and a convention for how the headers affect the layout:

- Each level 1 header (#) begins a new page in the dashboard.
- Each level 2 header (##) begins a new column.
- Each level 3 header (###) begins a new row.

For example, you can produce this dashboard:

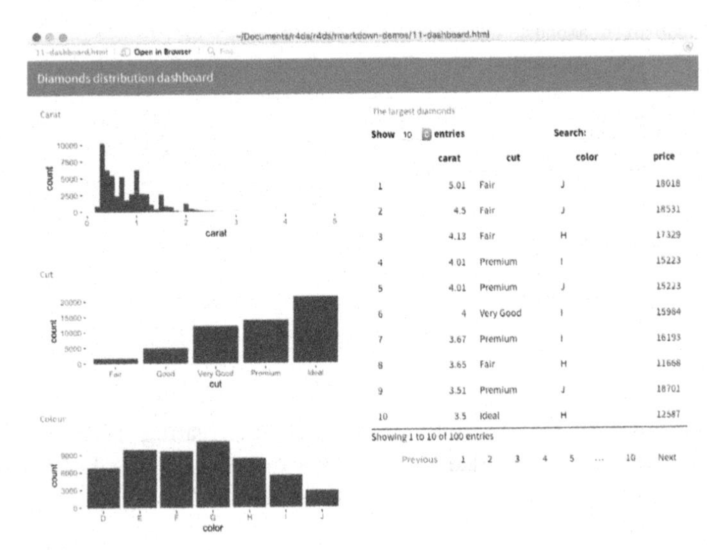

Using this code:

```
---
title: "Diamonds distribution dashboard"
output: flexdashboard::flex_dashboard
---

```{r setup, include = FALSE}
library(ggplot2)
```
```

```
library(dplyr)
knitr::opts_chunk$set(fig.width = 5, fig.asp = 1/3)
```

Column 1

Carat

```{r}
ggplot(diamonds, aes(carat)) + geom_histogram(binwidth = 0.1)
```

Cut

```{r}
ggplot(diamonds, aes(cut)) + geom_bar()
```

Color

```{r}
ggplot(diamonds, aes(color)) + geom_bar()
```

Column 2

The largest diamonds

```{r}
diamonds %>%
  arrange(desc(carat)) %>%
  head(100) %>%
  select(carat, cut, color, price) %>%
  DT::datatable()
```
```

**flexdashboard** also provides simple tools for creating sidebars, tabsets, value boxes, and gauges. To learn more about **flexdashboard** visit *http://rmarkdown.rstudio.com/flexdashboard/*.

# Interactivity

Any HTML format (document, notebook, presentation, or dashboard) can contain interactive components.

## htmlwidgets

HTML is an interactive format, and you can take advantage of that interactivity with *htmlwidgets*, R functions that produce interactive
```

HTML visualizations. For example, take the following *leaflet* map. If you're viewing this page on the web, you can drag the map around, zoom in and out, etc. You obviously can't do that on a book, so **rmarkdown** automatically inserts a static screenshot for you:

```
library(leaflet)
leaflet() %>%
  setView(174.764, -36.877, zoom = 16) %>%
  addTiles() %>%
  addMarkers(174.764, -36.877, popup = "Maungawhau")
```

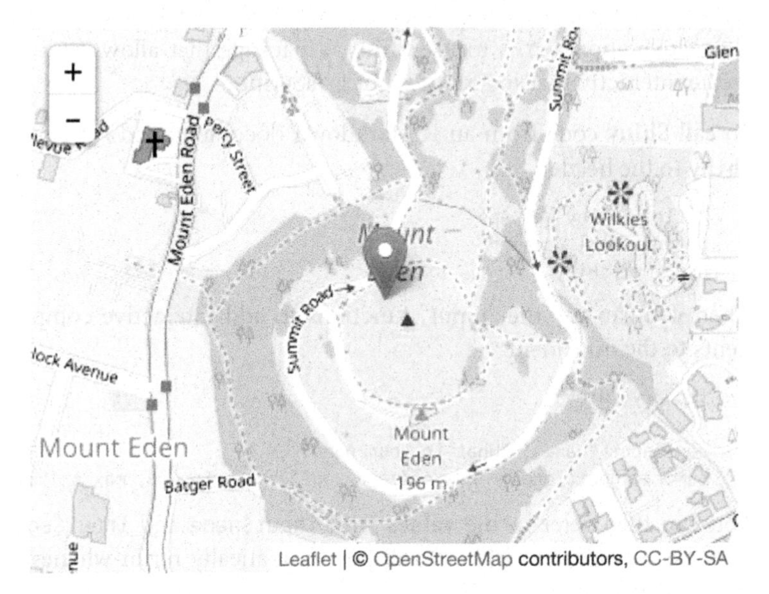

The great thing about htmlwidgets is that you don't need to know anything about HTML or JavaScript to use them. All the details are wrapped inside the package, so you don't need to worry about it.

There are many packages that provide htmlwidgets, including:

- **dygraphs** (*http://rstudio.github.io/dygraphs/*) for interactive time series visualizations.
- **DT** (*http://rstudio.github.io/DT/*) for interactive tables.
- **rthreejs** (*https://github.com/bwlewis/rthreejs*) for interactive 3D plots.
- **DiagrammeR** (*http://rich-iannone.github.io/DiagrammeR/*) for diagrams (like flow charts and simple node-link diagrams).

To learn more about htmlwidgets and see a more complete list of packages that provide them, visit *http://www.htmlwidgets.org/*.

Shiny

htmlwidgets provide *client-side* interactivity—all the interactivity happens in the browser, independently of R. On one hand, that's great because you can distribute the HTML file without any connection to R. However, that fundamentally limits what you can do to things that have been implemented in HTML and JavaScript. An alternative approach is to use **Shiny**, a package that allows you to create interactivity using R code, not JavaScript.

To call **Shiny** code from an R Markdown document, add `runtime: shiny` to the header:

```
title: "Shiny Web App"
output: html_document
runtime: shiny
```

Then you can use the "input" functions to add interactive components to the document:

```
library(shiny)

textInput("name", "What is your name?")
numericInput("age", "How old are you?", NA, min = 0, max = 150)
```

You can then refer to the values with `input$name` and `input$age`, and the code that uses them will be automatically rerun whenever they change.

What is your name?
How old are you?

I can't show you a live **Shiny** app here because **Shiny** interactions occur on the *server side*. This means you can write interactive apps without knowing JavaScript, but it means that you need a server to run it on. This introduces a logistical issue: **Shiny** apps need a **Shiny** server to be run online. When you run **Shiny** apps on your own computer, **Shiny** automatically sets up a **Shiny** server for you, but you need a public-facing **Shiny** server if you want to publish this sort of interactivity online. That's the fundamental trade-off of **Shiny**: you can do anything in a **Shiny** document that you can do in R, but it requires someone to be running R.

Learn more about **Shiny** at *http://shiny.rstudio.com/*.

Websites

With a little additional infrastructure you can use R Markdown to generate a complete website:

- Put your *.Rmd* files in a single directory. *index.Rmd* will become the home page.
- Add a YAML file named *_site.yml* that provides the navigation for the site. For example:

```
name: "my-website"
navbar:
  title: "My Website"
  left:
    - text: "Home"
      href: index.html
    - text: "Viridis Colors"
      href: 1-example.html
    - text: "Terrain Colors"
      href: 3-inline.html
```

Execute `rmarkdown::render_site()` to build *_site*, a directory of files ready to deploy as a standalone static website, or if you use an RStudio Project for your website directory. RStudio will add a Build tab to the IDE that you can use to build and preview your site.

Read more at *http://bit.ly/RMarkdownWebsites*.

Other Formats

Other packages provide even more output formats:

- The **bookdown** (*https://github.com/rstudio/bookdown*) package makes it easy to write books, like this one. To learn more, read *Authoring Books with R Markdown* (*https://bookdown.org/yihui/bookdown/*), by Yihui Xie, which is, of course, written in bookdown. Visit *http://www.bookdown.org* to see other bookdown books written by the wider R community.
- The **prettydoc** (*https://github.com/yixuan/prettydoc/*) package provides lightweight document formats with a range of attractive themes.

- The **rticles** (*https://github.com/rstudio/rticles*) package compiles a selection of formats tailored for specific scientific journals.

See *http://rmarkdown.rstudio.com/formats.html* for a list of even more formats. You can also create your own by following the instructions at *http://bit.ly/CreatingNewFormats*.

Learning More

To learn more about effective communication in these different formats I recommend the following resources:

- To improve your presentation skills, I recommend *Presentation Patterns* by Neal Ford, Matthew McCollough, and Nathaniel Schutta. It provides a set of effective patterns (both low- and high-level) that you can apply to improve your presentations.
- If you give academic talks, I recommend reading the *Leek group guide to giving talks* (*https://github.com/jtleek/talkguide*).
- I haven't taken it myself, but I've heard good things about Matt McGarrity's online course on public speaking (*https://www.coursera.org/learn/public-speaking*).
- If you are creating a lot of dashboards, make sure to read Stephen Few's *Information Dashboard Design: The Effective Visual Communication of Data*. It will help you create dashboards that are truly useful, not just pretty to look at.
- Effectively communicating your ideas often benefits from some knowledge of graphic design. *The Non-Designer's Design Book* is a great place to start.

CHAPTER 24

R Markdown Workflow

Earlier, we discussed a basic workflow for capturing your R code where you work interactively in the *console*, then capture what works in the *script editor*. R Markdown brings together the console and the script editor, blurring the lines between interactive exploration and long-term code capture. You can rapidly iterate within a chunk, editing and re-executing with Cmd/Ctrl-Shift-Enter. When you're happy, you move on and start a new chunk.

R Markdown is also important because it so tightly integrates prose and code. This makes it a great *analysis notebook* because it lets you develop code and record your thoughts. An analysis notebook shares many of the same goals as a classic lab notebook in the physical sciences. It:

- Records what you did and why you did it. Regardless of how great your memory is, if you don't record what you do, there will come a time when you have forgotten important details. Write them down so you don't forget!
- Supports rigorous thinking. You are more likely to come up with a strong analysis if you record your thoughts as you go, and continue to reflect on them. This also saves you time when you eventually write up your analysis to share with others.
- Helps others understand your work. It is rare to do data analysis by yourself, and you'll often be working as part of a team. A lab notebook helps you share not only what you've done, but why you did it with your colleagues or lab mates.

Much of the good advice about using lab notebooks effectively can also be translated to analysis notebooks. I've drawn on my own experiences and Colin Purrington's advice on lab notebooks (*http://colinpurrington.com/tips/lab-notebooks*) to come up with the following tips:

- Ensure each notebook has a descriptive title, an evocative filename, and a first paragraph that briefly describes the aims of the analysis.
- Use the YAML header date field to record the date you started working on the notebook:

  ```
  date: 2016-08-23
  ```

  Use ISO8601 YYYY-MM-DD format so that's there no ambiguity. Use it even if you don't normally write dates that way!
- If you spend a lot of time on an analysis idea and it turns out to be a dead end, don't delete it! Write up a brief note about why it failed and leave it in the notebook. That will help you avoid going down the same dead end when you come back to the analysis in the future.
- Generally, you're better off doing data entry outside of R. But if you do need to record a small snippet of data, clearly lay it out using `tibble::tribble()`.
- If you discover an error in a data file, never modify it directly, but instead write code to correct the value. Explain why you made the fix.
- Before you finish for the day, make sure you can knit the notebook (if you're using caching, make sure to clear the caches). That will let you fix any problems while the code is still fresh in your mind.
- If you want your code to be reproducible in the long run (i.e., so you can come back to run it next month or next year), you'll need to track the versions of the packages that your code uses. A rigorous approach is to use **packrat** (*http://rstudio.github.io/packrat/*), which stores packages in your project directory, or **checkpoint** (*https://github.com/RevolutionAnalytics/checkpoint*), which will reinstall packages available on a specified date. A quick and dirty hack is to include a chunk that runs `sessionInfo()`—that won't let you easily re-create your packages as they are today, but at least you'll know what they were.

- You are going to create many, many, many analysis notebooks over the course of your career. How are you going to organize them so you can find them again in the future? I recommend storing them in individual projects, and coming up with a good naming scheme.

Index

Symbols

A

B

C

D

E

F

N

O

P

S

T

U

V

About the Authors

Hadley Wickham is Chief Scientist at RStudio and a member of the R Foundation. He builds tools (both computational and cognitive) that make data science easier, faster, and more fun. His work includes packages for data science (the tidyverse: ggplot2, dplyr, tidyr, purrr, readr, ...), and principled software development (roxygen2, testthat, devtools). He is also a writer, educator, and frequent speaker promoting the use of R for data science. Learn more on his website, *http://hadley.nz*.

Garrett Grolemund is a statistician, teacher, and R developer who works for RStudio. He wrote the well-known lubridate R package and is the author of *Hands-On Programming with R* (O'Reilly).

Garrett is a popular R instructor at *DataCamp.com* and *oreilly.com/safari*, and has been invited to teach R and Data Science at many companies, including Google, eBay, Roche, and more. At RStudio, Garrett develops webinars, workshops, and an acclaimed series of cheat sheets for R.

Colophon

The animal on the cover of *R for Data Science* is the kakapo (*Strigops habroptilus*). Also known as the owl parrot, the kakapo is a large flightless bird native to New Zealand. Adult kakapos can grow up to 64 centimeters in height and 4 kilograms in weight. Their feathers are generally yellow and green, although there is significant variation between individuals. Kakapos are nocturnal and use their robust sense of smell to navigate at night. Although they cannot fly, kakapos have strong legs that enable them to run and climb much better than most birds.

The name kakapo comes from the language of the native Maori people of New Zealand. Kakapos were an important part of Maori culture, both as a food source and as a part of Maori mythology. Kakapo skin and feathers were also used to make cloaks and capes.

Due to the introduction of predators to New Zealand during European colonization, kakapos are now critically endangered, with less than 200 individuals currently living. The government of New Zealand has been actively attempting to revive the kakapo population by providing special conservation zones on three predator-free islands.

Many of the animals on O'Reilly covers are endangered; all of them are important to the world. To learn more about how you can help, go to *animals.oreilly.com*.

The cover image is from *Wood's Animate Creations*. The cover fonts are URW Typewriter and Guardian Sans. The text font is Adobe Minion Pro; the heading font is Adobe Myriad Condensed; and the code font is Dalton Maag's Ubuntu Mono.

Lightning Source UK Ltd.
Milton Keynes UK
UKOW06f2042111017
310830UK00001B/1/P